THE SCOTTISH BUILDING REGULATIONS

EXPLAINED and ILLUSTRATED

THE SCOTTISH BUILDING REGULATIONS

EXPLAINED and ILLUSTRATED

SECOND EDITION

W.N. Hamilton, P. Kennedy,
A.R.M. Kilpatrick, A.C. MacPherson
and R.K. McLaughlin

Blackwell
Science

Blackwell Science Ltd
Editorial Offices:
Osney Mead, Oxford OX2 0EL
25 John Street, London WC1N 2BL
23 Ainslie Place, Edinburgh EH3 6AJ
238 Main Street, Cambridge
Massachusetts 02142, USA
54 University Street, Carlton
Victoria 3053, Australia

Other Editorial Offices:
Arnette Blackwell SA
224, Boulevard Saint Germain
75007 Paris, France

Blackwell Wissenschafts-Verlag GmbH
Kurfürstendamm 57
10707 Berlin, Germany

Zehetnergasse 6
A-1140 Wien
Austria

First published 1993
Reprinted with updates 1994
Second edition published 1996

Set in 10pt Times
by Archetype, Stow-in-the-Wold
Printed and bound in Great Britain by
Hartnolls Ltd, Bodmin, Cornwall

DISTRIBUTORS

Marston Book Services Ltd
PO Box 269
Abingdon
Oxon OX14 4YN
(*Orders*: Tel: 01235 465500
Fax: 01235 465555)

USA
Blackwell Science, Inc
238 Main Street
Cambridge, MA 02142
(*Orders*: Tel: 800 215-1000
617 876-7000
Fax: 617 492-5263)

Canada
Copp Clark, Ltd
2775 Matheson Blvd East
Mississauga, Ontario
Canada, L4W 4P7
(*Orders*: Tel: 800 263-4374
905 238-6074)

Australia
Blackwell Science Pty Ltd
54 University Street
Carlton, Victoria 3053
(*Orders*: Tel: 03 9347 0300
Fax: 03 9347 5001)

A catalogue record for this title is available from the British Library

ISBN 0-632-04115-3

Library of Congress
Cataloging-in-Publication Data
The Scottish building regulations : explained and illustrated / W.N. Hamilton . . . [et al.].
— 2nd ed.
p. cm.
Includes bibliographical references and index.
ISBN 0-632-04115-3
1. Building laws—Scotland. I. Hamilton, W. N.
KDC448.S26 1996
343.411'07869—dc20
[344.11037869] 93-34575
CIP

CONTENTS

PREFACE

Following the success of the hardback edition of this book, first published in 1993 and then reprinted with updates in 1994, the publishers have now, after consultation with the authors, decided to introduce this current paperback edition.

The aims remain as before – in essence, to provide a simplified guide to the Scottish Building Regulations and their associated Technical Standards set out in a manner which gives ease of reference to each section of these documents. As with previous editions this is preceded by sections on the Building (Scotland) Act, administration of building control and legal liability related to building control.

Since the first edition appeared there have been two major sets of Amendments to the Regulations/Technical Standards, these being in July 1993 and June 1994. In some cases these changes have been driven by European harmonisation (e.g. revisions to Part B – Fitness of materials). Part T which deals with provisions for the disabled has undergone several revisions. Part S which deals with stairs, (etc.), has undergone significant change. All these changes are incorporated here.

With regard to Part M (Drainage and sanitary facilities) the Scottish Office has decided to dispense with the series of rules within the Technical Standards and to use a policy of reference to the appropriate British Standards. The opportunity has been taken to add illustrative examples while reviewing Part M in this edition.

With regard to several other Parts of the Technical Standards (e.g. Parts A, C, D, E, K, N, P, and Q) there have been minor amendments and these also have been incorporated in this edition of the book.

It was also felt useful to include a new Appendix (which links with the section of the book dealing with the 1959 Act and its regulations) which illustrates how some of the building control prescribed forms would normally be completed.

Regulation of building control is never static and further revisions are likely, in particular related to Parts A, D, E, J, K and T. Some comment has also been included in relation to proposed changes in these areas.

The authors would also like to take this opportunity to thank Julia Burden and Janet Prescott of Blackwell Science for their courteous perseverance in urging completion of this latest edition.

May 1996

I Legal and Administrative Matters

1. BUILDING CONTROL

An overview

History

Building control in Scotland has been in existence since around AD1119. The establishment of Royal Burghs gave rise to a dean of guild court in each Burgh, which had jurisdiction in matters of building control and nuisance. Many such burghs had been established by 1707 and others with these dean of guild building control functions were established in the nineteenth and early twentieth centuries. A more detailed historical account of the development of building control is provided in Appendix 1.

Dean of guild courts

The Guest Committee

By the mid 1950s public interest concerned itself with investigating all aspects of local government. In 1957 the Guest Committee Report on building legislation was published. This report made recommendations with regard to building control and the setting down of building standards to apply uniformly throughout Scotland.

Guest

The Building (Scotland) Act 1959

The outcome of the Guest Report was the passing of the above Act which, as amended, is still the governing statute on the subject of building control in Scotland.

The Act

The Act established new machinery for examining building proposals and for inspecting buildings under construction in relation to a national code of building standards to be made by the Secretary of State by way of regulations.

This new machinery replaced the arrangements at that time which resided within the Burgh Police (Scotland) Acts and other Acts. It required the setting up of building authorities. In many cases the existing dean of guild courts were subsumed into the building authorities. Local bye-laws were replaced by a uniform system of buildings standards regulations applying throughout Scotland. Provision was made for the relaxation of these standards for particular cases.

Burgh Police (Scotland) Acts

Building operation regulations were to be made for the protection of the public while building work is going on. Warrants would be required from building authorities for the construction, demolition or change of use of buildings. Before buildings could be occupied, a certificate of completion would be required from the building authorities. The certificate would only be provided when the authorities were satisfied that the conditions of the warrant had been fulfilled.

Operations Regulations

Provisions of the Act

The Act also provided powers for the local authorities to deal with construction work undertaken without warrant and with buildings whose life has expired. Further powers are provided to require buildings in certain situations to conform to the standards. There is also provision which allows building authorities to deal with dangerous buildings and which includes if necessary the demolition of the building.

There is also provision for appeals against building authority decisions or decisions of the Secretary of State and there are provisions for the inspection of building work.

The Act, with minor alterations throughout the years, has continued to provide the Secretary of State (with the backup of the Scottish Office Development Department's Construction and Building Control Group and through consultation with the Building Standards Advisory Committee) with the powers to make regulations concerning building control matters. These regulations are published by Her Majesty's Stationery Office in Statutory Instruments.

Building Standards Regulations

These regulations provide standards which can be reasonably attained in buildings of the classes to which these regulations relate in such matters as security, health, safety, amenity, convenience and welfare of the persons who will inhabit or frequent such buildings as well as the safety of the public generally. Conservation of fuel and power is a further important consideration within these standards.

Introduction of the new arrangements

The national system of building control provided for in the 1959 Act, was introduced in 1964, giving the local authorities the powers and the responsibility for enforcement. A new structure of local government came into operation as a result of the Local Government (Scotland) Act 1973. As indicated earlier, dean of guild courts were abolished and their building control functions were transferred in most cases to the new district councils and islands councils created at that time. However in two regional councils, the Borders and Dumfries and Galloway, building control has been a regional responsibility. Highland Region was also empowered to centralise building control within its jurisdiction but chose to devolve it to its district councils. In April 1996 the new unitary authorities were in place with consequent re-arrangement again of building control.

Local Government (Scotland) Act 1973

Matters of consultation

The Secretary of State in the framing of Building Standards Regulations is required to consult the Building Standards Advisory Committee which is highly representative of the whole of the building and construction industry. The Secretary of State may also consult with such other bodies as appear to be representative of the interests concerned.

Building Standards Advisory Committee

Users of the Regulations and those affected by them can also make their views known with respect to matters such as being too wide in scope or too onerous in application. By these various consultative procedures it is intended to generate sensible and practicable regulations and to minimise the possibility of the promulgation of unsuitable regulations.

Building Control Forum

A further development in the consultation process was the introduction in 1985 of the Building Control Forum. This consists of representatives from the Scottish Office and the Convention of Scottish Local Authorities (COSLA) to promote common interpretation and administration of the Building Standards Regulations. Thus senior officials from the Scottish Office meet (three or four times a year) senior local authority officials to resolve and smooth difficulties which arise in the operation of the Scottish system of building control. Changes of practice rather than changes in legislation are most often found to be the best way forward. The minutes of these meetings and occasional information papers agreed by the Forum are distributed to all building control authorities to increase dissemination of these matters.

The Act and its regulations

As will be seen in section 2, which deals with the Building (Scotland) Act in some detail, there are several sets of regulations which arise from the Act. The current provision is:

(1) the Building Standards (Scotland) Regulations 1990 and the associated Technical Standards 1991 which set the relevant standards for a wide range of aspects such as fitness of materials, means of escape, thermal and sound insulation and so on; **Standards Regs**

(2) the Building Standards (Scotland) (Relaxation) by Local Authority Regulations 1991 which deal with the extent of delegation of powers by the Secretary of State to local authorities to relax the building standards regulations mentioned in (1); **Relaxations Regs**

(3) the Building Operations (Scotland) Regulations 1975 which make provisions for the safety of the public during construction or demolition work, in respect of matters such as erection of protective works, clearing of footpaths, securing or closing entry to sites and so on; **Operations Regs**

(4) the Building Procedures (Scotland) Regulations 1981 and its 1991/95 Amendments which deal with matters such as how to make application to the local authority for warrant or for direction (relaxation), how to make applications and/or appeals to the Secretary of State, general procedures of local authorities, and so on; **Procedures Regs**

(5) the Building Forms (Scotland) Regulations 1991 which prescribe the forms which should be used for applications, orders, warrants, directions, notices, and so on. **Forms Regs**

Fees

In particular the 1991 No. 2 Amendment to the Procedures Regulations deals with the table of fees to be charged by local authorities for the administration of the building control function.

The framework of this legislation is shown diagrammatically in Figure 1.1.

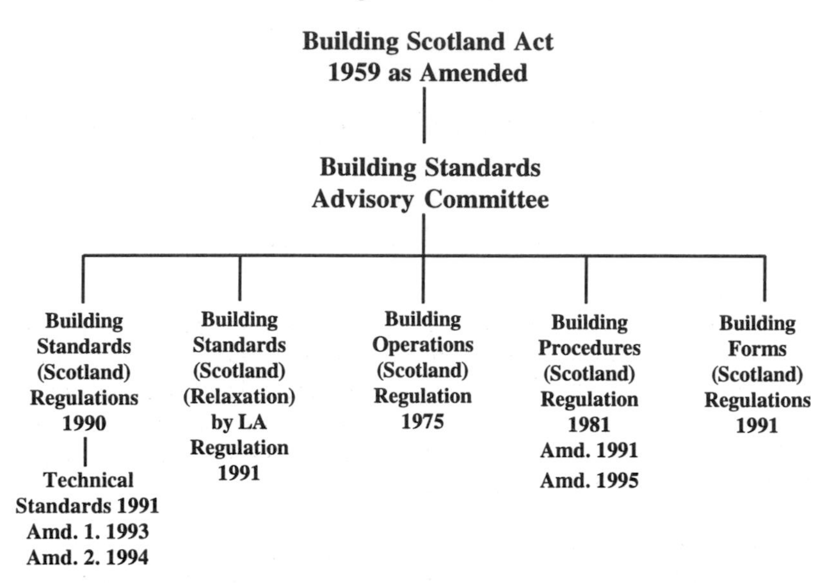

Figure 1.1 The 1959 Act and its regulations.

Associated Acts

Housing (Scotland) Act 1974

Other areas of legislation can impinge on building control matters and collectively it is useful to identify these as associated Acts. For instance the Housing (Scotland) Act 1974 deals with such matters as repair grants and grants for the provision of fire escapes and may well be administered by a building control department.

Roads (Scotland) Act 1984

The Roads (Scotland) Act 1984, (by way of further illustration), deals with the occupation of parts of roads for the deposit of building materials, which is likely to be a matter of liaison between roads departments and building control departments.

These and the other associated acts are covered in more detail along with the Building (Scotland) Act 1959 in the next section of this book.

E.C. Directives

European Community Construction Products Directive

An important change from outwith the Scottish system of building control lies in the implementation of the EC Directive. The purpose of the directive is to ensure that the regulations made from time to time by the Secretary of State do not inadvertently cause barriers to trade within the European Community.

The Department of Trade and Industry's publications on the subject of European Community Directives include information related to construction products. There is further comment on this area within the section of this book which deals with Part B of the Regulations (fitness of materials).

References

Guest Committee (1957) *Report of the Committee on Building Legislation in Scotland*, (HMSO).

2. THE 1959 ACT AND ITS REGULATIONS

Introduction

All building control responsibilities and powers spring from the Building (Scotland) Act 1959 (as amended). The scope covers all types of buildings and deals with matters such as safety and health in and around buildings and energy conservation. As a result of the Act, building authorities, (i.e. building control departments) responsible for ensuring that minimum acceptable buildings standards would be met, were established.

Scope

Earlier, it was indicated that several sets of regulations came into place as a result of the Act. This section deals with the Act and its resulting regulations in some detail.

The 1959 Act (as amended)

The Act was amended in the Building (Scotland) Act 1970 and as a result of various provisions within the Health and Safety at Work Act 1974, the Town and Country Planning (Scotland) Act 1972, the Local Government (Scotland) Act 1973 and the Roads (Scotland) Act 1984.

The Act is in four Parts, the first dealing with Buildings Authorities, the second with Standards and Operations, the third with Dangerous Buildings and the fourth with supplementary matters including Schedules.

Structure of Act

Part 1 – Buildings Authorities

Within Part 1 of the Act several items have been repealed and those which remain relate to the matters which follow.

(1) Provision of powers to make regulations to be discharged from time to time by the creation of statutory instruments, s. 2(4). The current range of regulations deal with Forms, Procedure Standards, Relaxation by Local Authorities and Operations as indicated earlier in Figure 1.1.

These regulations deal, among other matters, with warrants, directions, appeals and procedures of local authorities building control departments in maintaining acceptable standards for buildings, and these will be looked at in more detail later.

(2) Provisions for joint building control and planning applications are allowed to be combined within one document, as per s. 2(5 and 6). To date no such powers have been adopted.

Part II – Building Standards and Building Operations

This Part deals with various matters including creation of standards, relaxations, appeals, powers to approve building types, operations, demolitions, occupations of roads, and certificates of completion.

Building standards

Powers of Secretary of State

Section 3 deals with the powers of the Secretary of State (through the Scottish Office Development Department and after consultation with the Building Standards Advisory Committee), to make building standards regulations.

Section 3(1) allows prescription of standards in terms of performance (e.g. sound tests, drain tests), types of material (e.g. thermal insulants), methods of construction (e.g. stairs), or otherwise. The concept of relating the standards to particular classes of building (e.g. sound insulation standards only relate to dwellings) is included here. The Fourth Schedule of the Act lists all the matters to be covered (which subsequently appear as Parts of the Building Standards Regulations such as say Part M Drainage and Sanitary Facilities).

The concept is also raised of different classes of building being linked to different (and appropriate for that class) standards. In the latest regulations these classes are now known as Purpose Groups.

The current provision is 1990 No 2179 (s. 187) Building Standards (Scotland) Regulations 1990. These are likely to be of most interest to lawyers whereas the blue covered ring binder entitled *Technical Standards* is for the use of designers, builders, building control officers and other persons with an interest in the design and construction of buildings.

Standards are prescribed which have a reasonable expectation of being attained, s. 3(2). Prescription is allowed, s. 3(3), of different standards in relation to buildings of different classes. Special provisions may also be made for short life buildings (ten years or less). Provision is also set out here for the acceptance of materials or methods of construction which are deemed to satisfy the standards.

Exempted buildings are dealt with as per s. 3(4), these being buildings other than dwellings or offices belonging to the United Kingdom Atomic Energy Authority or any buildings of other classes which are specified in the regulations as being exempt.

Classification of buildings is dealt with as per s. 3(5), classification being by size, design, purpose, description, location or any other appropriate characteristic.

Consultation is covered as per s. 3(6) and requires that the Secretary of State shall consult with the Building Standards Advisory Committee and other relevant bodies before making regulations.

Powers to repeal any previous legislation which contains provisions now covered by the Standards Regulations are dealt with as per s. 3(7). Such powers were used, for example, to repeal the Thermal Insulation Industrial Buildings Act 1957.

Relaxation of building standards regulations

Section 4 deals with powers and procedures related to the relaxation of buildings standards regulations.

Secretary of State and Relaxations

The Secretary of State has powers in s. 4(1) to dispense with or to relax the provisions relating to either an application made in relation to a particular building for a relaxation or dispensation (see Forms Regulations, later), or an application made in relation to a class of building. He may also provide of his own accord a direction to dispense with or relax provisions in relation to particular classes of buildings.

Powers for local authorities

The Secretary of State has powers in s. 4(2) to make regulations which allow local authorities to exercise power in the dispensation and relaxation of certain provisions in the standards. Current powers reside in the 1991 No 158(s. 13) Building Standards (Relaxation by Local Authorities) (Scotland) Regulations 1991.

In the previous relaxation regulations (1985), powers of dispensation and relaxation had been limited to alterations, extensions or changes of use of buildings. Such powers have been extended to include all new buildings as well, with the exception of enclosed shopping centres (buildings which contain shops having frontages to an arcade, mall or other covered circulation area) and cases referred to the Secretary of State (applications for warrant under section 6A of this Act – see later).

No directions can be given, s. 4(3) for specific cases which are spelled out in the relaxation regulations. A current example of this is shopping centres.

In granting a dispensation or a relaxation, this is given as a direction by the Secretary of State and may or may not include conditions, s. 4(4).

Forms 1 and 2, Forms Regs

The mechanism for application for a dispensation or a relaxation is the use of Form 1 of the 1991 No 160 (s. 15) Building (Forms) (Scotland) Regulations 1991. Form 2 of these regulations is the form for the Relaxation Direction decision of the local authority (or in some cases, the Secretary of State) provided in response to application on Form 1 (see Appendix 7).

Class relaxations

Class dispensations or relaxations, s. 4(5), will cease to be effective at the end of any specified period, when this is indicated as part of the direction. They may also be revoked by subsequent directions by the Secretary of State.

The cut off time in applying these measures is related to the time when warrant applications are lodged with local authorities, s. 4(5A).

Directions relating to class dispensations or relaxations by the Secretary of State, s. 4(6), are made after consultation with the Building Standards Advisory Committee.

Notification, s. 4(7), must be made to local authorities of the making of class dispensation or relaxation directions by the Secretary of State.

Fees

Payment (or if the Secretary of State so decides, remission) of fees by persons making application for class directions is covered by s. 4(7A).

Procedure Regs

Powers for the Secretary of State to create the procedures regulations, which deal with applications for warrants, directions and appeals, are given in s. 4(8). The current provision is 1981 No 1499(s. 152): The Building (Procedure) (Scotland) Regulations 1981, along with the 1991 No 159(s. 14) and the 1995 No 1572(s. 112) Amendments. This latter amendment deals with the scale of building control fees which may be charged (see Appendix 2).

In the case of class directions the regulations shall be read in the context of and subject to directions in force, s. 4(9).

Appeals against decisions

Section 4A deals with appeals against decisions related to dispensations and relaxations.

Applicants may appeal to the Secretary of State, s. 4A(1) regarding refusals of dispensations or relaxations or the imposition of conditions by the local authority which are considered to be unacceptable.

Appeals, Form 3

Application is made on Form 3 (see Appendix 7) and within 28 days of notification of the local authority decision (Procedures Regulations). An Appeal, s. 4A(2) can also be filed (Form 3) where the local authority has failed to provide a decision within two months, (Procedures Regulations) or such longer time as agreed in writing between the applicant and the local authority.

The Secretary of State has powers, s. 4A(3) to either uphold the local authority decision or to provide his own direction with regard to the application.

Approval of building types

Type Approvals

Section 4B deals with the powers of the Secretary of State to approve particular types of building. Its purpose is to avoid repetitive assessment of buildings of identical construction. This is most likely to relate to standard building types within Purpose Group 1 (dwellings), Part A of the Technical Standards.

Powers for dispensation by the Secretary of State for any particular building type to particular provisions of the standards regulations are given, s. 4B (1) either arising from an application or of his own volition.

Applications must be made in the prescribed manner, s. 4B(2).

Certification

The Secretary of State may, on his approval, issue a certificate, s. 4B(3), specifying the building type, standards to which the certificate relates and the class or classes of case to which the certificate refers.

The certificate may have a specified period of operation as per s. 4B(4). Buildings conforming to the type covered by the certificate shall in the appropriate context be deemed to conform with the standards, s. 4B(5), and the Secretary of State has powers to vary the certificate either as a result of an application or of his own volition, s. 4B(6).

Anyone making an application to the Secretary of State for a certificate is required to pay a fee although the Secretary of State has powers to give remission of the fee, s. 4B(7).

Section 4B(8, 9, and 10) deal with matters associated with revocation of certificates while s. 6(11) provides the Secretary of State with considerable scope in framing certificates or their revocations.

Building operations regulations

Operations Regs

Powers are given to the Secretary of State as per s. 5(1) to provide for the safety of the public during construction, repair, maintenance or demolition of buildings by making operations regulations. The current provision is the 1975 No 549 (s. 74) Building Operations (Scotland) Regulations 1975.

These regulations provide for the separation of a building or a building site from roads or other places used by the public by the erection of protective works, by the clearance of footpaths and by the securing or closing of entries and exits from building sites.

Section 8 of the Act which dealt with application to the local authority for permission to occupy parts of roads (including pavements) when involved with building work has been replaced by s. 58 of the Roads (Scotland) Act 1984 which deals with the same issues but requires such applications to be made to the roads authorities.

Application of building standards regulations via warrants

Warrant system

Section 6 is designed to ensure that building work will comply with the standards, through exercising a system requiring applications to be made giving details in plans, specifications, etc. as to how the work of construction is to be carried out.

Warrants

Form 4

Section 6(1) requires that a warrant be obtained for all construction, demolition or changes of use with the exceptions of those fixtures not requiring a warrant which are detailed in Schedule 2 to regulation 4, within Part A of the Technical Standards (e.g. replacement of windows does not require a warrant).

Granting of warrant by L.A. Form 5

The local authority is required to grant a warrant, s. 6(2), when satisfied that all activity conforms with the requirements of the operations and standards regulations. All activity will be subject to any conditions specified in the warrant, s. 6(3), and must be in accordance with these standards, including any dispensation/ relaxation directions (see Appendix 7).

Staged warrants

Section 6(3A) provides powers to the local authorities to grant a warrant subject to further information being provided in respect of the prescribed stages which are set out in regulation 5 of the Building (Procedure) (Scotland) Regulations 1981 as follows:

- construction of foundations;
- construction of substructure;
- construction of underground drainage system;
- construction of superstructure, excluding stages specified below;
- construction of external wall cladding or internal walls or linings;
- construction of roof;
- installation of lift, escalator, or electrical, ventilation, heating or plumbing system.

Limited warrants

These are dealt with in s. 6(4 and 5) and relate to buildings with a limited life not exceeding ten years. Such warrants are issued on the condition that the building will be demolished before the end of the stated period although there is also provision to make an application for extension of the period.

Form 13

Demolition warrants

Form 6

A warrant for demolition is subject to the condition in s. 6(6) that the work will be carried out within a certain time period which is usually related to the size and nature of the work.

Alterations to warrants

Form 7

During the lifetime of the warrant and prior to issue of any certificate of completion, application may be made to the local authority for an amendment of the terms as per s. 6(7 and 8). This would normally be accepted if there is no worsening of the standards as per the original warrant.

Work required or authorised by local authorities

No warrant application is required in circumstances where construction or demolition work is either required or authorised by the local authority as per s. 6(9). An example would be work relating to dangerous buildings.

Projects outwith normal scope of L.A.

Secretary of State and warrants

Sections 6A and 6B were inserted as a result of the Building (Scotland) Act 1970 and are designed to deal, in the main, with projects which are outwith the normal scope of the work administered by the local authority and which as a result may require fresh deliberation and decision making at the Scottish Office.

The Secretary of State may give directions, s. 6A(1), which require warrant applications or applications for amendments to warrants to be made to him. Such directions may be, s. 6A(2), to a particular local authority or a general direction to all local authorities. This then requires all local authorities to refer any applications which fall within the framework of the direction, s. 6A(3), to the Secretary of State. The local authority may however refuse any application which has not been made in the prescribed manner.

Interaction of Scottish Office and L.A.

On receipt of such applications, the Secretary of State may decide in any particular case, by issue of notice, that the application is a matter to be dealt with in the normal way by the local authority, s. 6A(4). On the other hand he may give notice that he proposes to consider and make decisions about specific matters in the application. Thereafter the remainder of the application will be decided by the local authority.

The matters he may consider are specified in s. 6A(5) and consist of:

(1) for new building work, whether or not the proposals in any warrant application or alteration to warrant application for a building will conform to the standards;

(2) for alterations or extensions to buildings, whether or not the existing building conformed to the standards, whether or not it will conform on completion of the work and in cases where originally it did not conform, whether or not it will conform even less on completion of the work;

(3) for changes of use of buildings, whether or not the building will then comply with certain parts of the regulations as specified by the Secretary of State.

In cases of changes of use, certain requirements can become more stringent and others may arise which did not apply to the original building. This is likely to be the case with regard to fire resistance (more stringent) and sound insulation (new requirement) in a case where say a large stone villa is to be converted into a number of flats.

Powers to impose more onerous require-ments

In cases of major or more unusual projects, s. 6A(4c), the Secretary of State has powers to impose more onerous requirements to the application involving consultation with the Building Standards Advisory Committee and others, s. 6A(7).

Generally, once the Scottish Office has reached decisions over issues and specified these in the notice(s) to the local authorities, each local authority then processes applications in the normal manner.

Limited life buildings

With regard to applications for extension of the period specified in the warrant for buildings having a limited life (e.g. the prefabricated housing stock of the immediate post-war period, the 'prefabs') to local authorities, the Secretary of

State may give a direction, s. 6A(8), that such applications must be referred to him. On direction from the Scottish Office, the local authority then processes such warrant applications in the normal way.

Powers of Scottish Office

The Secretary of State has powers to dispense with or relax standards in relation to applications referred to the Scottish Office, s. 6B(1), either by application or by his own accord notwithstanding that such powers reside under the relaxation by local authorities regulations in the hands of the local authorities.

It is not competent to appeal to a sheriff against a local authority decision to refuse warrant (construction, change of use, extension of period of life of building) where the refusal is based on determination by the Scottish Office, s. 6B(2).

In order to expedite these matters of referred applications, the Secretary of State has powers, s. 6B(3), to make procedure regulations, which are embodied in those same procedures regulations (Part III of the 1981 Regulations) already identified when discussing s. 4(8), earlier.

Within Part III of these procedure regulations, regulation II identifies the appropriate application of Part III as being applications for warrant for construction or for extension of life of buildings which are subjects of referral notices from the Secretary of State.

Regulation 12 of the procedure regulations specifies that referred warrant applications should be sent to the Scottish Office together with plans, local authority observations and confirmation that application is in the prescribed manner. In the case of limited life buildings, a copy of any previous warrant should also be sent.

Under regulation 13 of the procedure regulations, the applicant must be informed of any referral to the Secretary of State and also of the contents of any resulting notice.

Consultation

Where the Secretary of State intends imposing more onerous requirements than the standards, on any application, he requires (regulation 14) to consult the applicant, the local authority, the Building Standards Advisory Committee and any other relevant interested person.

Where he proposes to relax the standards either of his own accord or under application from a local authority, he must notify the applicant, the local authority and any other interested party of his intention and consider any representations received by him within 21 days for issue of notification of his intentions.

Secretary of State and hearings

The Secretary of State may initiate a hearing (regulation 16) for the applicant, the local authority and any other interested parties to be heard by a person appointed by him for that purpose.

The local authority must notify the applicant and any other interested party of the decisions of the Secretary of State and the reasons for reaching such decisions.

Occupation of parts of roads

Section 8 of the Act, which dealt with issues of occupation of parts of roads adjacent to construction sites, has been repealed and for the most part re-enacted within s. 58 of the Roads (Scotland) Act 1984.

Building Control and roads authorities

The effect of this change has been to transfer to local roads authorities the former powers of building authorities which had resided in s. 8 of the 1959 Act to control occupation of parts of roads in connection with building operations on adjacent buildings. Section 58 of the Roads Act deals with control of occupation and deposition of materials on roads and s. 85 with control of the use of builders skips on roads.

It is an offence, s. 58(1) and s. 58(3), to deposit materials or to erect staging or scaffolding which projects over a part of a road (public or private) for any purposes connected with building, without written permission from the roads authority. Occupation, in the terms of such permission, s. 58(2), does not constitute an offence. Where a person is charged under s. 58(1) with such an offence, s. 58(4), an available defence is that all reasonable precautions were taken to avoid commission of such an offence. Where a person is charged under s. 58(3), that is, a person other than the occupier of the land, s. 58(5), available defences are to prove that the offence:

(1) took place on instructions, or by the authority, of his employer or of the person conducting the operations;
(2) was due to a mistake;
(3) was due to reliance on information supplied to him, and that he was unaware that he was depositing the materials otherwise than under the terms of a written permission.

Unless leave of court is given, such a person charged with the offence, must, s. 58(6), within seven days of the hearing, supply the prosecutor with the identity of the person on whose instructions or authority the offence was carried out.

Certificates of completion

Section 9 of the Act deals with the issues associated with the granting (or refusing) of a certificate of completion at the end of the building work associated with the warrant.

Form 8

Application may be made to the local authority, s. 9(1), on Form 8 of the 1991 Forms Regulations, for a certificate of completion. Within a prescribed period of 14 days after receipt of the application (regulation 6 of the procedures regulations), the local authority is required to grant such a certificate (Form 10) or notify the applicant of their refusal to do so (see Appendix 7).

The local authority requires to grant such a certificate where the building complies, s. 9(2), so far as they are able to ascertain after taking all reasonable steps to find out, with the conditions of the warrant.

Form 9

With regard to any electrical installation in the building, s. 9(3), and before a certificate of completion can be granted, the applicant must also provide a certificate (Form 9) completed by the person who installed the electrical system certifying that the installation is in accordance with the standards and the relevant conditions of the warrant (see Appendix 7).

Reasonable cause for failure to produce such a certificate will be acceptable if it is to the satisfaction of the local authority (e.g. worthy of consideration would be where an electrician of good reputation died immediately after the completion, including checks, of an electrical installation).

Section 9(3A) deals with all other installations in a building in like manner, while s. 9(4) makes it an offence to provide false or misleading information on a certificate.

It is an offence to occupy or use a building before granting of the certificate of completion by the local authority, s. 9(5), unless solely for the purpose of construction of the building.

Temporary occupation
Form 11
Form 12

Powers are available to the local authority under s. 9(6) to grant written permission, on application by the warrant holder, for temporary occupation of the building, in exceptional circumstances, prior to the issue of the certificate of completion. Such permission would be for a specified period and further temporary occupation could only be granted on further application.

Appeal to the sheriff

Before dealing with matters of enforcement under s. 10, 11, 13 etc., it may be useful to deal here with matters on which appeals can be made to the sheriff. Section 16 deals with appeals and there are seven circumstances, s. 16(1), in which appeals can be made:

(1) refusals by local authorities to grant warrant;
(2) refusals by local authorities to extend the life of limited life buildings;
(3) refusals by local authorities to issue certificates of completion;
(4) orders made by local authorities under section 10;
(5) orders made by local authorities under section 11;
(6) orders made by local authorities under section 13;
(7) charging orders made by local authorities.

A person wishing to appeal must give notice of the appeal within 21 days of being served a decision under (1), (2), (3) or an order under (4), (5), (6) and (7).

The appeals procedure shall be determined by the Court of Session, s. 16(2), which may also make provisions in relation to expenses and as to the sitting with the sheriff of technical advisers (expert witnesses).

Appeals mechanisms

The sheriff may, under (1), (2), (3) either confirm the local authority decision, s. 16(3), or substitute some other decision as seems to him proper.

He may under (4), (5), (6), (7) vary or quash the order or replace it with one which seems to him more equitable.

In all cases, the determination of the sheriff is binding on all parties and shall be final.

The sheriff may if he so wishes state a case on a point of law to the Court of Session if he desires to seek clarification. The Court of Session in turn may require, in certain circumstances, that the sheriff state a case to the Court on a point of law.

An interesting point of law arose with respect to s. 16(3) in the case of *Waddell and others* v. *Dumfries and Galloway Regional Council* (1979). An appeal was made to the sheriff principal on a matter of expenses and the appellants, while accepting the determination of the sheriff, argued that awarding of the expenses was a procedural matter divorced from the determination of the appeal. The sheriff principal held that the sheriff's determination of the appeal, including the matter of expenses, was final and binding on both parties.

Any order [(4), (5), (6), (7)] shall not become operative, s. 16(4), until:

(1) where an appeal has not been made, the 21 days allowed for appeals has passed;
(2) where an appeal has been made, the time at which the sheriff determines the appeal or the appeal is abandoned.

The provisions of s. 16 (5) give powers to the sheriff to require from an appellant a sum of money to cover the expenses of the appeal.

By way of illustration of the appeals procedure, consider a case where construction work on a building has been carried out without warrant and in a situation where the work in the opinion of the local authority does not meet the standards required by the regulations. After trying persuasion, notices, etc., the local authority finally serves an s. 10 notice on the owner requiring certain works to be carried out within a certain time.

The owner on the other hand thinks that the local authority are over reaching themselves and are being unreasonable. He decides to appeal to the sheriff. Figure 2.1 shows a flow chart of the possibilities which can then arise.

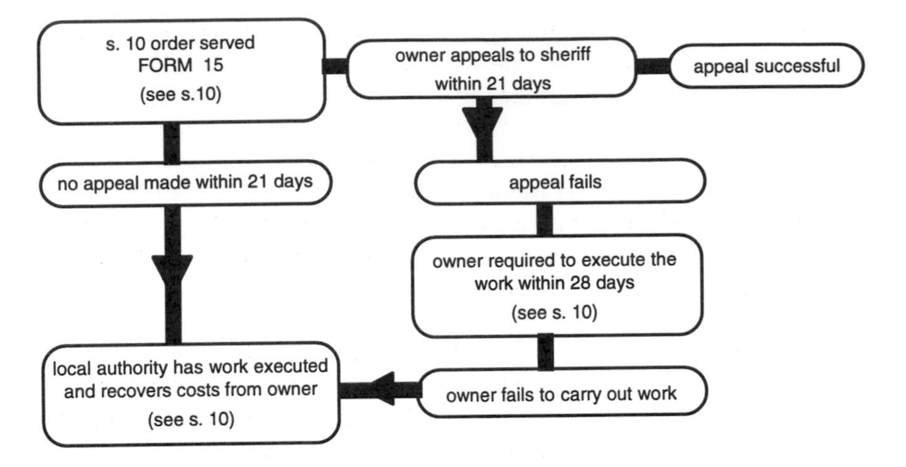

Figure 2.1 Possibilities resulting from appeal.

Enforcement – Section 10

Need to conform

Section 10 provides powers to local authorities to have building work removed or altered to conform with the standards. Whereas s. 6 provides the means of ensuring that those involved with building work understand the requirements necessary to comply with the standards via warrant, s. 10 provides powers for enforcement of these standards.

Enforcing mechanisms

Form 14

Where it appears to the local authority that any building to which the standards apply has been or is being constructed without a warrant, s. 10(1), or in contravention of conditions set out in the warrant, an s. 10 notice (Form 14) can be served on the person responsible. Such a notice may also be served in circumstances where a building is still in existence after the period limited for it by warrant (see Appendix 7).

The notice can be served on the person who caused the building to be erected but it can also be served on the successor to his interest. An important effect of this is that it places an onus on prospective buyers of properties to ensure that they make proper enquiries regarding conformity of such properties with standards before concluding any sale.

Such a notice requires a response (regulation 43 of the procedure regulations sets out the procedures for both the person alleged to be offending and the local authority, in these matters) within the time specified (which must not be less than 21 days from service of the notice) showing cause why remedial measures should not be taken to ensure conformation with the standards.

To show cause, it is necessary to lodge a written statement showing why the operations specified in the notice should not be carried out and give a statement of the grounds on which the alleged offender intends to rely.

In response the local authority must provide information as to date (not less than seven days after receipt of the statement), time and place of a hearing on the matter. The local authority may require a supplementary statement.

Whether the person or his representative appears, or not, at the hearing the local authority must take account of the statement lodged. Where the local authority is satisfied that cause has not been shown s. 10(1B), an order (Form 15) can be executed requiring specified operations to be carried out. The period must be at least 28 days after the notice becomes operative.

Provisions exist for retrospective approval in s. 10(1A). Where work has been carried out without warrant or in contravention of the conditions of a warrant, an application can be made to either the local authority or the Secretary of State under s. 4 of this Act for a relaxation of standards. This application must be made within the period of the notice.

This effectively extends the period of notice which then expires 21 days after the direction applied for is granted by either the local authority or the Secretary of State, or is refused by the Secretary of State. Where the direction is refused or granted subject to condition by the local authority, the period of notice is deemed to be extended for two different circumstances. The first is where no appeal is made, in which case the extension is the period of time in which appeals can be made to the Secretary of State. The second is where an appeal has been made to the Secretary of State under s. 4A in which case the extension is the period of 21 days from the date on which the appeal has been determined or abandoned.

Form 15

With regard to a local authority order on Form 15, if the person receiving the order is aggrieved, an appeal may be made to the sheriff, s. 16(1), within 21 days of the making of the order as mentioned earlier.

Subject to the references to other legislation such as the Ancient Monuments Act 1931 (etc.) and to persons such as trustees (etc.) given in s. 17, if the person still fails to carry out the requirements set out in the s. 10 order (Form 15), he shall be guilty of an offence and the local authority may then execute the operations s. 10(2), which will satisfy the requirements of the notice.

The remedy is available to the local authority, s. 17(3), to enter the building and carry out the work to secure compliance where the person is not in occupation, on condition that reasonable notice is given to the occupier.

Expenses incurred by the local authority in meeting the requirements of the order are recoverable from the person concerned, s. 10(2). Such expenses, where there is more than one interested party for the building, may on application to the sheriff, s. 17(6), be apportioned. If the sheriff's decision is not acceptable, an appeal can be made to the Inner House of the Court of Session in Edinburgh and if necessary, the House of Lords.

On acceptance of default or failure of appeal, the person is liable to the penalties which then come into force and which are set out in s. 19. He may also be liable to damage proceedings in the Civil Court, s. 19A, to the extent that the breach of the s. 10 notice has caused damage. If necessary, the local authority can protect its interest, s. 10(3), by a secured annuity at 6% per annum over 30 years by way of a charging order under Schedule 6 of the Act.

Form 26

Form 15

The order set out in Form 15 may only be served, s. 10(4), in accordance with the rules set out in s. 16(4) relating to times for appeals being made, determined or abandoned, which were mentioned earlier.

The local authority may include in an s. 10 notice (Form 14), the requirement that work be suspended, s. 10(5), until the matters raised in the notice have been determined or until one month from the date on which the notice was served.

The local authority may continue the existence of a limited life building, s. 10(6), as long as it does not contravene the standards.

Fixtures and their manner of fitting are covered by s. 10(7) with regard to their conformity with the standards. Schedule 2 to regulation 4, within Part A of the Technical Standards, provides a list of fixtures which do not require a warrant.

The flow diagram in Figure 2.2 outlines the s. 10 procedures.

Enforcement – Section 11

Section 11 deals with the powers of local authorities to require buildings to conform to the standards but within clearly defined limits.

The provisions of s. 11, s. 11(1), relate to buildings to which the standards apply. Where the building does not conform to the standards, the local authority has powers to pursue compliance. In making the judgement as to whether or not to pursue compliance, the local authority must take account of how practicable and reasonable it is to pursue compliance. One aspect of this judgement must be the cost of the work.

However, the provisions of s. 11 will only apply in certain cases, since close scrutiny of the Building Standards (Scotland) Regulations 1990 reveals exemptions from requirements to conform in certain circumstances for certain types of buildings.

An example of this is to be found in regulation 21 which deals (along with regulations 19 and 20) with the resistance to transmission of sound. Regulation 21 states that 'regulations 19 and 20 shall not be subject to specification in a notice served under section 11 of the Act'. Regulations 19 and 20 deal with the requirement to provide adequate resistance to airborne and impact sound transmission in dwellings.

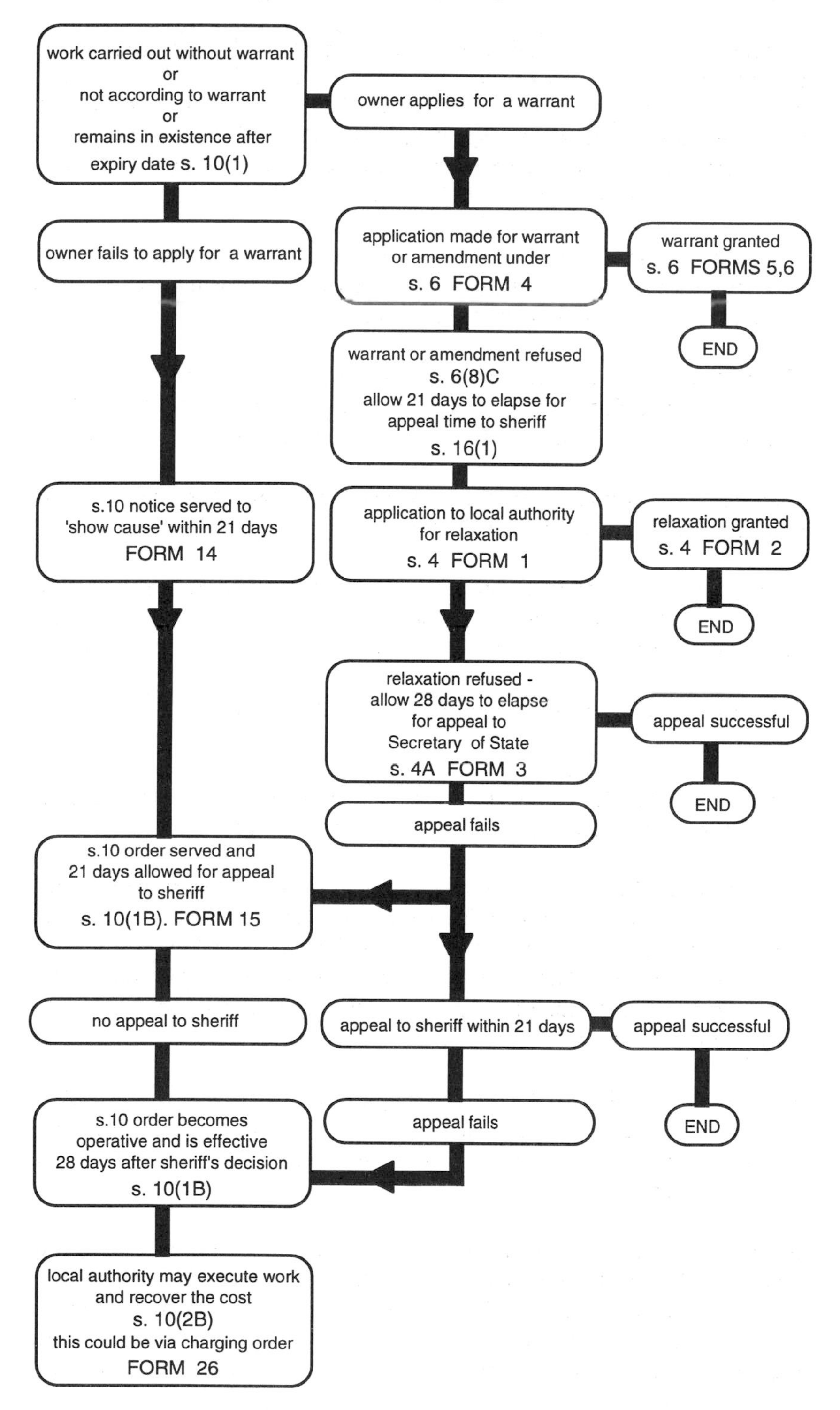

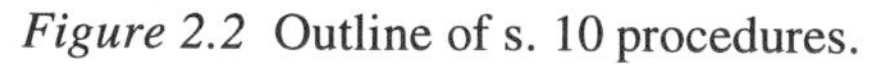

Figure 2.2 Outline of s. 10 procedures.

The intention of regulations 19, 20 and 21 are that the strictures of adequate sound insulation are to be applied to new buildings and to alterations and changes of use but not to buildings already in existence. The presumption (in terms of reasonability and practicability) is that these latter buildings had conformed with the requirements of the warrant at the time they were built.

Similar provisions (on the same basis of reasonability and practicability) reside within the following regulations: 12(2), 13(2), 23(4), 25(3), 29(3), 30(3), 32(3), and 33(6).

When an s. 11 notice is to be served it must be after disposal of any relevant warrant application, including appeal to the sheriff where the period of grace has passed, s. 16(4), or where the appeal has been made timeously until it is determined or abandoned, in the case of the refusal of a warrant.

Form 16

Form 16 of the forms regulations is used for serving of notice and the local authority must specify the provision, s. 11(2), to which they consider the building ought to conform. The notice must specify the period (not less than 28 days) allowed for response and the response must show cause why the building should not conform to the provisions specified in the notice.

Retro-spective approval

As was the case in relation to s. 10 notices, provisions exist for retrospective approval, in this case within s. 11(3). Thus where the building does not conform to the provisions specified in the notice an application may be made for a direction (dispensation or relaxation of standards) under s. 4 of the Act. Such an application must be made (and the local authority notified) within the period of the s. 11 notice.

This effectively extends the period of notice which then expires 28 days after the direction applied for is granted or refused. Where the direction is refused or granted subject to conditions by the local authority, the period of notice may be deemed to be extended for two different circumstances. The first is where no appeal is made (s. 4A of the Act), in which case the extension is the period of time in which appeals can be made to the Secretary of State. The second case is where an appeal has been made to the Secretary of State under s. 4A, in which case the extension is the period of 28 days from the date on which the appeal has been determined or abandoned.

Form 17

If within the period indicated in the notice (as extended, as the case may be, in terms of applications for directions or appeals in relation to decisions), the owner fails to show cause why the building should not conform to the satisfaction of the local authority, the local authority may then order him, s. 11(4), by means of a notice issued on Form 17 of the 1991 Forms Regulations, to make the building conform within the period specified, being not less than 28 days.

Subject to the references to other legislation such as the Ancient Monuments Act 1931 (etc.) and to persons such as trustees (etc.) given in s. 17, if the person still fails to carry out the requirements set out in the s. 11 order (Form 17), he shall be guilty of an offence and the local authority may then execute the operations required to make the building conform, s. 11(5). Any expenses incurred by the local authority would then be recoverable.

Settling expenses by use of charging order

These expenses may be settled, s. 11(6), in the ordinary manner or if necessary by the procedures for a charging order, already discussed under s. 10 and which are indicated in the Sixth Schedule of the Act.

Section 11(7) of the Act states that no s. 11 notice shall specify any provision of the standards not subject to such notice. This is a clear warning to local authorities to be sure of their ground before issuing an s. 11 notice.

Any s. 11 order shall only become operative, s. 11(8), after the period for dealing with any application for a warrant in respect of the work identified in the order (or for any appeals procedures) has passed.

Application of s. 11 powers

As with s. 10 powers, the concept of the local authority and/or the Secretary of State, deciding that requirements are reasonably practicable, before applying s. 11 powers, s. 11(9), is stated. Such decisions must involve consideration, among other matters, of the expense of the work.

The flow diagram in Figure 2.3 outlines the s. 11 procedures.

The Building Standards Advisory Committee

Building Standards Advisory Committee

Section 12 deals with the Building Standards Advisory Committee. This is a committee appointed by the Secretary of State after consultation with the range of specialist opinion representing relevant interests. The remit of the committee, s. 12(1), is to:

(1) advise the Secretary of State about building standards regulations;
(2) review the operation of these standard regulations and make recommendations to the Secretary of State regarding their development;
(3) advise the Secretary of State on any matters which he refers to it.

Matters of constitution and procedure, s. 12(2), for this committee may be dealt with by regulations made by the Secretary of State. This process is assisted by the committee itself reporting to him at intervals of not more than five years on matters related to the exercising of its functions.

Matters of reimbursement of expenses and allowances to members of this committee are dealt with in s. 12(3) of the Act.

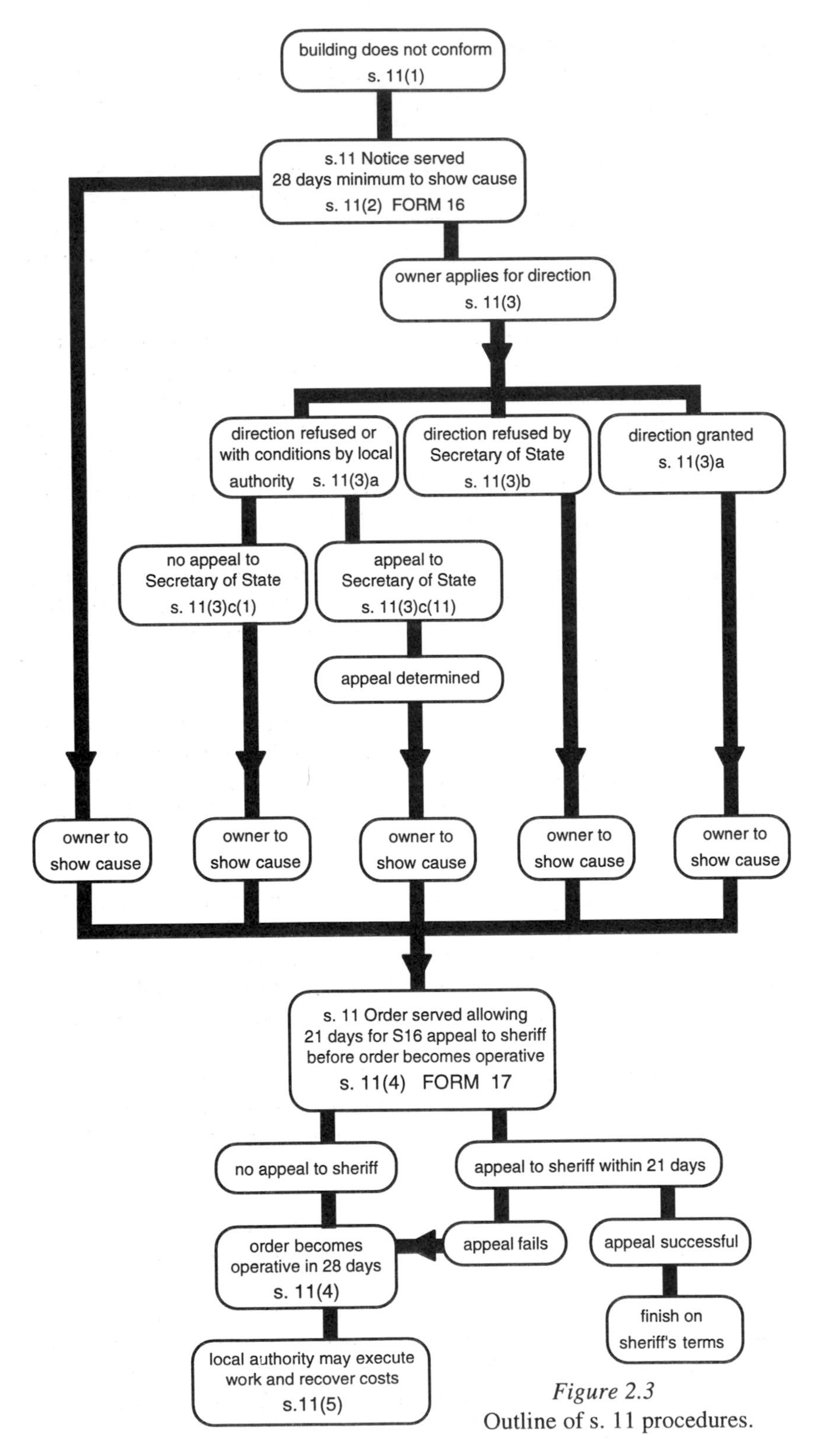

Figure 2.3
Outline of s. 11 procedures.

Part III – Dangerous Buildings

This part of the Act deals with the powers of local authorities in relation to buildings which are found to be dangerous.

Where it appears to the local authority, s. 13(1), that a building is in a dangerous condition for a variety of reasons (powers of inspection are available under s. 18) including danger to inhabitants, danger to people who frequent the building, danger to users of adjacent buildings and danger to the public generally, prompt action is then required of the local authority.

Form 21

In such circumstances, under s.13(1)(a) it is required of the local authority that they remove all the occupants of the building deemed dangerous (and if necessary, any adjacent building). This is done by the serving of the appropriate notice issued on Form 21 of the Forms Regulations. Note that s. 25 of the Act deals with the statutory position in relation to the serving of notices. This s. 13 notice would state a specified time for removal which is fixed by the local authority, having regard to all the circumstances.

If any person fails to comply with the notice then the local authority may apply to the sheriff for a warrant for its enforcement. The procedures which then follow are given within schedule 7 of the Act which deals with the evacuation of dangerous buildings.

L.A. and duty of care

Under s. 13(1)(b), the local authority has a duty to protect the public and persons and property on adjacent land. Thus the local authority has the power to carry out any operations needed (if necessary, demolition) to prevent access to the building and/or to any road or public place adjacent to the building which appears to be dangerous.

Under s. 13(1)(c), the local authority has a duty to serve a notice (Form 18 of the Forms Regulations) on the owner of the building requiring him within seven days of service of the notice to begin such works (and also to specify, being not less than 21 days after the seven days, the required completion date) as are necessary to make the building safe. Section 13(1)(c) procedures on Form 18 are used in situations where the building is considered dangerous, but not of imminent danger, and work would be specified to repair, secure or demolish. The period of time specified by the local authority must be reasonable or else it is open to be extended by the sheriff as was the case in *Stevenson* v. *Midlothian County Council.*

In relation to s. 13(1)(b), an interesting point of law was debated in *City of Edinburgh District Council* v. *Cooperative Wholesale Society* (1986).

In 1974 the then Edinburgh Corporation had served three s. 13 notices on Messrs John Jackson (Kirkintilloch) Ltd requiring demolition work at 10/12 West Bowling Street, Edinburgh. Following failure to do so, this work had been done by the Corporation and the cost of the work being £4871.14, this was the sum sued for by the Corporation.

Counsel for the CWS maintained that s. 13(1)(b) did not entitle the Corporation to have the work carried out to remove the danger. He asked the sheriff principal to construe the words 'or otherwise' on the ejusdem generis rule. If this had been upheld it would have meant that the only justification for a local authority demolishing any part of a building was as a means of preventing access to a dangerous building.

The sheriff principal did not accept that the interpretation of s. 13(1)(b) could be so narrow. His interpretation was that paragraphs (a) and (b) of s. 13(1) conferred powers on the local authority to act swiftly in situations where immediate action was called for, whereas paragraph (c) related to circumstances of less imminent danger since the statutory procedures allowed a period of at least 28 days for the works to be carried out.

Clearly any decision by the officers of the local authority to demolish, would not be taken lightly and only after taking opinion from a structural engineer. Such opinion would be essential under the law of evidence as corroboration by an expert witness of the actions of the building control officers.

Subsections 13(2), 13(3) and 13(4) deal with matters which arise consequentially from service of notice regarding removal of danger which is dealt with in s. 13(1)(c).

Form 19 Section 13(2) deals with the situation where the owner of the alleged dangerous building has not responded to the Form 18 notice which required a start on the specified work within seven days and completion on not less than a further 21 days. In such circumstances the local authority may issue a notice of intention to make an order (Form 19 of the Forms Regulations), after having given the owner and any other interested party an opportunity to be heard, requiring any objections and grounds to be lodged with the local authority within seven days of serving of Form 19. This further protracts the administrative procedures.

The procedures related to orders for dangerous buildings are to be found within regulation 44 of the procedures regulations. These procedures are illustrated in the flow diagram of Figure 2.4 along with the other matters associated with s. 13(1)(c) and s. 13(2, 3, 4 and 5).

Section 13(3) reminds us that such an order can become effective, since the appeals procedures of s. 16(4) relate to such orders, only after the period of time in which appeals can be made, or, in cases where appeals are made, only after the period of time until the appeal is either determined or abandoned.

Where the owner remains intransigent and flouts the s. 13(2) order of Form 19, the local authority, s. 13(4), may execute the operations which are specified in the order including if necessary the demolition of the building.

Form 26 Any expenses incurred by the local authority with respect to work which it has carried out on dangerous buildings, s. 13(5), are recoverable from the owner.

This can be done, if necessary, by using the provisions of schedule 6 to generate a charging order, Form 26, which can be served on the owner of the building. However, in carrying out the work the local authority must take account of the provisions of s. 17 (see later) which relate to buildings of historic interest, etc.

The issue of the charging order is an important one since it burdens the building as well as the site. The legal position was clarified in the case of *Howard* v. *Hamilton District Council* (1985) where the owner of a building asked for the charging order to be quashed on the grounds that it had been imposed by the local authority on a previous owner. The owner was interpreted as the 'owner for the time being' which means the original and all subsequent owners. Thus the plea failed.

Section 13(6) ensures that the provisions of schedule 7 shall be effective in securing the removal of occupants from buildings where there is immediate danger, or from dangerous buildings which are the subject of an s. 13 order.

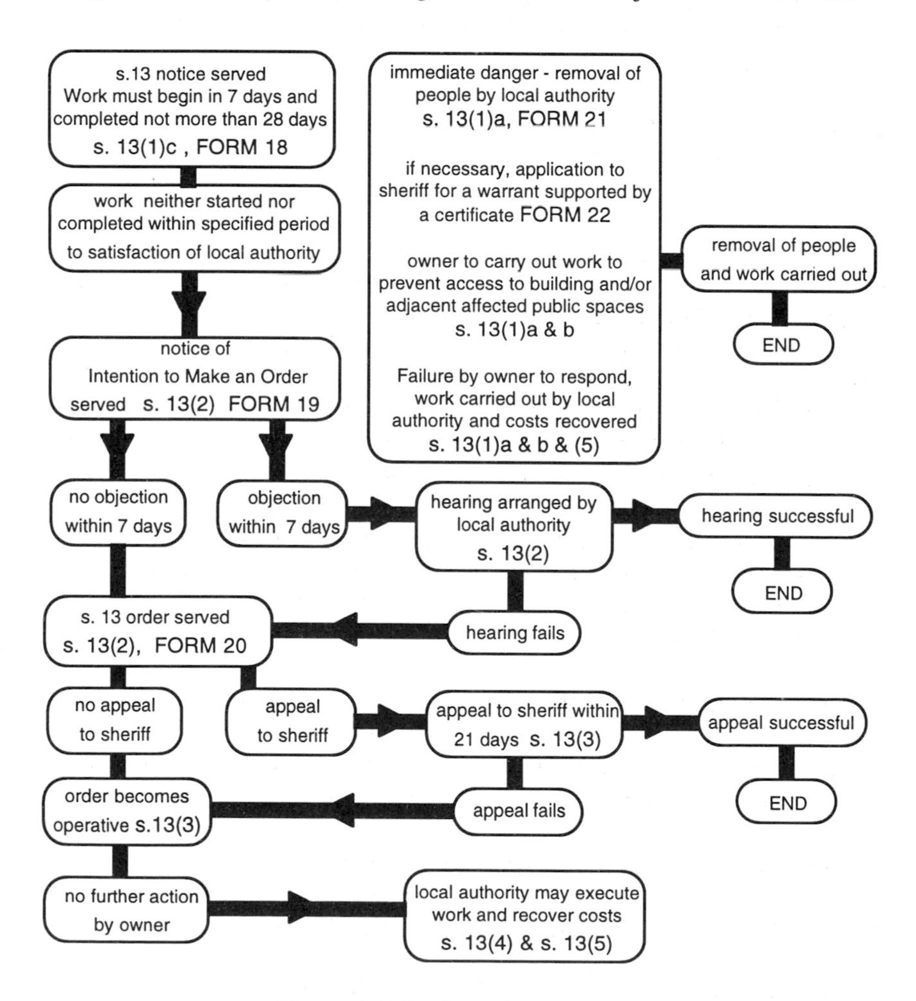

Figure 2.4 Outline of s. 13 procedures.

Evacuation from dangerous buildings

Schedule 7 deals with evacuation from dangerous buildings and begins with the requirement that the occupants will remove themselves from the building within the period specified on the written notice served by the local authority, Form 21.

On failure of occupants to remove themselves, the local authority may then apply to the sheriff for a warrant for removal of the occupants of the building. Such an application by the local authority requires to be supported by a certificate, Form 22, delivered to the sheriff by an officer of the authority.

The sheriff may require service of an additional notice before granting the warrant for ejection, usually not more than seven days after receipt of the application from the local authority.

The sheriff's decision is final in these matters.

Any person so ejected from tenancy within the said building shall be considered not to have the tenancy terminated but only interrupted until the building is made safe.

Use of s.13 procedure

An interesting case relating to proper use by the local authority of the s. 13 procedures outlined in Figure 2.4 occurs in *GUS Property Management* v. *City of Glasgow District Council* (1989) where the local authority having failed to secure demolition of a building via these procedures attempted to effect the same outcome through the use of other legislation. The attempt was not successful.

It is also important that the requirements set out by the local authority in the s. 13 notice match those set out by them in any subsequent s. 13 order, otherwise such an order is invalid as was the case in *Pegg* v. *City of Glasgow District Council* (1988). This case also dealt with the issue of ownership in multi-ownership buildings such as tenements or blocks of flats.

The local authority has powers, s. 14, to sell off any materials arising from its demolition of a building in order to defray the cost of the work.

In circumstances where the local authority has demolished a building and the owner cannot be found, s. 15(1), the local authority may be authorised by the Secretary of State to procure compulsory purchase of the property and the site. The provisions which apply for such a purchase, s. 15(2), are the Acquisition of Land (Authorisation Procedure) (Scotland) Act 1947.

Compulsory purchase by L.A.

The local authority is then entitled to deduct, s. 15(3), from any compensation payable to the owner, arising from such a compulsory purchase, any expenses arising from work done by the local authority to make the building safe, including demolition.

Procedures for appeals to the sheriff by the owner of the building and for the seven defined categories of grievance, s. 16, were already dealt with prior to dealing with s. 10.

Supplementary matters associated with notices or orders under s. 10, 11 or 13, are dealt with in s. 17.

With regard to buildings which are listed, historic, ancient monuments or the subject of preservation notices, s. 17(2), any requirements in notices or orders under s. 10, 11 or 13 must be consistent with provisions in the following legislation:

(1) Ancient Monuments Act 1931, s. 6(1) and s. 6(4)
(2) Town and Country Planning Act 1972 s. 52 and s. 56
(3) Historic Buildings and Ancient Monuments Act 1953 s. 10 and s. 11

In circumstances where the owner of a building has received an s. 10, 11 or 13 notice but is not himself the occupier of the building, s. 17(3), he may, having given reasonable notice to the occupiers, enter the building to carry out the work specified.

Recovery by L.A. of cost of works

Situations can arise where a local authority may seek to recover, from a person, expenses incurred by the authority in carrying out the works specified in a notice or order. If that person can prove, s. 17(4):

(1) his interest in the building is limited to that of trustee, tutor, curator, judicial factor, liquidator and;
(2) his funds or assets are insufficient to meet any demand;

then his liability will be limited to the funds and assets which he has available.

More specifically, with regard to orders under s. 10 and 11, it is a defence in law to prove that failure to comply with the said orders, s. 17(5), was because:

(1) his interest in the building was limited to that of trustee, tutor, curator, judicial factor, liquidator and;
(2) his funds or assets are insufficient to meet the demand.

Apportionment

Where more than one person has responsibility for the building and the cost of any specified work should be shared equitably, the person receiving the demand for repayment of expenses to the local authority, may apply to the sheriff for apportionment, s. 17(6). The sheriff may then make an order for such apportionment.

The same provisions apply to an apportionment appeal, s. 17(7), as have already been discussed within the general provisions for appeals at s. 16(3).

Inspection of buildings and tests

Matters of entry by L.A.

Section 18 deals with matters of entry of persons, authorised by the local authority, into buildings for various inspections and tests.

An authorised person may at all reasonable times be allowed entry, s. 18(1), for the following purposes:

(1) inspection of buildings in course of construction;
(2) inspection of finished buildings yet to receive a completion certificate;
(3) spotting sites of buildings for which application for directions under s. 4 has been made;
(4) inspecting sites of buildings for which application for a warrant under s. 6 has been made;
(5) applying any reasonable tests to determine the quality and strength of any material being used or intended for use;
(6) inspecting buildings where there is reason to believe that a change of use is either proposed or has taken place;
(7) inspecting buildings which there is reason to believe are dangerous;
(8) inspection of buildings where there is reason to believe that the exercise of powers by the local authority under s. 11 may be necessary;
(9) execution of operations under s. 10, 11 or 13.

Such authorised persons may carry out any of the above activities as are required.

Section 18(2) has been repealed.

Form 24 & 25

With regard to items (6), (8) and (9), such authorised persons, s. 18(3), must give three days notice, Form 24, to the occupier and also to the owner if he is known, stating the intention and purpose of entry to the premises. The authority for entry is completed by the local authority on Form 25.

In cases of urgency, or of refusal of entry, a warrant for entry may be provided to the local authority by a justice of the peace, s. 18(4). This allows an officer of the local authority to enter the building, if need be, by force. The justice of the peace would require first to be satisfied that there are reasonable grounds for entry, and that entry has been refused or the case is indeed one of urgency.

The person authorised to enter the building either under Form 25 by the local authority or by warrant from the justice of the peace, s. 18(5), may take with him any other person as appears to him to be necessary. When entry is to an unoccupied building such a person is required to ensure that on leaving the building it is left secure against any unauthorised entry.

Any warrant, s. 18(6), which is granted under section 18 continues to be in force for a period of one month.

Any person who wilfully obstructs any authorised person entering the premises for the matters indicated in s. 18(1) is guilty of an offence against the Act, s. 18(7), and is then liable on summary conviction to a fine.

Care during authorised entry

Authorised persons who enter factories or workplaces must be careful not to reveal to any other person any manufacturing process information or trade secrets unless required in performance of their duties. Such disclosures, s. 18(8), are considered to be an offence against the Act.

Any reference in s. 18 to premises includes a reference to sites and buildings, s. 18(9), and any reference to a justice of the peace includes a reference to a sheriff.

The local authority may require:

(1) anyone who has made application for a direction under s. 4;
(2) anyone who has made application for a warrant under s. 6;
(3) anyone constructing a building under an s. 6 warrant;
(4) anyone who has constructed a building under an s. 6 warrant;

Specified tests

to have carried out at their expense, tests of specified materials.

There is a proviso, however, where the local authority on application being made to them, may if they think fit, meet part or all of the expense of any such test. One illustration of this proviso to s. 18 (10) was circumstances where a local authority was prepared to meet the cost of a sound insulation test on a dwelling (need to conform to Part H) to gain familiarity with the test procedures as well as information about the particular specification.

Penalties for offences against the Act

Offences against the Act

Section 19 makes provisions for penalties related to offences against the Act which are pursued to summary conviction. Where offences relate to s. 18(7) (obstruction of authorised persons) or to s. 25(3) (failure to provide information or provision of false information), the convicted person is liable to a fine not exceeding £50, s. 19(1).

For all other offences against the Act, s. 19(2), on summary conviction, the convicted person is liable to a fine not exceeding £400 and in the case where the offence continues, a further £50 for each day of further offence.

Civil liability

Section 19A, which deals with civil liability, was inserted into the Act as a result of the Health and Safety at Work Act 1974.

Although remedies at law already exist for dealing with damage, it is intended that a possible means of pursuing damage is by the use of s. 19A for various breaches indicated in s. 19A. In relation to an action brought under s. 19A, any defences prescribed within s. 19A are available.

The breaches specified, s. 19A(2), are:

(1) failure to comply with the terms and conditions of a warrant;

(2) contravention of any provision within the buildings operations regulations;
(3) constructing a building without warrant which does not conform to the building standards regulations;
(4) change of use without warrant where the building does not conform to the building standards regulations.

Pursuit of liability

Liability cannot be pursued, s. 19A(3), where a building was erected prior to the date on which s. 19A(1) came into force, unless the breach has arisen as a result of change of use, extension, alteration, repair, maintenance, fitting or demolition.

Defining damage

Section 19A(4) takes account of other remedies at law which are not prejudiced by the existence of s. 19A.

Damage is taken to include, s. 19A(5), death or injury (which in turn includes disease and any impairment of a person's physical or mental condition).

Fees

Provision is made, s. 20(1), for the charging of fees by local authorities in exercising their building control function. The fees prescribed are set out (see Appendix 2) in the Building (Procedure) (Scotland) Amendment Regulations 1995. Interpretation of the table of fees by local authorities has been challenged, notably in the case of *Dunbarton County Council* v. *George Wimpey & Co* (1968) in relation to how to charge for multiple housing units. The local authority interpretation was to deal with each unit as an individual application whereas Wimpey's interpretation was one of multiple applications related to the warrant. The Wimpey interpretation was sustained.

Interpretation of the table of fees

Section 20(2) and s. 21 have been repealed.

Information

Provision is made in s. 22 for local authorities to submit information as and when required by the Secretary of State.

Inquiries

Public Inquiries

Provision is made in s. 23 for the Secretary of State to hold public inquiries, s. 23(1), as and when he thinks fit, and s. 23(2) indicates that subsections (2) and (8) of s. 210 of the Local Government (Scotland) Act 1973 would apply to such inquiries.

The Forms Regulations

Provision is made in s. 24(1) for the Secretary of State to make regulations prescribing the form of any notice or other document. He uses these powers to make the Building (Forms) (Scotland) Regulations, the current set being brought into force in 1991.

These forms would normally be used, s. 24(2), in all cases. Section 24(3) indicates that any powers to make such regulations are by statutory instrument.

Service of Notices and Orders

Within s. 192 of the Local Government (Scotland) Act 1973 are provisions relating to the service of notices, orders and other documents. These provisions, s. 25(1), also apply to the service of notices, orders or documents under the Building (Scotland) Act.

Identifying recipient

One of the difficulties encountered by building control departments when serving notices is knowing who the recipient is. Provision is made in s. 25(2) for local authorities to obtain necessary information from occupiers or those who receive rent from the premises in question.

It is an offence against the Act, s. 25(3), for any person to fail to give information or to give false information with respect to these matters.

Crown rights

Section 26 presently exempts a crown building from the application of the Act and from any regulations, order or notice under it, subject to certain limited exceptions. The basic exemption is subject to three provisos which detail circumstances in which the Crown exemption is qualified. The first proviso, s. 26(1)(a), allows for operations carried out on a Crown building by a private person to be subject to the necessity of obtaining a warrant under s. 6(1) unless it is being carried out by or on behalf of the Crown Estate Commissioners or a government department or special approval has been obtained from the appropriate government department. The second proviso, s. 26(1)(b), covers the situation where the Crown has a minor interest in a building and the appropriate department is of the opinion that the Crown interest should not stand in the way of local authority action under s. 10, 11 or Part III of the Act. The appropriate department can issue a direction allowing action against the owner to be started while the Crown still occupies part of the building. The third proviso, s. 26(1)(c), provides that where operations are being carried out on a Crown building by private persons or firms, the building operations regulations must be complied with by the private persons or firms.

Subsection 2 excludes Crown buildings and land occupied by a Crown building from the powers of entry contained in the Act except with the consent of the appropriate government department.

Effect of HSWA

Subsections (2A) to (2F) were added by the Health and Safety at Work etc. Act 1974 but have not yet been brought into force. Under these provisions Crown buildings will become subject to the requirements of the building standards regulations. They will not, however, be subject to requirement to obtain a warrant from a building authority under s. 6 of the Act nor to any of the enforcement powers which the Act confers on building authorities.

Subsection (2A) applies the building standards regulations to all Crown buildings unless they fall within a general class of buildings specified in the regulations as being exempt or where some special exemption of a particular class of Crown building has been included in the regulations.

Subsection (2B) requires all Crown buildings to which the building standards regulations apply to be constructed in accordance with those regulations. This requirement is a condition of warrant in the case of a non-Crown building and since the warrant procedure will not apply to the Crown it is necessary to specify this requirement.

Changes to existing buildings

Subsection (2C) requires that in a case of an extension to or alteration of a Crown building, the building as altered or extended, as a whole, comply with the building standards regulations to at least the same extent as the original building. In the case of a change of use the building is required to conform to any additional or more onerous requirements which become applicable as a result of the change of use.

Subsection (2D) applies, with suitable adaptions, the provisions of s. 19A which were inserted by the Health and Safety at Work etc. Act 1974 but have not yet been brought into force. Under this subsection the Crown would be liable for damage caused by a failure to comply with subsections (2B), (2C) or the building operations regulations.

Relaxations

Subsection (2E) allows the Secretary of State to use his powers under s. 4 of the Act to give a direction relaxing the requirements of the building standards regulations in relation to a Crown building.

Subsection (2F) provides for a minor adaption of the operation of s. 4B in relation to work on a Crown building. The subsection allows for a type approval which has expired, been varied or revoked to continue to operate where appropriate in relation to a Crown building on which construction work has begun before the relevant date.

Section 3 defines 'Crown building' and 'appropriate authority' and allows for the Treasury to determine the appropriate authority in cases of dispute.

Changes in Crown immunity

Crown buildings have been expected to conform to building standards regulations although not legally required to do so, (the Royal Prerogative). There is currently a general policy of transferring Crown properties out of Crown immunity. A recent example of this has been the loss of Crown immunity of Health Board buildings which are now under the general provisions of the Act other than those buildings in the planning or construction phase. These will continue to be exempt during this transitional phase, which basically means until such buildings are completed and occupied.

The exemption of Crown Property has deep historical and constitutional roots and it cannot be expected that removal of exemption will be easily achieved. It does, however, appear likely that major steps in this direction will be taken in the near future, bringing into the scope of the Scottish Act a number of buildings (and in England and Wales also) which have considerable architectural significance, such as St Andrew's House in Edinburgh.

Section 27 has been repealed.

Financial provisions

Cost of making provisions

Money is provided by Parliament, as indicated in s. 28, to defray the expenses incurred by the Secretary of State as a result of these provisions.

Interpretation

Matters of interpretation (including definitions) are covered by s. 29 of the Act.

In s. 29(1) 'building' and 'construct' are defined in like manner to that indicated in the list of definitions given in Appendix 3.

'Building operations regulations', 'Building Standards Advisory Committee', 'building standards regulations' and 'certificate of completion' have the meanings assigned to them in s. 5, s. 12, s. 3 and s. 9 respectively.

'Change of use' relates to changes in the class of a building which either brings it within the standards or within more onerous standards. An example would be where an individual dwelling is subdivided into several flats in which case more onerous sound insulation and fire resistance standards would then apply.

'Contravene' in relation to any provision of the building standards includes failure to comply with that provision.

'Enactment' is a legal term which includes an order, regulation or other instrument which is brought into effect by virtue of the Act being in place.

'Government department' includes a Minister of the Crown.

'Local Act' is another legal term which includes a decree-arbitral, a provisional order or any other instrument ratified or confirmed by Parliament.

'Local authority' means islands or district councils except in the case of the Highland, Borders and Dumfries and Galloway regions where it means the region itself. This will require re-interpretation as a result of the bringing into being of the unitary authorities.

'Operations' includes those in relation to enclosing and preparing the site.

'Prescribed' is very precise and means prescribed by the Secretary of State by regulations made under this Act.

'Road' includes street and any pavement, footpath, drain, ditch or verge at the side of the road or street.

'Warrant' has two meanings. The first is as assigned to it under s. 6 in terms of requiring a warrant for building activity. The second is a warrant for entry to inspect under s. 18 and also under s. 26(2).

Definitive building

Section 29(2) informs us that building can be taken to include a prospective building (one at the design, planning, drawing, specification stage) and in relation to alterations, extensions or changes of use of an existing building it would only include those parts pertinent to such changes.

Defined ownership

Section 29(3) deals with the issue of ownership of land or buildings and such person would include any person who is able to sell and convey land or buildings to the promoters of an undertaking under the Lands Clauses Act.

'Public road' is taken to mean one maintainable by the Secretary of State, (effectively the highway authority), while 'private road' is taken to mean one not so maintainable.

Section 29(5) deals with the dynamic nature of legislation and the term enactment shall be taken to mean its up to date position.

Sections 29(6) and (7) have been repealed.

LA jurisdiction

Section 29(8) recognises those situations where a building may straddle the boundary between local authorities. The local authorities have powers to agree which shall have jurisdiction and in the event of failure to agree, the decision shall be made by the Secretary of State.

Section 29(9) has been repealed.

Local Act provisions (bye-laws)

Position of bye-laws

Section 30(1) indicates that the Act supersedes any local act provisions.

Section 30(2) allows the Secretary of State some discretion in regard to s. 30(1) if he decides to make an order regarding certain provisions in local acts in relation to corresponding provision in the Act itself.

Section 30(3) gives the Secretary of State powers to repeal by order any provisions within local acts.

Section 30(4) indicates that the powers in s. 30(3) shall be brought into place by statutory instrument.

Section 31 has been repealed.

Section 32 gives the Building (Scotland) Act its short title, namely the Act and also its extent and commencement.

The Schedules
Following the Act there are ten schedules of which six (the first, second, fifth, eighth, ninth and tenth) have been repealed.

The third schedule deals with matters which may be provided for by regulations under s. 2(4) such as notices, hearings, decisions from local authorities, maintenance of local authority records, duration and validity of warrants, etc.

The fourth schedule, which deals with matters in regard to which building standards regulations may be made, has been covered previously listing the matters such as preparation of sites, strength and stability, resistance to moisture, etc.

The sixth schedule has also been covered previously and deals with the recovery of expenses by local authorities via charging orders.

The seventh schedule which has also been covered previously deals with the evacuation of dangerous buildings.

This concludes the Act.

Associated Acts
There is a range of legislation other than the Act itself which impinges on the work of building control officers. Collectively it is useful to call these parliamentary acts the Associated Acts. Some of this legislation is covered briefly as follows:

The Homes Insulation Act 1978
This act makes provision for local authority grants towards the thermal insulation of dwellings with the Secretary of State having powers to alter the percentage or money sum or both. It has been administered by building control departments dealing with issues such as draughtproofing and loft insulation.

LA grants for energy conservation

Current government privatisation strategy has led to such work being passed to an organisation outwith the local authority known as Energy Action (Scotland).

The Road (Scotland) Act 1984
When dealing earlier with the Building (Scotland) Act it was indicated that s. 8 which covered the issue of occupation of parts of roads for deposition of materials had been repealed and that the operative provisions resided within the Roads (Scotland) Act along with other matters of concern to the building control officer such as procedures for placement of skips.

There is thus a considerable liaison between building control departments and highways authorities or roads departments.

The Licensing (Scotland) Act 1976
Applications for the granting of a new licence or a provisional licence in relation to licensed premises will only be entertained when supported by certification as

to the suitability of the premises in the matters of:

Interaction with the LA departments

(1) planning;
(2) building control;
(3) food hygiene.

Clearly there is an interface here between planning, building control and environmental health departments.

From the building control point of view:

(1) where there is an application for a new licence the certificate should state that an s. 6 warrant and an s. 9 certificate of completion have either been granted or not required or an s. 6 warrant for a change of use of the premises has either been granted or not required;

(2) where there is an application for a provisional licence the certificate should state that an s. 6 warrant has been granted and either a warrant for change of use under s. 6 has been granted or on completion of the construction in accordance with the warrant will be granted.

The building control department liaises with the legal department and the licensing court. The work involves inspection of premises, and items of key importance include the provision of alarm systems, emergency lighting and adequate means of escape. In the case of provisional licence applications where building work is involved, the provisional licence may be granted on the basis of plans and specifications checked by the building control department. Final approval can only follow on completion of the work, after inspection by the building control officer.

The Fire Precautions Act 1971

Role of fire officer

The building control officer requires to liaise with the fire authority and should be familiar with the Fire Precautions Act. The building standards regulations are applied to all classes of buildings, but for buildings other than dwellings, and as a consequence of the Fire Precautions Act, the fire officer may insist on more stringent measures than are necessary to satisfy the building standards regulations.

The Sewerage (Scotland) Act 1968

The building control officer requires to liaise with the sewerage authority and should be familiar with the Sewerage (Scotland) Act. An area of common consultation would be the nature of the connection of new buildings into local authority sewers.

The Water (Scotland) Act 1946

The building control officer requires to liaise with the regional council water authority and should be familiar with the Water (Scotland) Act. In like fashion to the sewerage authority, an area of common consultation is the connection of water supply to new buildings. Another area of consultation is with regard to the renewal of lead piping.

Town and Country Planning Acts
There is a considerable interface between the building control officer and the planning officer and the building control officer should have some familiarity with the Town and Country Planning Acts. It is normal practice that in addition to the copy of building proposals held by the building control department a further copy is supplied to the planning department to ensure that in addition to satisfying the building standards regulations the proposals satisfy local planning requirements.

The Housing (Scotland) Act 1974
The building control officer may be involved with improvement grants, repairs grants or grants for the replacement of lead piping. In addition disablement grants may be made, for instance, to provide ground floor bathrooms which are appropriate for the disabled. This will also involve liaison with the social work department.

Grants for home improv-ements

Civic Government (Scotland) Act 1982
Building control departments require to undertake inspections of a variety of premises (such as places of entertainment, places with raised structures, houses in multiple occupation) in relation to this Act in order to report to the relevant district council or local authority.

The naming of streets is covered by this Act and in many local authority organisations this is dealt with by building control.

Safety at Sports Grounds Act 1975
There is currently a great deal of work (generated by the findings of the Taylor Report) in upgrading football stadia. Much of this is work associated with building warrant which is the direct responsibility of the building control department.

However, in considerable measure it has been regional authority work associated with requirements to be satisfied before a Safety Certificate can be issued under the Safety at Sports Grounds Act 1975. Building control departments were consulted by the regional authorities on building control issues associated with such safety matters.

In April 1996 the new unitary authorities came into operation and a fresh look is currently being taken at these operational arrangements

Other legislation
The building control officer should be aware of other legislation which interfaces with buildings but is more likely to come into the domain of environmental health. This would include:

(1) the Noise Insulation Regulations which make provision for grants for effective window insulation around airports and motorways;
(2) the Control of Pollution Act 1974, which makes provision for raising noise nuisance complaints (which might in some cases be a consequence of inadequate sound insulation of walls and/or floors in dwellings);
(3) the Offices, Shops and Railway Premises Act 1963;
(4) the Factories Act 1961.

References

The Building (Scotland) Act 1959 (revised to October 1977)
The Building Standards (Scotland) Regulations 1990
The Building Standards (Relaxation by Local Authorities) (Scotland) Regulations 1985 and 1991
The Building (Procedure) (Scotland) Regulations 1981 and subsequent Amendment Regulations 1991 and 1995
The Building Operations (Scotland) Regulations 1975
The Building (Forms) (Scotland) Regulations 1991
The Roads (Scotland) Act 1984
Civic Government (Scotland) Act 1982
Safety at Sports Grounds Act 1975
Scottish Building Legislation, CIRIA Special Publication 34, which reviews all relevant legislation up to August 1984.

3. ADMINISTRATION OF BUILDING CONTROL

Introduction

As was seen earlier the three principal documents which are instrumental in the prescription of building control in Scotland and provide for its enforcement are:

(1) the Building (Scotland) Act 1959 (plus amendments) which is the enabling Act of Parliament;
(2) the Building Standards (Scotland) Regulations 1990 which is the Statutory Instrument to the Act describing the scope of the Regulations (including exemptions) the classification of buildings, the meaning of defined terms and stating the building standards in terms of general performance requirements;
(3) the Technical Standards 1990 which is a Scottish Office publication.

Technical standards

All building construction and demolition must comply with the requirements of the Regulations except where a specific exemption is granted. The Technical Standards describe the specific minimum requirements of the Regulations and it is knowledge and understanding of the principle and detail of these requirements which is necessary in practice. The layout and content of the Technical Standard will be described here with detailed discussion of specific requirements in the technical section of this book.

Responsibility for enforcement

As was indicated earlier when dealing with the Act, the responsibility for the enforcement of the regulations is devolved to local authorities under s. 10 and s. 11 of the Act and is usually delegated to a building control department. Local authority approval must be obtained for all building and demolition work before it can begin, except work that is exempted in the regulations, and for completed buildings before they can be used.

Approval of proposed work

Approval of proposed building work is granted by the issue of a building warrant and approval of the satisfactory execution of the work by the issue of a completion certificate. The basic stages and procedures which must be followed to comply with statutory requirements and to obtain a building warrant and completion certificate will be discussed here. A wider discussion of the legal aspects and responsibilities of the statutory control of building is covered by other sections of this book.

Using the Technical Standards

The Technical Standards to the Regulations are set out in Parts A to T together with appendices, later in this book. Each of the Parts of B to T refers to an aspect of building performance (e.g. Part B Fitness of Materials, Part T Facilities for the Disabled) and sets out the minimum relevant standards for one or more of regulations 10 to 33. Part A covers regulations 3 to 9 which are necessary in the application of Parts B to T. The relationship between the different Parts and the main functional elements of a building are summarised in Table 3.1.

Table 3.1 Structure of the Regulations

Theme	Part
How to apply the Regulation	A
Materials	B
Structure/fabric:	
Walls, floors, roofs	C,D,G,H,J,
Stairs	C,D,S,
Protected zones	C,D,E
Services	F,H,J,K,M,N,R
Miscellaneous	P,Q,T

For ease of use all Parts of the Technical Standards are divided into four sections with different coloured pages for each section. This arrangement will now be outlined.

Section 1: Contents (white page)
The contents always include an introduction, the relevant regulation(s), the standards to be met and the provision deemed-to-satisfy the standards.

Section 2: Introduction and regulation (grey pages)
The introduction outlines the aims of the regulation(s) and the associated Part and provides general information on the way in which the standards attempt to achieve these aims.

Section 3: Standards (white pages)
This section sets out the technical standards themselves. The first clause in the section describes the scope of application of the standards (e.g. all buildings) and subsequent clauses set out in detail the minimum standard to be met for the full range of building types and relevant situations in order to achieve compliance with the regulation(s).

Section 4: Provisions deemed-to-satisfy the standards (yellow pages)
The provisions deemed to satisfy the standards describe specific ways by which to achieve compliance with the standards. The method given in the provisions may be prescriptive in terms of materials and/or construction (e.g. specifying that cement mortar can be taken as meeting the requirements of providing 'suitable fire-stopping' for service openings in structural elements, D2.18); or, may describe a procedure by which compliance can be demonstrated (e.g. methods of calculating maximum sizes of openings in walls, Small Buildings

Guide to Part C; or, may refer to BSI, or other relevant, standards or codes of practice. Not all of the clauses in S3 have associated provisions deemed-to-satisfy the standards; where these do occur they are indicated by an asterisk against the clause number in S3.

To satisfy the requirements of the Technical Standards it is necessary either to meet the criteria set out in the provisions deemed-to-satisfy or to demonstrate by some other means that the standard is achieved. The latter approach is intended to allow for innovation in design and construction practice but given that the differences in interpretation between local authorities already exist there may be considerable difficulty in ensuring uniformity of acceptance criteria by this route.

Working with building control

As seen earlier, the statutory responsibility for the enforcement of the regulations rests with the local authority and is normally delegated to the building control department. Traditionally the building control officers vet the design and specification of proposed work before issuing a building warrant, carry out inspections and tests during construction and following a final inspection are responsible for issuing the completion certificate to allow occupation of the building. The introduction in November 1992 of self-certification of structural design carried out by chartered engineers removed one aspect of the vetting of proposals from the control of the local authority. Proposals to extend the responsibility for compliance with other regulations to professions in the industry coupled with moves towards allowing private sector 'approved inspectors' to carry out building control functions mean that the traditional activity of building control departments is likely to change significantly over the next few years. In this context some building control departments are attempting to provide an advisory service prior to formal submission of an application for a building warrant, whereas others simply vet proposals against the requirements of the regulations. These types of differences between building control departments should be understood if an effective working relationship is to be established for the duration of a building project. Although there are some variations in the structure and administration of building control departments in different local authorities, the process of applying for a building warrant through to obtaining a completion certificate can be separated into four common stages outlined below.

Approved inspectors

Stage 1: Application for a building warrant

The application for a building warrant is made on a warrant application form which must be accompanied by sufficient information about the proposed design and construction to establish whether or not it complies with the regulations. The level and depth of the submission, particularly with reference to the number and scale of the drawings, is dependent on the complexity of the building. Guidance on this is given in the Building Procedures Regulations. Normally three sets of the drawings must be submitted with the application, one of which must be on plastic, or traditionally, linen. The plastic or linen copy is kept as a permanent record by the local authority. The workload generated by a project, and hence

Submission of drawings

the fee payable, is linked to the estimated cost of the proposed works using the fee scales mentioned earlier. Where direct labour is involved the cost of materials alone has in the past been accepted as the basis for the fee. On receipt of the warrant application form, the drawings and the fee, the application is logged in the Building Register and a warrant number allocated to the project. The person named on the warrant application form as the agent will receive all correspondence from the building control department.

Stage 2: Vetting for the application

Warrant and/or points letter

Following the vetting of the submission by the building control department either a building warrant is granted or a points letter is sent to the agent. If the building warrant is granted the warrant form is issued and one stamped set of drawings returned. The warrant will remain valid for three years. If the warrant is not granted, a points letter, normally accompanied by an unaltered set of drawings, is returned. The letter identifies those points on which the building does not comply with the regulations; it does not suggest how to achieve compliance. If it is not possible to achieve compliance, or it is not reasonably practical (e.g. in refurbishment of a property) it will be necessary to agree a compromise with the building control department. This would usually involve providing a trade off. It should be noted that there is no requirement to approve any proposal that does not comply with the regulations. Following the issuing of a points letter a full set of drawings, taking account of the points raised, must be re-submitted to the building control department. In some cases the department may be willing to accept amendments to the submitted drawings.

Stage 3: Inspection during construction

Notification for inspection purposes

The building control department will inspect the building during construction to ensure that the work is carried out in accordance with the drawings and specifications on which the warrant was issued. The building control department must receive at least seven days notification of the intention to start work on the site. In general, visits are planned to allow monitoring of critical aspects of the construction but are at the discretion of the building control officer. There are three statutory inspections for which the building control department must be given formal notice. The minimum period of notice required is shown for each of these inspections, but it is advisable to keep the building control officer on the project informed of progress in order to minimise the time which can be lost waiting for the inspection to be carried out. The statutory notifications and inspections are:

(1) start of work	7 days;
(2) open foundation inspection	24 hours;
(3) starting pouring foundations	24 hours;
(4) open drain test	24 hours.

Pro-formas for the notification of the building control department are usually issued with the building warrant.

Approval must be obtained from the building control department for any amendments to the design and specification against which the warrant was issued. The amendment form must be submitted with relevant drawings for vetting by the department.

Stage 4: Application for completion certificate

Final inspection

14 DAYS.

On receipt of the application for a completion certificate the building control department is required to carry out a final inspection within seven days, although alternative arrangements can be agreed with the agent. The final inspection ensures that the building complies with the approved drawings, including any approval amendments, and involves a final drain test and an emergency lighting test where appropriate. This inspection is often carried out with the fire prevention officer who is responsible for issuing the fire certificate for the building. On the satisfactory completion of the final inspection the completion certificate is issued. It is an offence to occupy or use a building, or part of a building, for which a completion certificate has not been issued against a building warrant.

4. LEGAL LIABILITIES

Introduction

Following an outline of the Scottish court system and the provisions for dealing with liability claims in Scottish courts, the remainder of this chapter considers the implications in terms of liability for various parties, such as builders, architects, building control officers and those others involved in the processes of construction in Scotland.

Court system

The civil system in Scotland

Courts are classified as civil, criminal or of special jurisdiction, but our main concern here is with the civil courts. These civil courts concern themselves with the rights and duties of citizens towards each other.

The civil courts in the Scottish system are:

(1) the sheriff court;
(2) the Outer House of the Court of Session;
(3) the Inner House of the Court of Session;
(4) the House of Lords.

Sheriff-doms

The sheriff courts deal with a large volume of civil business, within the six sheriffdoms. Each sheriffdom has about 14 full-time resident sheriffs (the sheriffdom of Glasgow and Strathkelvin has about 21) and they operate within a system of sheriff court districts. Appeals may be taken from the sheriff courts to the Inner House of the Court of Session and thereafter, if necessary, the House of Lords.

Nature of delict and reparation

Delict

Delict (or in England, tort) is a legal wrong which causes loss or injury to the person or property of another. This may take the form of a deliberate act, but more commonly arises as a result of negligence. Strictly speaking, negligence is a quasi-delict because there is an absence of intent so that the damage is the consequence of carelessness rather than malice. The law imposes a general duty on all to take reasonable care to avoid loss, damage or injury to others and failure to observe that duty constitutes negligence. But however the loss or injury arises, where there is fault (culpa) there will be liability. The obligation to compensate is known as reparation.

However there are occasions where there need be no fault (culpa) on the part of a party who nevertheless has a statutory liability or one in strict law. This would be the case, for example, where a driver for building firm X was involved in a crash resulting in damage for some other party Y. The manager of X, although not directly at fault, would be liable as long as the driver had been about the firm's business.

The reasonable man

The usual defence against blame in cases of liability is that all was done that a reasonable man could be expected to do in the circumstances to prevent the loss or injury. This would be the case for any defender, be he builder, architect, building control officer, demolition or structural engineer or whoever.

Such was the case in:
Morrisons Associated Companies Ltd v. *James Rome & Son Ltd* (1964) (SLT) where the builder was not found liable in negligence when a building collapsed after the builder had supported it in accordance with the recognised practice at that time. In other words the builder had acted as a reasonable man.

On the other hand, in the case of:
Wright v. *British Transport Commission* (OH) (1962) (SLC) where a contractor demolished a chimney head in a situation where the street had not been cleared and a person was injured in the vicinity of the operation, the contractor was found to be negligent. In other words the behaviour was not that of a reasonable man.

Reasonable man and strict law

It is interesting to note that the defence of reasonableness is not acceptable in strict law, which may be statutory (e.g. Acts of Parliament) or in the nature of vicarious liability.

An illustration of this occurs in the case of:
Millar v. *Galashiels Gas Co. Ltd* (1949) SC (HL) the background to which was the death of a workman through failure of a brake mechanism of a hoist in his employer's factory.

The statute in question was the Factories Act 1937 which made provision that every hoist had to be properly maintained, with 'maintained' being defined as 'in an efficient state, in efficient working order, and in good repair'.

In court it was demonstrated that the employers had taken every practicable step to ensure that the mechanism worked properly and that the hoist was safe. The failure was unexplained and could not have been anticipated and for this reason the defence case failed.

Appeal was made to the Court of Session in Edinburgh, and thereafter to the House of Lords. The House of Lords in affirming the judgment of the Court of Session held that the duty imposed by the statute was strict and that the fact that the brake mechanism had failed was sufficient to establish that the employers were in breach of their statutory duty.

Vicarious liability

There are three kinds of vicarious liability. The first has already been covered with the case of the lorry driver, this being the liability of the employer. The second is the case of partnerships where any negligence on the part of a partner can hold the firm (the other partners) liable.

The third case is that of the agent. Any third party has the right to believe that an agent has the powers normally associated with that position unless otherwise informed (verbally in front of witnesses or by advertisement in the Edinburgh Gazette). In other words the principal is liable for the actions of its agents.

Not every perceived wrong will give rise to an action for reparation if the court considers that no legal wrong exists. By way of illustration consider the situation where Builders Merchant P is operating profitably until Builders Merchant Q opens in adjacent premises. P might consider Q's behaviour as harmful conduct but the court would consider the matter to be one of fair competition rather than a legal wrong.

Duty of care

Duty of care is owed to those persons whom the defender, as a reasonable man, would anticipate could be at risk of loss or injury, if there were negligence on his part.

For instance, a building control officer who failed to check out structural calculations for a part of a building which later collapsed would leave his employers (i.e. the local authority) liable (strict law or vicarious liability) for his alleged negligence.

Standard of care

In deciding whether or not negligence had occurred, the court would require to assess the standard of care of the defender. The standard would be that of the average building control officer, neither outstanding nor less than mediocre. In other words, he would be required to demonstrate the average standard of care expected from members of that profession.

A case which illustrates this point in relation to alleged negligence by an architect was that of:
Wagner Associates v. *Joseph Dunn (Bottlers)* (OH) (1986) (SLT) where the situation was that a combination of circumstances led to movement in a wall requiring it to be demolished. It was found that the architects had not exercised duty of care in that they had not warned their clients of the risks associated with that particular design of wall, in which case the clients would probably have opted for a safer design.

Higher standards of care

It is worth noting that there are certain circumstances, for instance in the protection of children and the elderly, where the appropriate standard of care may be considered to be much higher. Consideration of potential liability for a builder or demolisher would arise in relation to potentially dangerous areas such as building or demolition sites, for instance. Matters of adequately securing plant and fencing off the site would require special care.

The substance of delict
This is embodied in the maxim *damnum injuria datum*. The three essential

elements in delict are:

Nature of delict

(1)	*damnum*	loss or an injury to the pursuer;
(2)	*injuria*	conduct which amounts to a legal wrong, alleged to have been the fault of the defender, and recognised as such by court;
(3)	*datum*	the loss or injury must have been caused by the legal wrong, in other words there must be a causative link between damnum and injuria.

Damnum must entail financial (patrimonial) loss or physical injury and/or suffering. Suffering alone (as could be the case where a construction worker dies as a result of an industrial accident) can be the basis for a claim for damages as solatium for mental anguish.

In the case of *injuria*, the court will decide (as illustrated earlier by the circumstances of builders merchants Q and P) whether or not the wrong is such as would allow the court to consider an action for damages. If so, it is a legal wrong (delict).

There are situations where a legal wrong takes place without loss (*injuria sine damno*) such as careless practices on a construction site which fortunately do not lead to injury.

In order to demonstrate that delict exists in any given situation it must be demonstrated that a breach of a legal duty has occurred, such a duty arising either from statute, for example the Building (Scotland) Act, and any of its associated regulations, or from common law.

Causal link

Delictual liability may only be demonstrated where *damnum* and *injuria* are shown to have occurred and there is a causal link (*datum*) between the two.

Lack of causal link

A case which illustrates lack of causation is:
McWilliams v. *Sir William Arrol & Co. Ltd and Lithgows Ltd* (1962) SC (HL).
A steel erector was killed when he fell from a tower in a shipyard in Port Glasgow. The widow raised an action for damages against her late husband's employers on the grounds that they had failed to provide him with a safety belt. Although it was found that the employers were in breach of duty, it was also proved that if a safety belt had been provided, the steel erector would not have worn it. The employers were held not liable since although there had been *damnum* and *injuria*, the causal link (*datum*) to prove delict, was absent from the case.

A similar failure to make causal connection took place in:
Galek's Curator Bonis v. *Thomson* (OH) (1991) (SLT) in relation to a scaffolding accident.

Liability in relation to the Act and its regulations

Liability and the act

The 1959 Act and its regulations, as outlined earlier, contain numerous statutory duties for various groups of people intending to carry out, or in the course of carrying out, building works. Breach of the provisions of this legislation is an offence punishable on conviction by the level of fine indicated in s. 19(1) and s. 19(2) of the Act.

Breach of statutory duty

Breach of duty

Breach of duty imposed by building control legislation may result in civil liability giving rise to a claim for damages by someone who suffers loss or damage as a result of such breach. The liability is thus for damages arising from the breach of statutory duty.

Civil liability

Civil liability is covered by s. 19A of the Act where provisions are made for specified breaches, so far as they cause damage, to be actionable. Damage is defined in s. 19A as including the death of, or injury to , any person (including any disease and any impairment of a persons's physical or mental condition).

The Act goes on to state at s. 19A(4), significantly, that nothing in the civil liability section of the Act shall prejudice any right of action in civil law which exists apart from the provisions of the section.

Section 19A has not yet been tested in the courts since, to date, the usual common law remedies have seemed to be sufficient.

Position of local authorities

Liability and local authorities

The Act imposes a range of statutory duties on local authorities in the exercising of the building control function. For example, within s. 6(2) of the Act it states that local authorities shall grant building warrants where applications to them are made in the prescribed manner, where all conduct will be in accordance with the building operations regulations and where they are satisfied that the building when it is constructed, will conform to the building standards regulations.

More generally, in s. 2(4) of the Act, the role of the local authority is indicated since it states that the Secretary of State may by regulations make provision with respect to the procedure of local authorities in the exercise of their jurisdiction and functions under the Act. In this way the powers and the duties of local authorities are defined in statutory law.

This raises the important matters of the scope of the duties and of the liabilities of those people whose work is in building control.

Scope of the act

The foreword to the Act introduces the theme that it is 'an Act to make new provisions for safety, health and other matters in respect of the construction of buildings and for safety in respect of the conduct of building operations'. Clearly the local authorities have a duty of care with respect to health and safety.

They also have a duty with respect to conservation of fuel and power as indicated in s. 3(2) of the Act. What is not so clear in terms of the scope of their duties is the extent of the 'other matters'.

The position in Scotland has been clarified to some extent by events in England, in the recent case of:

Murphy

Murphy v. *Brentwood District Council* (1990). The situation here was that the plaintiff's semi-detached house had been built in 1969, being one of a number of houses on an estate. The district council building control department had given approval for its construction. Ground conditions on the site were such as might have given rise to differential settlement and because of this a special raft foundation had been designed. The proposals for this had been considered and approved by an independent firm of consulting engineers acting on the district council's behalf.

It was found by the trial judge that the design of the raft foundation was defective and that the consulting engineers had been negligent in approving the design.

The plaintiff had purchased the house in question in February 1970, and in 1977 discovered serious cracks, which he had repaired. In 1981, insurers acting on behalf of the plaintiff commissioned another firm of consulting engineers to investigate further serious cracks which had appeared on internal walls.

As a result of this investigation it was discovered that the raft foundation had subsided differentially and this had caused cracking and distortion. In 1985 a gas leak had occurred which might well have been caused by distortion of the floor slab. There was also cracking of, and leakage from, a soil pipe connected from the house to a main drain.

The plaintiff's neighbour had also suffered settlement damage to his house but had not been able to afford any contribution towards the cost of remedial work to the building. Repairing the plaintiff's house alone was not practicable and as a result he sold the house to a builder who was aware of the structural defects.

The insurers settled his claim for subsidence damage for £35,000 being the loss of value sustained on the sale of the house. This was the sum at issue in the case.

Anns

The trial judge found for the plaintiff holding that the defendants (Brentwood District Council) were liable in negligence under the principle of an earlier case, *Anns* v. *London Borough of Merton* (1977) and that the reduction in the value of the plaintiff's house by reason of the state of its foundations was recoverable as damages. Brentwood District Council then took the matter to the Court of Appeal which upheld the decision of the trial judge.

House of Lords and Anns

The district council then carried the Appeal to the House of Lords. This provided an opportunity for the House of Lords to reconsider the *Anns* case. In the *Anns* case the House of Lords had held that a local authority, in exercising its statutory functions of building control, could be liable in tort (in Scotland, delict) to occupiers of a building for the cost of remedying defects which amounted to a present or imminent danger to health and/or safety, where such defects were the result of negligent failure of the council to ensure that the building had been erected in conformity with building regulations.

The effect of *Anns* was generally to hold building control authorities (and indeed, builders) liable not only for personal injury or damage to other property resulting from defective construction but also for damage to the building itself (such as arose in the *Murphy* case with regard to the design of the raft foundation). Liability, thus interpreted, meant that it included economic loss.

There were wide misgivings about the *Anns* interpretation of liability and since 1978 the English courts had begun to limit the operation of the so-called *Anns* doctrine particularly by limiting the class of people to whom the duty was owed. On consideration of the *Murphy* v. *Brentwood District Council* case (with the background of *Anns*) the House of Lords held that *Anns* (and therefore *Murphy*) had been wrongly decided and that a local authority in the exercise of its building control functions is not under a duty to protect property owners from pure economic loss.

Judgment on building control liability

It is understood that in Scotland, liability of building control authorities in the matter of discharging their function would be interpreted in Scottish courts in line with the judgment of the House of Lords in the case of *Murphy* v. *Brentwood District Council.* In other words, such authorities, in exercising their statutory functions, owe no duty of care to owners or occupiers of buildings to safeguard them against pure economic loss.

Disclaimers

Another aspect of duty of care is the function of disclaimers. The influence of the disclaimer is illustrated in the case of:
Hadden v. *City of Glasgow District Council* (OH) (1986). Here a husband and wife intending to purchase a flat applied to the district council for a home loan. Before offering a loan the district council had carried out a survey of the property for their own purposes.

On the basis of that survey the district council offered to make a loan to the prospective purchasers. The prospective purchasers were aware that a survey was being carried out but they were also informed that the offer of a loan did not imply that the council was satisfied with the state of the property. The prospective purchasers were further informed that the state of the property was a matter for them to satisfy themselves about and that the district council had no responsibility thereto.

Economic loss

The offer of the loan was accepted and missives were entered into. Thereafter the district council's director of building control issued a notice requiring the execution of extensive building repairs. This led to the purchasers suffering an economic loss which they sought to recover from the district council. The basis of their claim was that the employee of the district council who had surveyed their flat had been negligent in failing to notice the defects in the building and as a consequence this negligence had led to the purchasers' loss, since, but for the negligence, the district council would not have offered and the purchasers would not have accepted the loan. The district council argued that the purchasers' case was irrelevant.

It was held that the purchasers having had notice of the council's provisions and conditions before accepting the loan offered (written evidence that can be produced in court) and those conditions amounting to a disclaimer of a duty of care, there was no duty of care and the action was dismissed.

Position of the builder

Liability and the builder

There is a great deal of case law related to contractual liability of the builder in relation to client, architect, sub-contractors and so on. These are matters which are considered to be outside the scope of this book and are well covered elsewhere.

In addition to the contractual liability to his client through the terms of the contract between them, the builder is liable both under the Health and Safety at Work Act 1974 and also the Construction (Working Places) Regulations 1966 for the provision of a safe working environment. For instance in the case of *Manford* v. *George Leslie* (Sh.Ct.) (1987) (SCLR), it was held that the builder had fulfilled his duty in the provision of a safe means of access to a place of work in a situation where a worker was injured in taking a shorter but less safe route.

Again in the case of *McChlery* v. *J W Haran* (OH) (1987) (SLT) a glazier fell from a step ladder while repairing a window. He alleged that the step ladder was required to be footed under the provisions of the Construction (Working Places) Regulations 1966, at regulation 32(3). It was held that under the said regulation the provisions do not apply to stepladders.

Delict at common law

The builder can also be liable in delict at common law to third parties who suffer physical damage to person or property as was illustrated earlier in the case of *Wright* v. *British Transport Commission* (1962) in which a contractor had demolished a chimney head without having cleared the street.

The builder, like the building authorities, can also be held liable in relation to breaches of statutory obligations. As an instance, regulation 10 of the Building Standards (Scotland) Regulations 1990 requires that 'materials, fittings, components and other manufactured products used...shall be suitable for the purpose'. It is therefore necessary for the builder to ensure that items such as bricks, blocks, electrical sockets, hot water cylinders, etc. are of an appropriate standard.

Indeed, so far as the builder is concerned, most of the commonly used standard form building contracts contain an express provision which requires that the contractor (the builder) is required to comply with all relevant statutory obligations, which naturally includes the building standards regulations and all other regulations related to the Building (Scotland) Act 1959 as amended.

The National House Building Council

Purpose of NHBC

The NHBC's purpose is to protect the home buyer by promoting higher standards of house building, preventing defects, and providing a ten year warranty, known as Buildmark, to ensure that the home buyer does not suffer serious financial loss because a builder has failed to comply with NHBC's rules and technical requirements.

Builders warranty

It is obviously to a builder's advantage to be able to indicate to prospective purchasers that building operations will be carried out in conjunction with NHBC standards and that a warrant will be given stating that the property has been built :

(1) in accordance with the NHBC's requirements;
(2) in an efficient and workmanlike manner and of proper materials, to be fit for habitation.

Buildmark provides safeguards in addition to any contractual, common law or statutory rights that the purchaser may have against the builder.

Phases of Buildmark

Buildmark provides insurance up to £10,000 to safeguard deposits which could be lost through the builder's insolvency or to put right defects during the period of construction. In the first two years after the property has been built the builder must put right any defects which arise as a result of his not having kept to the NHBC standards for materials and workmanship. Where the builder fails to carry out such works, Buildmark provides the right to arbitration and to compensation if the builder is insolvent.

In the longer term, that is the third to the tenth year after construction, Buildmark insures the purchaser against any major damage arising from defect in the load bearing structure or in the render or other external cladding. This does not include ordinary repairs. Cover also extends to subsequent purchasers.

Other warranties

More recently the Municipal Mutual Insurance Company issued a similar scheme of warranty to that of NHBC, launching in England and Wales in 1989 and in Scotland in 1990. The scheme offers a 15 year warranty against major structural defects arising from poor workmanship, poor design or inadequate materials, in addition to the two year developers defect liability period.

Limitation on time for bringing actions

Healing time for personal injuries

There is a limitation on the time for bringing actions of delict in relation to the matter of personal injuries. The Prescription and Limitation (Scotland) Act 1973 provides that no action of damages in respect of personal injuries can be brought unless it has been commenced by service of an initial writ in the Sheriff Court (or summons in the Court of Session) within three years (which is generally considered to be a reasonable time to obtain a full medical discharge from injury) after the date on which the injuries were sustained.

This is a holding device and the action need not be immediately pursued. This further delay is usual where the doctor is unwilling to provide a clean bill of health. Extensions to the three year period are allowed where:

(1) the person is entitled to bring the action under legal disability (being under the age of majority or of unsound mind) and not in the custody of any guardian. In such cases the action can be brought within three years of his reaching 18 years of age or from the time when the unsoundness of mind ended;
(2) where decisive facts were not known to the pursuer until later.

The Law Reform (Miscellaneous Provisions) (Scotland) Act 1980 added another relaxation to allow an action to be brought after the expiry of the three year period if it seems equitable to do so.

References

Marshall, E.A.M., *General Principles of Scots Law*, 5th edition, Part III (delict), (W.Green/Sweet & Maxwell, Edinburgh), 1991

Walker, D.M., *Principles of Scottish Private Law*, 4th edition, Volume II, Book IV, Parts 6 & 7, (Clarendon Press, Oxford)

Gloag, W.M., & Henderson, R.C., *Introduction to the Law of Scotland*, 7th edition, (W. Green & Son Ltd., Edinburgh)

II Technical Matters

Example A1

The building in Figure A1 is classified as purpose group 1 (Dwelling) because the office is secondary to use as a house. Two assumptions should be noted:

A6.2

A Sch.3

(1) the office is used by an occupant of the house;

(2) the floor area of the office is 50 m^2 or less.

If either of these assumptions did not apply, the office and house would have to be treated as separate buildings and the regulations interpreted accordingly.

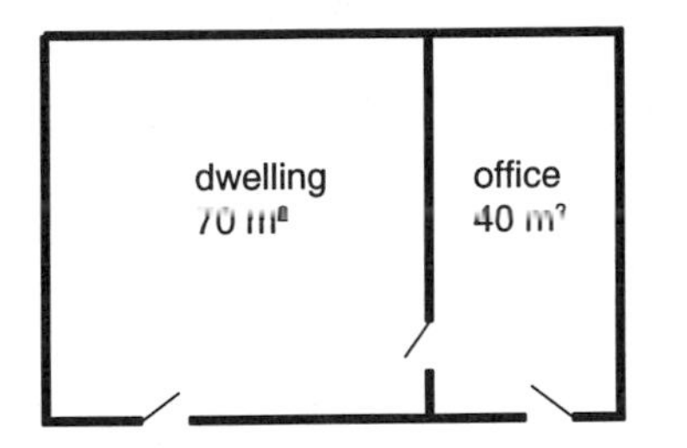

Figure A1 House with office annexe.

Example A2

In Figure A2 the factory shop is ancillary to the main use of the building, which is the manufacturing function of the factory, therefore the classification is purpose group 6 (Industrial). Although the specific limit to floor area only applies to ancillary uses of dwellings, the requirement that the use should be of 'a minor nature' allows for control in practice, as does the need to select the classification with the more onerous requirement for level of performance.

A6.2

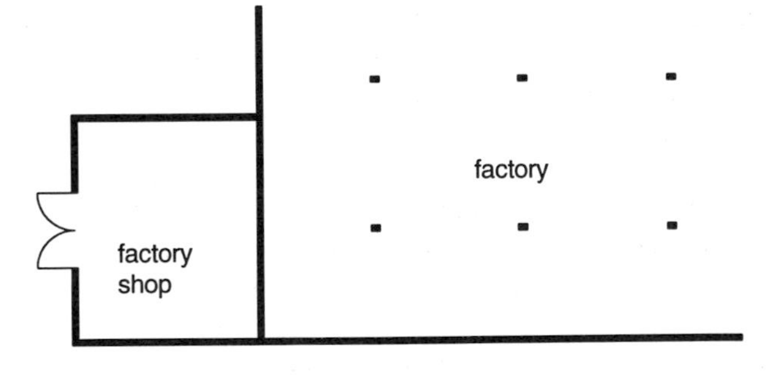

Figure A2 Factory and factory shop.

Example A3

The library, lecture theatres, foyer, administration offices and ward areas of a teaching hospital may be contained within one building such as that shown in Figure A3. The classification for each area taken in isolation is given in brackets. The overall classification of the building as a hospital (purpose group 2A) is obviously inappropriate because the other uses have a significant impact on the number of people, the activities and the contents in the different areas and hence on the required levels of performance. A simple solution in many buildings is to separate the different areas using compartment walls as this allows individual classification of each area. Structural fire precautions may require this anyway in order to limit the floor areas and volume of the building.

A6.3

D.2

A6.4 Compartment walls would not solve the problems for the foyer area where the most onerous level of performance required by any of the possible classifications would have to be applied for each regulation. Layouts of this type present particular difficulty for achieving compliance with the requirements for means of escape and fire-fighting in Part E.

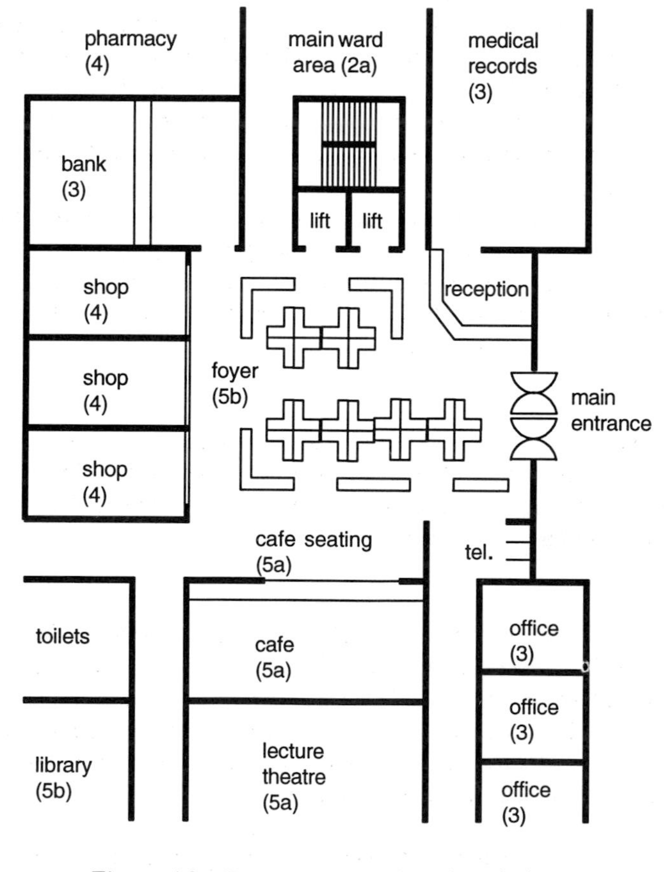

Figure A3 Entrance to teaching hospital.

DETERMINATION OF OCCUPANCY CAPACITY

BUILDING STANDARDS

Regulation

7 (1) For the purposes of these regulations, the occupancy capacity shall be:
a) in the case of a room or space described in column (1) of schedule 4 the number (or in the case of a fractional number the next lowest whole number) obtained by dividing the area of the room or space in square metres by the relevant occupancy load factor specified in column (2) of that schedule; or
b) in any other case the number of persons the room or space is likely to accommodate in use.

(2) The occupancy capacity of a building or storey which is divided into rooms or spaces shall be:
a) in the case of a storey or building of purpose group 2 which is not open to the public and which contains both day rooms and sleeping rooms, the sum of the occupancy capacities of either the day rooms or the sleeping rooms, whichever is the greater; or
b) in any other case the sum of the occupancy capacities of the rooms or spaces.

(3) Where a room or space is likely to be put to more than one use, the greater or greatest relevant occupancy capacity shall apply.

(4) In this regulation 'space' includes a roof or other area open to the external air to which there is access for a purpose other than the maintenance of the building.

Aim

The occupancy capacity of a room, storey or building is an estimate of the maximum number of people which could be present at any time and is used together with the classification in some parts of the Technical Standards to set the appropriate minimum standard required.

Background

The maximum number of people which can occupy any space within a building is basically dependent on what they are doing and the amount of furniture, fixtures or fittings in that space. For different types of spaces and activities the occupancy load factors, based on typical maximum numbers of people per square metre of floor area, have been determined and are given in the schedule to regulation 7 which is summarised on page A12.

The level of performance with respect to safety in case of fire, structural loading, ventilation and the provision of sanitary facilities required by the Technical Standards is dependent on the maximum number of people which could be in the building at any one time. A discotheque with a 700 m^2 floor area would have to provide escape routes which in an emergency could accommodate 1000 people, whereas a museum of the same size would be designed to evacuate only 140 people. While the requirement for the provision of escape routes is less onerous for the museum, the disadvantage is that the maximum number of people that can be allowed into the museum is also limited to 140. This limit could present a major problem should the museum wish to increase revenue by hosting a major exhibition that would attract large crowds.

ASch4

The occupancy capacity therefore not only determines, in certain cases, the required level of performance in the Technical Standards but also sets the safe limit for the number of people in a building at any one time.

Summary of the contents of the schedule to regulation 7

Description of room or space	Occupancy load factor
Standing spectators area	0.3
Amusement arcade, public bar, assembly hall	0.5
Dance hall, concourse, queuing area	0.7
Conference room, restaurant, betting office	1.0
Exhibition hall	1.5
Shops (Class 1):	
supermarkets, department stores, hairdressers etc.	2.0
Art gallery, factory production area, office(<60 m^2)	5.0
Kitchen, library, office(>60 m^2)	7.0
Shops (Class 2):	
wholesalers, furniture or other shops selling bulky goods	7.0
Bedroom or study bedroom	8.0
Bed-sitting room, billiards room	10.0
Car parks, warehouses	30.0

Application

The occupancy capacity of a room, or space, is calculated by dividing the floor area by the appropriate occupancy load factor from the schedule. The occupancy capacity of a storey or building is the sum of the occupancy capacities of the rooms, or spaces, within that storey or building. For rooms or spaces not specified in the schedule it is necessary to estimate by other means the maximum number of people likely to use the space at any one time. In most cases it is relatively straightforward. For theatres or cinemas the occupancy capacity can be based on the number of fixed seats and is therefore not difficult to estimate. In other cases, such as an open plan discotheque with bars and seating at different levels, the occupancy capacities for each area should be calculated separately and then aggregated. However, where areas overlap without a clearly established division, or where a room or space could be used for more than one activity, then the greater occupancy capacity will apply.

A7.1
A7.2
A7.1

To avoid double counting for institutional and residential buildings which have day rooms in addition to sleeping accommodation, the occupancy capacity should be based on either one or the other. However, if the public can visit the building, allowance will have to be made for the additional numbers. For the same reason corridors and toilets are not included in the calculation of occupancy capacity.

A7.3

A7.2

Particular care should be taken when refurbishing an existing building as apparently minor alterations to the structure can have a significant impact on the occupancy capacity and require considerable additional work to be undertaken. This is illustrated by the following example.

Example A4

The plan in Figure A4 shows a first floor restaurant and bar with snooker room all under the one ownership, although separated into different rooms. The occupancy capacity is 609, calculated as follows:

ASch4

Room	Floor area (m^2)	Occupancy load factor	Occupancy capacity
Kitchen	208	7	29
Restaurant	250	1	250
Snooker room	100	10	10
Bar	160	0.5	320
Total occupancy capacity			609

The owners decide to refurbish the premises and change the style to a cafe/bistro with a more open plan. This is to be achieved by removing the two non-structural walls (see Figure A4) and extending the area with tables and chairs. Although the total floor area remains the same, there are more tables and chairs, and the occupancy capacity increases to 1049. This is because the occupancy load factor applied to the 510 m^2 of bistro is 0.5, the figure for a bar. This is significantly more onerous than the figure for a restaurant and is used because in a bistro/bar people do not have to have a seat in order to use the bar. The cost, and difficulty, of providing adequate escape routes alone may well offset the benefit of accommodating more customers. In this situation it would not be possible to avoid meeting the requirements for the larger number of people even if there was no intention of ever allowing the premises to be overcrowded.

ASch4

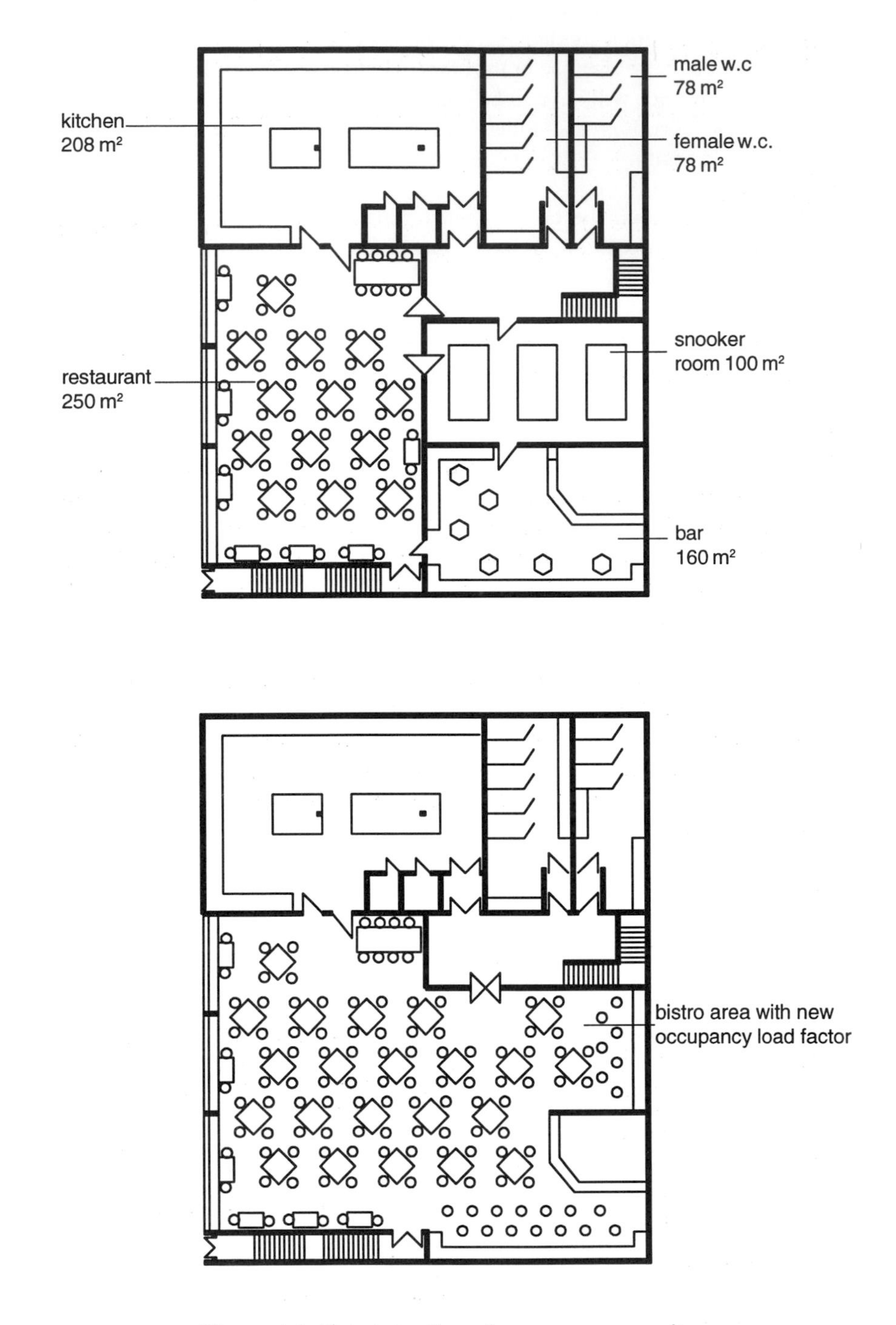

Figure A4 Determination of occupancy capacity.

MEASUREMENT OF BUILDINGS

BUILDING STANDARDS

Regulation

8 For the purposes of these regulations, measurements shall be made or calculated in accordance with schedule 5.

Aim

To provide a standard method for determining the lengths, areas and volumes of buildings for use in the Technical Standards.

Background

The particular requirements for both safety and quality of environment specified in the Technical Standards are, in some cases, based both on the classification of the building and on limiting values for height, area or volume. The height of the storey of a building, for instance, is used in the Small Buildings Guide of Part C to determine the minimum compressive strength of blockwork for a structural wall.

To ensure consistency in the determination of lengths, areas and volumes, the schedule to regulation 8 gives the rules to be applied when taking the measurements required. These rules are explained diagrammatically. Additional rules for the measurement of heights which are given in the Small Buildings Guide to Part C of the Technical Standards are illustrated in Part C rather than here.

Application

Measurement of areas

Area of room

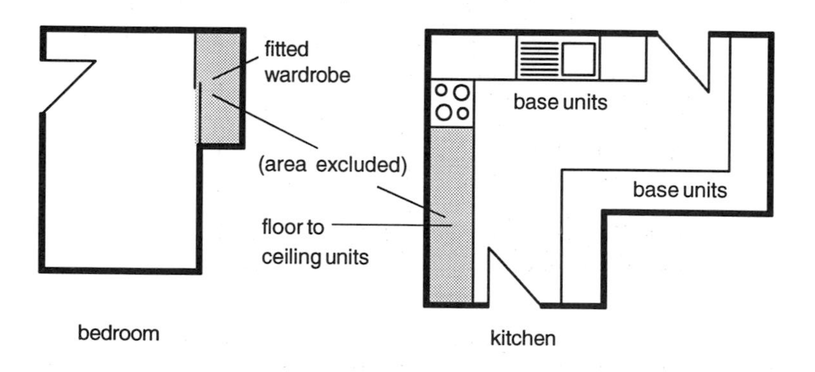

Area of storey/compartment

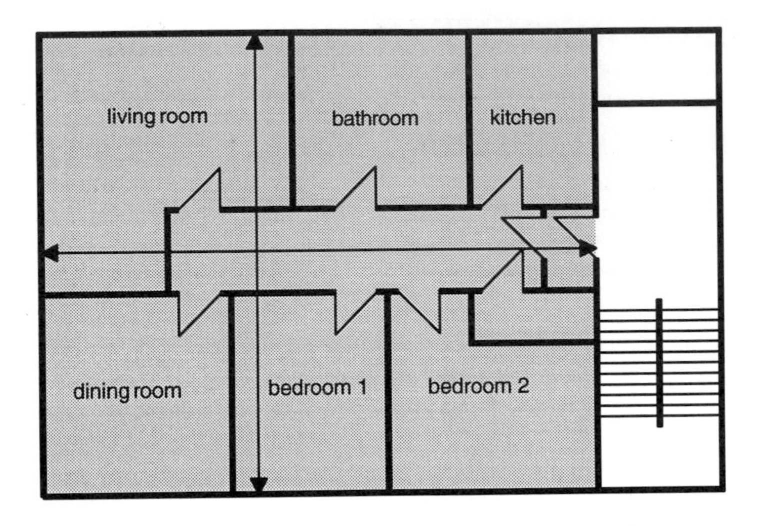

Measurement of cubic capacity

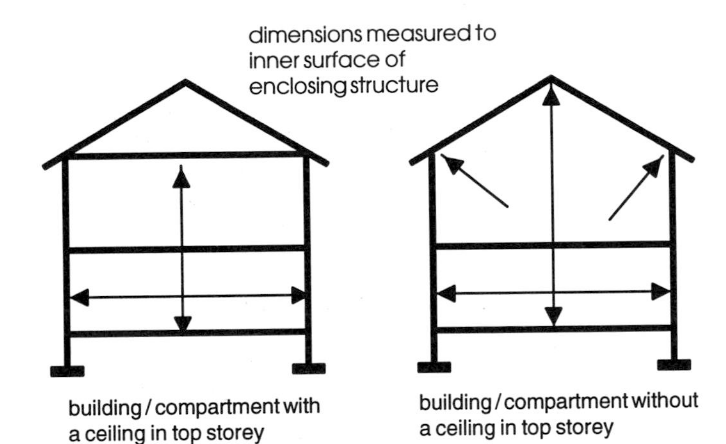

Measurement of height (level ground)

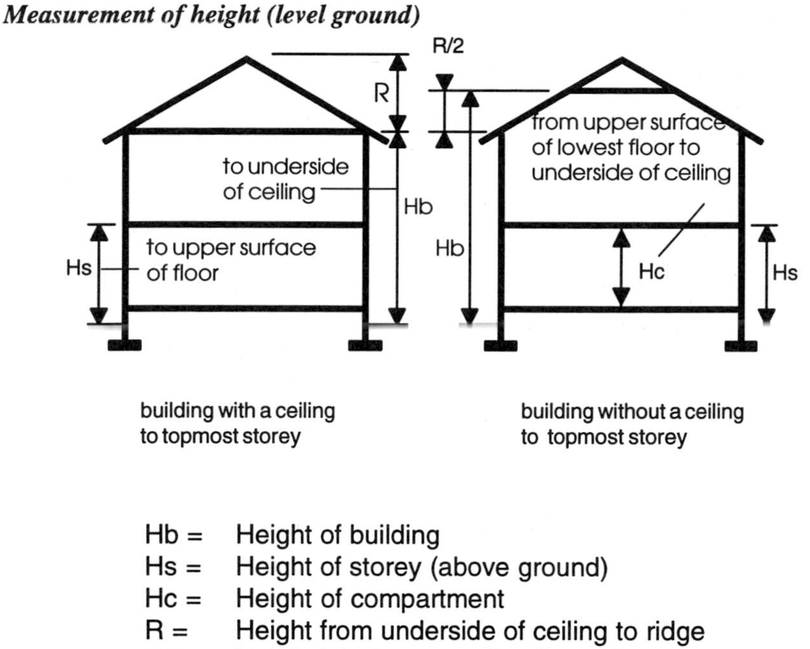

Hb = Height of building
Hs = Height of storey (above ground)
Hc = Height of compartment
R = Height from underside of ceiling to ridge
R/2 = Half height from top of wall to ridge.

Measurement of height (sloping ground)

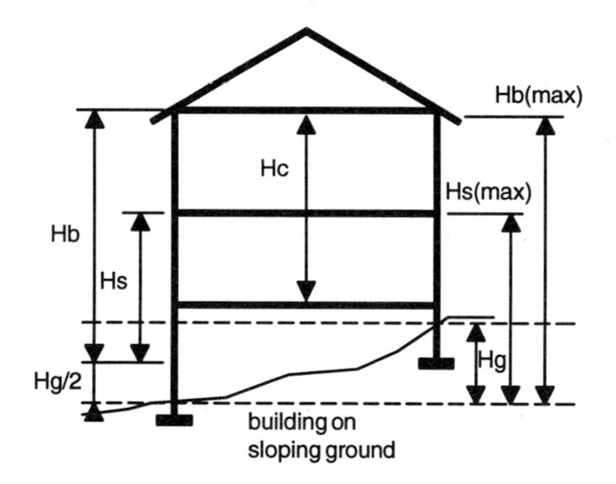

For Hg > 2.5 m
Hb = Height of building
Hs = Height of storey (above ground)
Hc = Height of a compartment
For Hg not > 2.5 m
Hb(max) = Height of building
Hs(max) = Height of storey (above ground)
Note: Hg difference in ground level on either side of building

NB. Additional rules for measurement of height are illustrated within Part C

COMPLIANCE WITH BUILDING STANDARDS

Regulation

9 (1) The requirements of these regulations 10 to 33 shall be satisfied only by compliance with the relevant standards.

(2) Without prejudice to any other method of complying with a relevant standard, conformity with provisions which are stated in the Technical Standards to be deemed to satisfy that standard shall constitute such compliance.

Aim

The regulation sets out the ways by which the requirements of the regulations can be satisfied and defines the relationship with the Technical Standards.

Background

The standards prescribed for buildings in the 1990 Regulations are statements of requirements which are supported by the Technical Standards produced by the Scottish Office. The Technical Standards are not statutory instruments in their own right. However 'relevant standard' is defined under regulation 2 as 'a standard set out in the Technical Standards' and therefore by virtue of clause (1) of regulation 9 the Technical Standards have statutory force. They provide the only means of satisfying the requirements of the regulations. The separation of the relevant standards from the statutory instrument allows changes to be more rapidly brought into force as amendments to the Technical Standards do not have to be approved by Parliament.

Application

To comply with the regulations the performance requirements set out in the Technical Standards must be satisfied either by:

(1) meeting the requirements which are stated in the Technical Standards to be deemed to satisfy the relevant standards (for those standards which have such provisions); or

(2) demonstrating compliance with the relevant standard by any other means.

Proposed Changes to Part A

These are in some measure of a harmonising nature with other UK regulations but also as a need to fine tune interpretative requirements and to consolidate content.

It is proposed to make adjustments to regulation 3 to remove the exemption from control from certain classes of building and extensions in situations where dangerous site conditions are present.

The reason for this is that some concern is being felt that control over ingress of dangerous substances into buildings via Part G might be compromised by existing exemptions under regulation 3.

FITNESS OF MATERIALS

PART B — Regulation 10

BUILDING STANDARDS

Regulation

10 Materials, fittings, components and other manufactured products used to meet a requirement of these regulations shall be suitable for the purpose for which they are so used and shall be used so as to comply with the requirements of these regulations.

Aim

Materials and components for use in the construction of buildings should be:

(1) selected on the basis of proven ability to meet the performance requirements specified in the regulations either for the materials and components themselves or for the elements of the building;
(2) correctly used or applied to ensure the potential performance is achieved;
(3) sufficiently durable to continue to meet the specified performance requirements for the life of the building, taking into account normal maintenance practice.

TECHNICAL STANDARDS

Scope

The requirements apply to all buildings, but only to materials, fittings, components and other manufactured products for which a specified performance is given in the regulations. **B1.2**

Requirements

Selection and use of materials and components

The selection and use of materials, components and fittings must be supported by evidence of suitability with respect to specified performance requirements. **B2.1** Where maintenance or periodic renewal is necessary to maintain satisfactory **B2.2** performance then the design and construction of the building must allow this to be reasonably practicable.

The workmanship and methods of construction, or installation, of materials and components must be in accordance with the methods and standards applied to assess their suitability.

Background

The performance based approach of this part of the Standards allows the evaluation and selection of new products, and new methods of use, against the same criteria as existing, or traditional, materials and products. This is intended both to assist innovation and to allow the use of materials and products imported from other European Community member states in compliance with the requirements of the European Commission Directive on the 'open market'. Imported materials and products must be accepted as suitable for use in a particular situation if:

(1) they comply with the standards of another member state which are deemed to be 'equivalent' to the standards for testing and specifying the same types of materials and products in the United Kingdom;
(2) they are labelled with a CE mark, and therefore comply with the Essential Requirements of the Construction Products Directive;
(3) they are covered by a European Technical Approval issued under the Construction Products Directive.

In practice a difficulty for building control officers will be whether or not different national standards and tests are 'equivalent' to those applied in the United Kingdom. Therefore until the Construction Products Directive is fully supported by harmonised standards and technical guidelines it is probable that manufacturers and specifiers will continue to rely on existing methods to demonstrate compliance with the Technical Standards.

Where materials or products carry the CE mark, are undamaged and are being used appropriately they can only be rejected by building control officers on the basis of failure to comply with the technical specification. The onus of proof in such cases will rest with the building control officer. However action can only be taken by the Trading Standards Authority and therefore an effective working relationship between Building Control and Trading Standards will be necessary.

COMPLIANCE

Introduction

B2.2 The suitability of materials, fittings and components can be established by any of the methods which follow, but it is important in all cases to ensure that the workmanship and methods of construction, or installation, used on site are the same as those specified, or used, to demonstrate suitability. In addition the design and construction must be such as to ensure that where maintenance, or periodic renewal, is necessary to maintain performance at a specified level the necessary tasks are reasonably practicable to carry out.

Provisions deemed-to-satisfy

Selection and use of materials and components

DTS B2.1 Compliance of materials, fittings, components or parts thereof can be demonstrated in a number of ways including their manufacture in accordance with:

(1) the standard (BS or otherwise) specified in the Technical Standards;
(2) a relevant standard or CP of a national standards institution (or equivalent) of any state in the European Economic Area;
(3) a relevant international standard recognised in any state within the EEA;

(4) a relevant specification acknowledged for use as a standard by a public authority of any state within the EEA;
(5) traditional procedures of manufacture of a state within the EEA where these are the subject of a written technical description sufficiently detailed to permit assessment of the materials, fittings, components or other products for the use specified;

and for products of a more innovative nature;

(6) a European Technical Approval issued in accordance with the Construction Products Directive (89/106/EEC); or

a specification sufficiently detailed to permit assessment.

In making assessment of fitness for purpose the Essential Requirements of Annex A of CPD (89/106/EEC) must be borne in mind, including consideration of Mechanical Resistance and Stability, Safety in Case of Fire, Hygiene, Health and Environment, Safety in Use, Protection Against Noise and Energy Economy and Heat Retention.

Assessment may involve testing, and where this is the case such tests should be carried out by an appropriate organisation, e.g. one accredited in accordance with BS 7501-BS 7503:1989 and BS 7511-BS 7514:1989 (or EN45000).

Products which comply with an EC Directive relating to a specific purpose do not have to comply with any specific standards or CP referred to in the Technical Standards, i.e. the EC Directive takes precedence. An example of this would be a self-contained smoke alarm manufactured in accordance with EC Directive 89/336/EEC which would be deemed to satisfy the requirements of the relevant regulation as it relates to the prevention of electromagnetic disturbances by and protection against disturbances to such smoke alarms.

STRUCTURE

PART C Regulation 11

BUILDING STANDARDS

Regulation

11 (1) Every building shall be so constructed that the combined loads to which the building may be subject are sustained and transmitted to the ground without impairing the stability of the building.

(2) Subject to paragraph (3) every building of five or more storeys shall be so constructed that in the event of damage occurring to any part of the structure of the building the extent of any resultant collapse will not be disproportionate to the cause of the damage.

(3) In calculating the number of storeys for the purpose of paragraph (2), no account shall be taken of any storey within a roof space where the slope of the roof does not exceed 70° to the horizontal.

Aim

The construction of a building must ensure that:

(1) the structure can safely carry all dead, imposed and wind loads;
(2) the foundations can transmit all loads to the ground without affecting the stability of the building by excessive settlement;
(3) collapse caused by damage to any part of the structure can be contained within a limited section of the building (this is focused on buildings of five or more storeys).

TECHNICAL STANDARDS

Scope

The requirements apply to all buildings.

Requirements

The following criteria must be taken into account when assessing the safety of the structure of a building:

(1) loading;
(2) properties of materials;
(3) structural analysis;
(4) details of construction;
(5) safety factors;
(6) workmanship;
(7) nature of the ground.

C3.1 Buildings of five or more storeys should be designed and built to withstand disproportionate collapse.

C2.1
C2.2 These requirements will be met by foundations and structures designed and constructed in accordance with the appropriate British Standards codes of practice.

The European Standards Organisation, CEN, have asked national standards organisations, (such as BSI), to make available the Pre-standard (ENV) Structural Eurocodes for trial use and comment prior to the development of the final European standards (EN).

In this context, DD ENV 1992-1-1; 1992 Eurocode 2; Part 1 together with DD ENV 1992-1-1; 1992 Eurocode 3; Part 1.1 *General Rules and Rules for Buildings in concrete and steel* have been thoroughly examined over a period of several years and are considered to provide appropriate guidance when used in conjunction with their National Application Documents for such buildings.

When ENVs are converted to ENs they are likely to be referenced in future Part Cs as offering an alternative means of compliance with the relevant standards in Part C.

For small buildings of traditional masonry construction simpler methods for ensuring compliance are set out in the Small Buildings Guide, now in its second edition, which is published separately from the Technical Standards. The guide is in five sections as follows:

Section 1: basic requirements for stability;
Section 2: strip foundations of plain concrete;
Section 3: thicknesses of walls in certain small buildings;
Section 4: proportions for masonry chimneys;
Section 5: sizes of certain timber floor and roof members.

Background
The British Standard codes of practice are presently the most relevant standards for the design and construction of safe buildings and continue to be specified within the provisions deemed to satisfy section C2.1 DTS of Part C.

Self-certification for structural design carried out by qualified professionals for building warrant purposes was introduced in 1992 and this system appears to be operating satisfactorily.

The scope for the design of structural elements by others is now effectively limited to traditional masonry constructions which lie within the scope of the Small Buildings Guide. The application of the guide is limited to a narrow range of carefully defined situations which are common.

Despite the intention to allow non-professional designers to use custom and practice, certain sections of the guide require the user to have experience in structural design.

For example, when sizing plain concrete strip foundations an assessment is required not only of sub-soil conditions but also of movement due to seasonal weather change and the existence of weaker sub-strata. Also, the footnote to the diagram showing permissible openings in buttressing walls states that such openings must not impair stability, which implies that other considerations must be taken into account. Such factors coupled with the introduction of self-certification may limit in practice the use of the guide.

COMPLIANCE

The standards and deemed-to-satisfy provisions require that loading should be determined, and the structure of buildings designed and constructed either:

(1) in accordance with the relevant British Standards; or
(2) in accordance with the provisions of the Small Buildings Guide (1994 edition) which applies to buildings of traditional masonry construction meeting the following criteria:
  (a) of purpose group 1 of not more than three storeys; or
  (b) annexes up to a height of 3 m to buildings of purpose group 1 including garages and outbuildings; or
  (c) single storey non-residential buildings of other purpose groups except that the sizes of certain timber floor and roof members in Section 5 of the guide shall only apply to dwellings of purpose sub-groups 1B and 1C.

The relevant British Standards cited in (1) above (discussion of which is beyond the scope of this book) are:

Calculation of loads

C2.2

(1) (a) for dead loads and imposed loads, other than imposed loads on a roof and wind loads, BS 6399: Part 1: 1984;
  (b) for imposed loads on a roof, other than wind loads, BS 6399: Part3: 1988;
  (c) for wind loads, CP 3: Chapter V: Part 2: 1972;
(2) for any building for agricultural use, BS 5502: Part 22: 1987

Construction

DTS C2.1

(1) foundations, general, BS 8004: 1986;
(2) structural work of reinforced, prestressed or plain concrete, BS 8110: Part 1, Part 2 and Part 3: 1985;
(3) structural work of composite steel and concrete construction, BS 5950: Part 3: s. 3.1: 1990 and BS 5950: Part 4: 1994;
(4) structural work of steel:
  (a) BS 449: Part 2: 1969; or

(b) BS 5950: Part 1: 1990
BS 5950: Part 2: 1992
BS 5950: Part 5: 1987
BS 5950: Part 8: 1990;

(5) structural work of aluminium, BS 8118: Part 1: 1991 and BS 8118: Part 2: 1991 (NB. for purposes of s. 7.2 of Part 1 of that code, the structure must be classified as a safe-life structure);
(6) structural work of masonry, BS 5628: Part 1: 1992, as read with BS 5628: Part 3: 1985;
(7) structural work of timber, BS 5268: Part 2: 1991 and BS 5268: Part 3: 1985.

DTS C3.1 ***Disproportionate collapse***

(1) for structural work of reinforced, prestressed or plain concrete, BS 8110: Part 1 and Part 2: 1985;
(2) for structural work of steel, BS 5950: Part 1: 1990, in which case the accidental loading referred to in paragraph 2.4.5.5 of that standard is to be chosen having particular regard to importance of the key element and the consequence of failure. The key element is always to be capable of withstanding 34 kN/m^2 applied from any direction;
(3) for structural work of masonry, BS 5628: Part 1: 1992.

Small Buildings Guide

The requirements of the five sections of the provisions deemed-to-satisfy the requirements of Part C for small buildings are as follows:

Section 1: Basic requirements for stability

The building can be considered to have adequate stability where:

(1) the walls are designed and restrained in accordance with the requirements of Section 3 and
(2) the roof structure is braced as recommended in BS 5628: Part 3: 1985 and anchored to the walls in accordance with the requirements of section 3. A framed roof with lipped returns and rigid sarkings will have sufficient stability without the need for additional bracing.

Section 2: Strip foundations of plain concrete

The minimum width of a strip foundation in plain concrete for given sub-soil type and condition, and wall loadings is set out in Table C1. Additional conditions which must be met to comply with the requirements of this section are:

Conditions related to the subsoil

(1) there should be no made ground, peat or wide variation in type of subsoil within the loaded area;
(2) there should be no weaker type of soil below the soil on which the foundation rests which could impair the stability of the structure;

(3) the possibility of soil movement due to seasonal weather change should be taken into account when designing the foundations.

Design provisions

(1) The cement in the concrete should comply with BS 12: 1991 and the aggregate with BS 882:1983.
The mix proportion should be either:
- grade ST1 to BS 5328: Part 2: 1991 or,
- 50 kg of cement to a maximum of 0.1 m³ of fine aggregate and 0.2 m³ of coarse aggregate. **DTS C2.1**

(2) Foundation dimensions

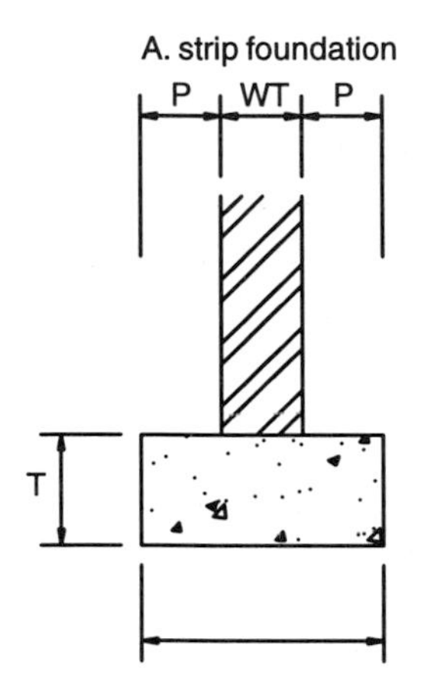

foundation width must not be less than the appropriate dimension in the table C1

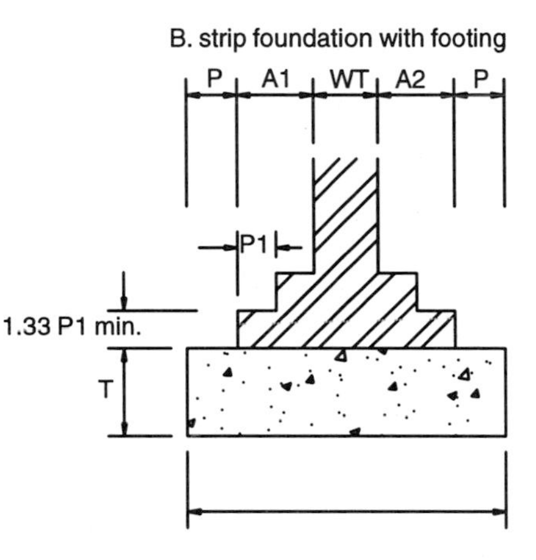

foundation width must not be less than the appropriate dimension in the table C1 plus offset dimensions A1 and A2

(3) Steps in foundations

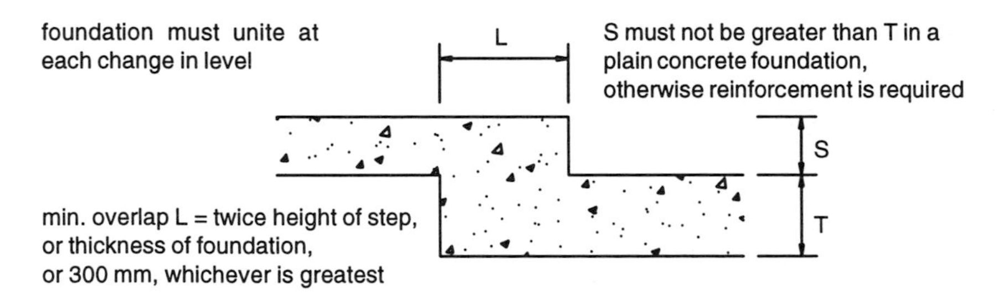

(4) Foundations for projecting piers and chimneys

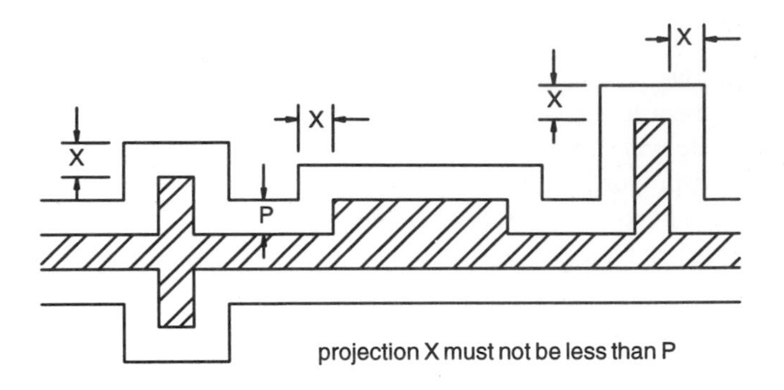

The use of the Table C1 to determine the minimum width of strip foundations is straightforward but the validity of the output depends on the accurate assessment of subsoil conditions. The guide outlines field tests for each soil type; however the evaluation of ground condition should be carried out by an experienced professional.

Although permissible under this section the use of unreinforced concrete foundations for long, low buildings in clay soils is inadvisable.

Table C1 Minimum width of strip foundations

		total load of load-bearing walling not more than (kN linear metre)					
		20	**30**	**40**	**50**	**60**	**70**
Type of subsoil	**Condition of subsoil**	**Minimum width of strip foundation (mm)**					
I rock	not inferior to sandstone, limestone or firm chalk	in each case equal to the width of wall					
II gravel sand sand/gravel mixtures	compact compact compact	250	300	400	500	600	650

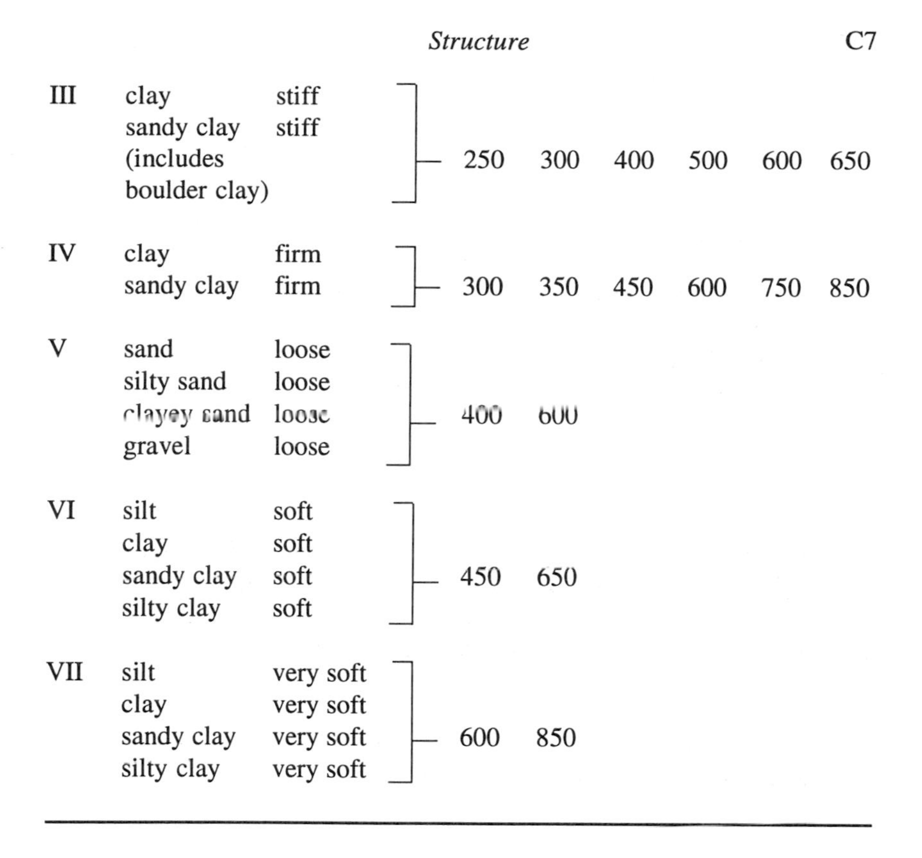

III	clay sandy clay (includes boulder clay)	stiff stiff	250	300	400	500	600	650
IV	clay sandy clay	firm firm	300	350	450	600	750	850
V	sand silty sand clayey sand gravel	loose loose loose loose	400	600				
VI	silt clay sandy clay silty clay	soft soft soft soft	450	650				
VII	silt clay sandy clay silty clay	very soft very soft very soft very soft	600	850				

Section 3: Thickness of walls in certain small buildings

The requirements for structural walls in terms of thickness, permissible number of openings, restraint for stability, materials and workmanship are set out in this section. The application of the section is limited by constraints relating to the building, the loading conditions and the wall as follows:

Limitations relating to the building

Size and proportion of residential buildings

For residential buildings of not more than three storeys:

(1) the maximum height of the building measured from the lowest finished ground level adjoining the building to the highest point of any wall or roof must not be greater than that specified in Table C2 and C3 correlating basic wind speed and site exposure;

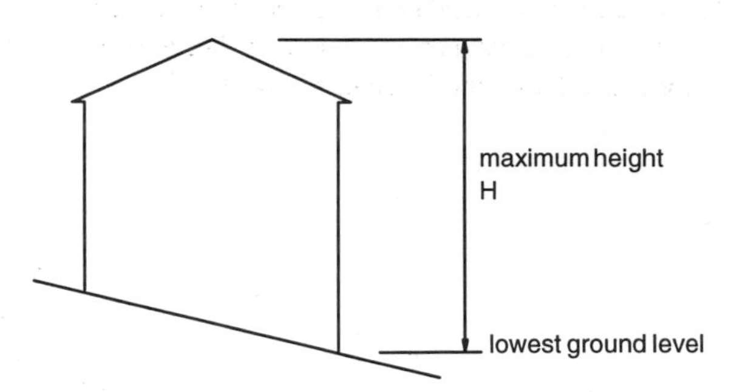

(2) the height of the building H, shall not exceed twice the least width of the building W;

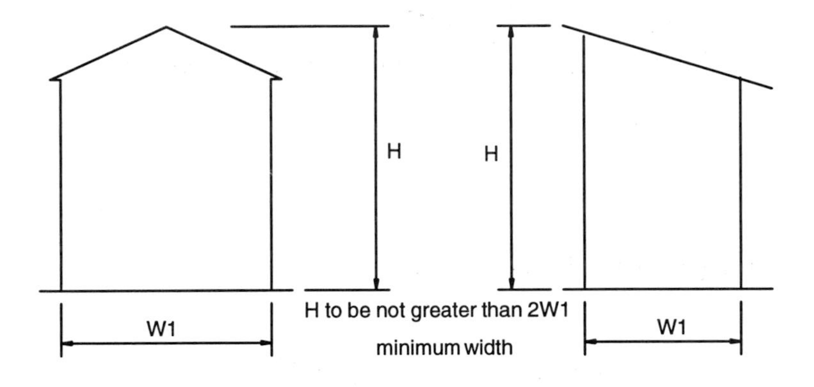

(3) the height of the wing H2 shall not exceed twice the least width of the wing W2 when the projection P exceeds twice the width W2.

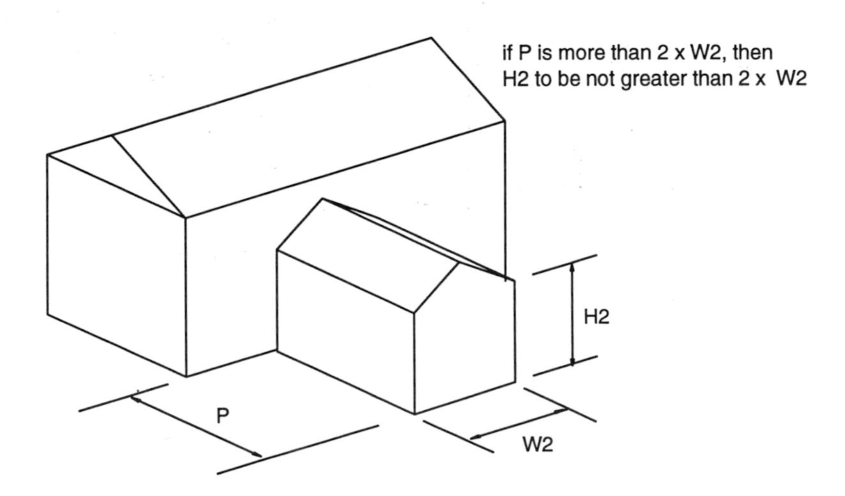

Size and proportion of non-residential buildings

For small single storey non-residential buildings the height H of the building shall not exceed 3 m and W shall not exceed 9 m.

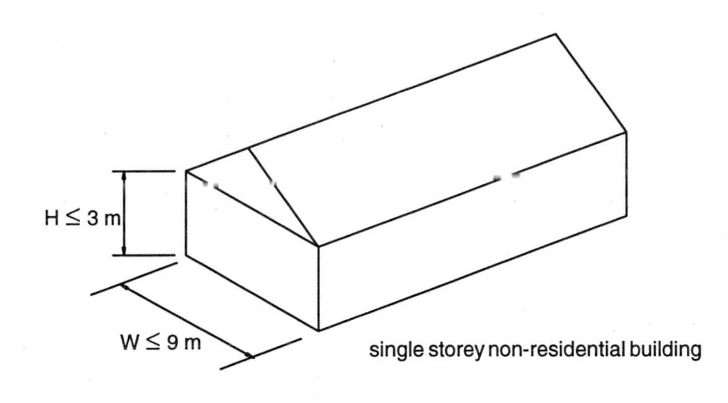

Size of annexe to residential building

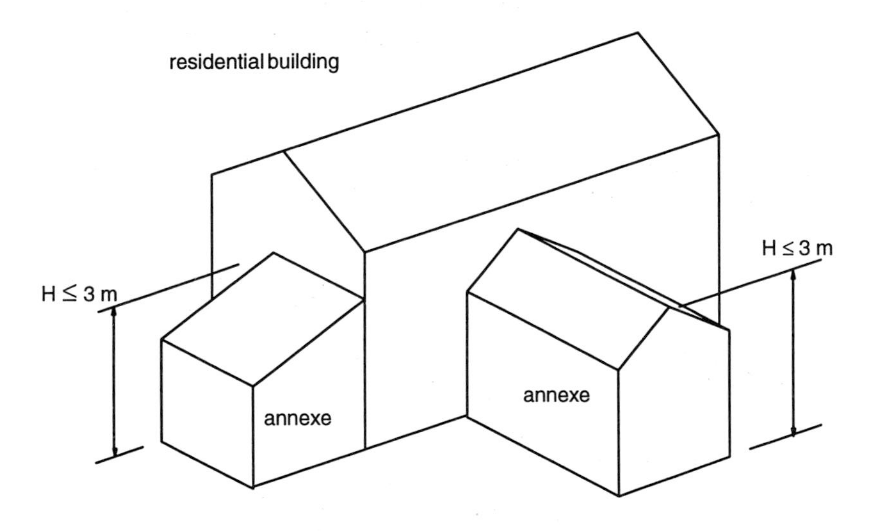

Table C2 Maximum heights of buildings on normal level sites

Basic wind speed m/s	Maximum building height (metres)			
	Location			
	Unprotected sites, open countryside	Open countryside with scattered windbreaks	Country with many windbreaks, small towns, city outskirts	Protected city centres
48	8.75	12.9	15	15
50	7.1	10	15	15
52	5.8	8.9	15	15
54	4.6	7.9	14.4	15
56	3.4	6.8	12.8	15

Note: Table applies to normal level or slightly sloping sites. For further guidelines see CP3: Chapter V: Part 2.

Table C3 Maximum heights of buildings on steeply sloping site

Basic wind speed m/s	Maximum building height (metres)			
	Location			
	Unprotected sites, open countryside	Open countryside with scattered windbreaks	Country with many windbreaks, small towns, city outskirts	Protected city centres
48	0	3	6.3	14
50	0	0	5	12
52	0	0	4	10
54	0	0	3	8.5
56	0	0	0	7

Note: Table applies to very exposed hill slopes and crests. For further guidance see CP3: Chapter V: Part 2.

Dead and imposed loads

Table C4 Maximum permissible loads

Element	Maximum loading	
Roofs	distributed load:	1.00 kN/m^2 up to 12 m span
		1.50 kN/m^2 up to 6 m span
Floors	distributed load:	2.00 kN/m^2
Ceilings	distributed load:	0.25 kN/m^2 together with
	concentrated load:	0.9 kN

Note: Total load at the base of the wall must not exceed 70 kN/m. There must be no lateral loads on the building apart from wind loads.

Limitations relating to the walls

Wall types

Residential buildings of up to three storeys of purpose group 1

external walls internal loadbearing walls compartment walls separating walls	(section applies to walls extending to full storey height)

Small single storey non-residential buildings and annexes to buildings of purpose group 1

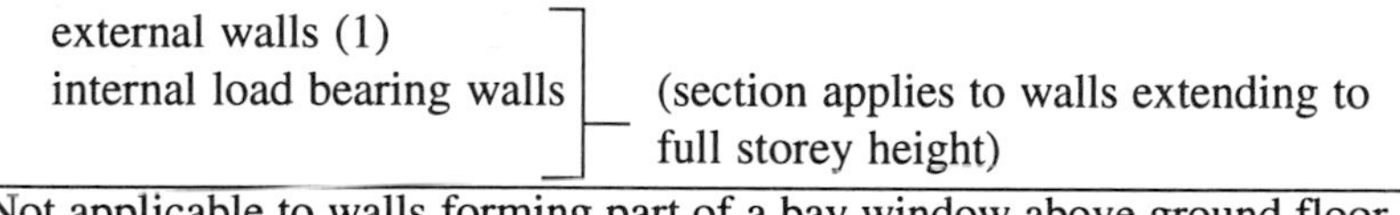

external walls (1) internal load bearing walls	(section applies to walls extending to full storey height)

Not applicable to walls forming part of a bay window above ground floor sill level.

Wall length and height

The maximum length of wall measured between centres of restraints is 12 m and the maximum height of walls is 12 m.

Differences in ground or construction level

The maximum difference in ground or construction level across a wall is four times the total wall thickness.

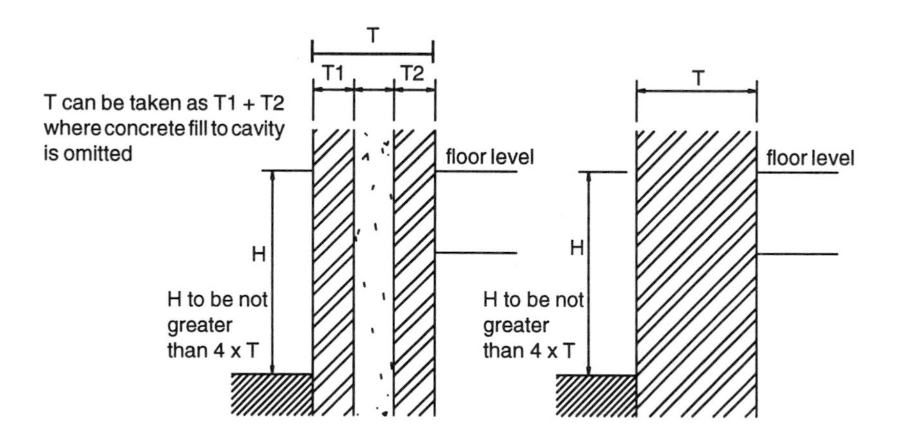

Limitations relating to the loading conditions

Maximum span of floors

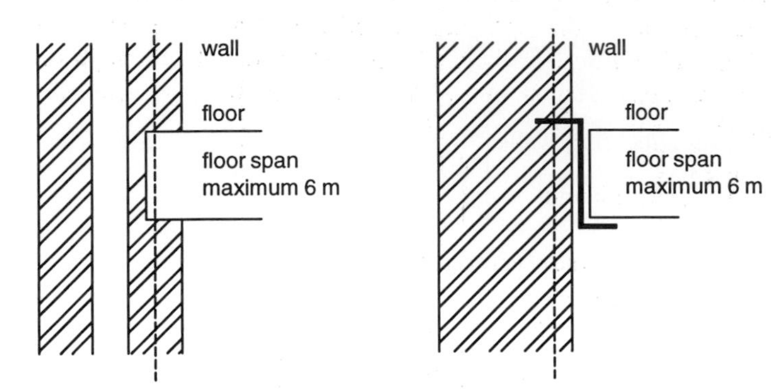

floor member bearing on a wall

floor member bearing on a joist hanger

Maximum floor area

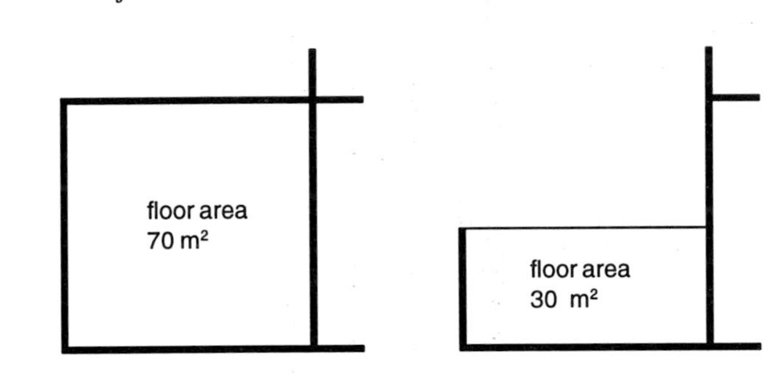

Lintel bearing length

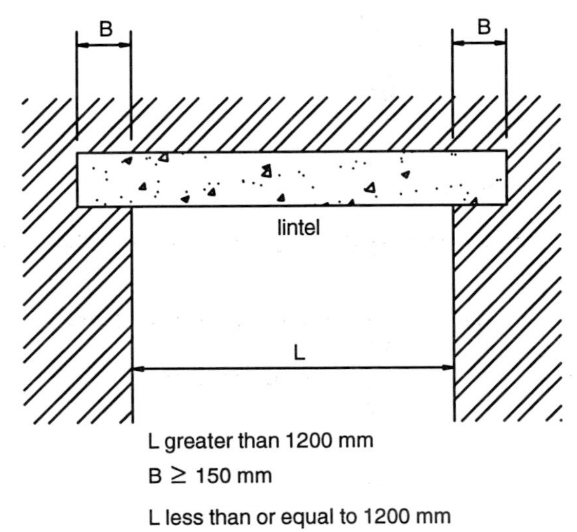

L greater than 1200 mm

B $\geq$ 150 mm

L less than or equal to 1200 mm

B $\geq$ 100 mm

for lintels carrying a concrete floor

B equal to the greater of 150 mm or L/10 (mm)

Determination of wall thickness

Wall thickness

The minimum wall thickness is dependent on the type of construction and the height and length of the wall, determined from Table C6. The method of measuring wall and storey height is illustrated below. Wall length should be measured between centres of restraints (e.g. buttressing walls, piers and chimneys).

Measuring wall height

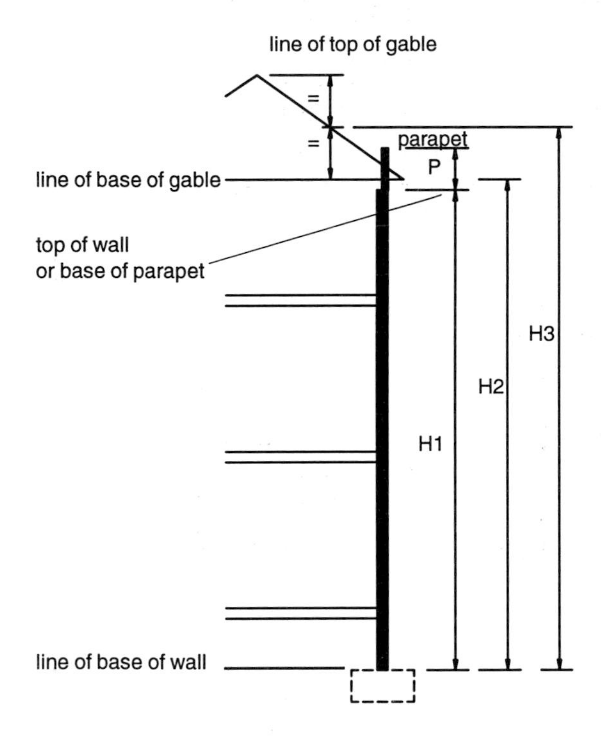

Wall height

H1 is the height of a wall that does not include a gable

H2 is the height of a compartment or a separating wall which may extend to the underside of the roof

H3 is the height for a wall (except a compartment or separating wall) which includes a gable

P if the parapet height is more than 1.2 m add the height to H1.

Measuring storey height

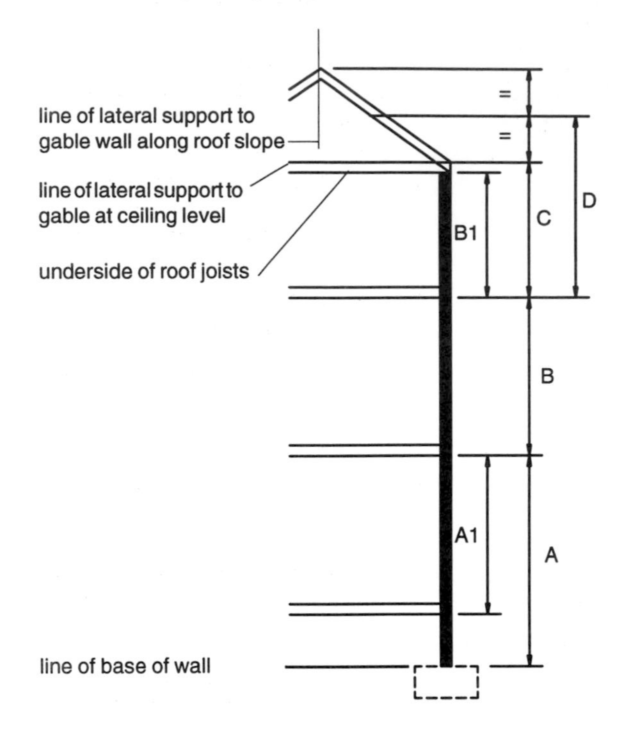

A is the ground storey height if the ground floor is a suspended timber floor or a structurally separate ground floor slab

A1 is the ground storey height if the ground floor is a suspended concrete floor bearing on the external wall

B is the intermediate storey height

B1 is the top storey height for walls which do not include a gable

C is the top storey height where lateral support is given to the gable at both ceiling level and along the roof slope

D is the top storey height for walls which includes a gable where lateral support is given to the gable only along the roof slope

Table C6 Minimum wall thickness

Type of construction	Wall height	Wall length	Minimum thickness
Solid walls in coursed brickwork or blockwork (external, compartment, separating walls)	1. All cases	All cases	1/16 of storey height
	2. ≤3.5 m	≤12 m	190 mm
	3. >3.5 m & ≤9 m	≤9 m	190 mm
		>9 m & ≤12 m	290 mm (1st storey) 190 mm(other storeys)
	4. >9 m & ≤12 m	≤9 m	290 mm (1st storey) 190 mm(other storeys)
		>9 m & ≤12 m	290 mm (2 storeys) 190 mm(other storeys)
Solid walls in uncoursed stone or vitrified material	The thickness determined as for 1-4 above multiplied by 1.33		
Cavity walls in coursed brickwork or blockwork	All cases	All cases	each leaf at least 90 mm cavity at least 50 mm.
For external, compartment and separating cavity walls	Total thickness determined as for 1-4 above should not be less than or equal to the thickness of both leaves + 10 mm.		
Internal loadbearing walls in brickwork or blockwork (not compartment or separating wall)	All cases	All cases	half thickness determined in 1-4 above less 5 mm
Lowest storey of three storey building carrying load from both upper storeys			140 mm
Single leaf external walls (single storey non-residential buildings)			90 mm (N.B. restraint criteria)

Notes:
Where bricks or blocks with modular dimension to BS 6750: 1986 are used wall thickness may be reduced by a maximum of the deviation from work size permitted by a British Standard relating to the equivalent sized bricks or blocks of the same material.
A wall must always be at least as thick as any part of a wall to which it gives vertical support.

Materials and workmanship

Compressive strength of bricks and block

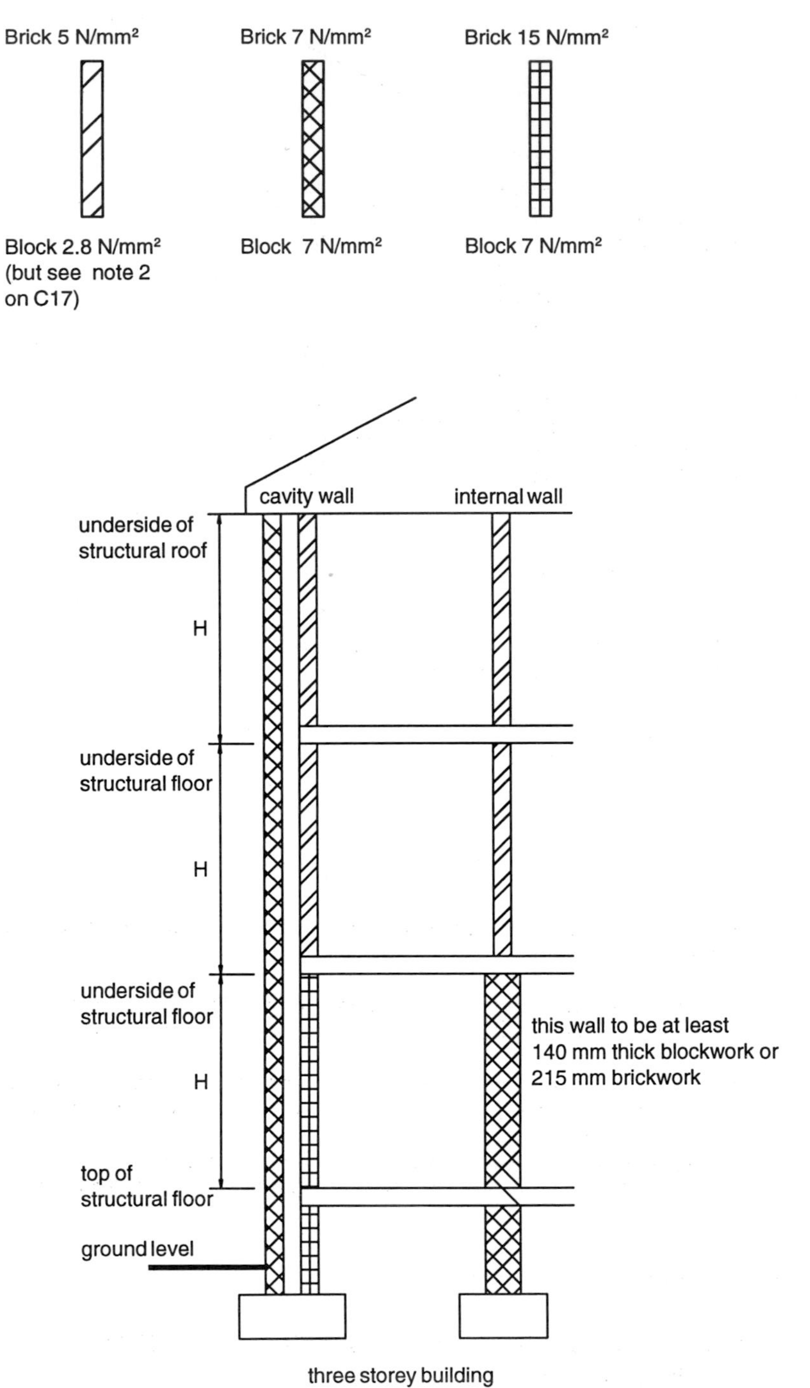

three storey building

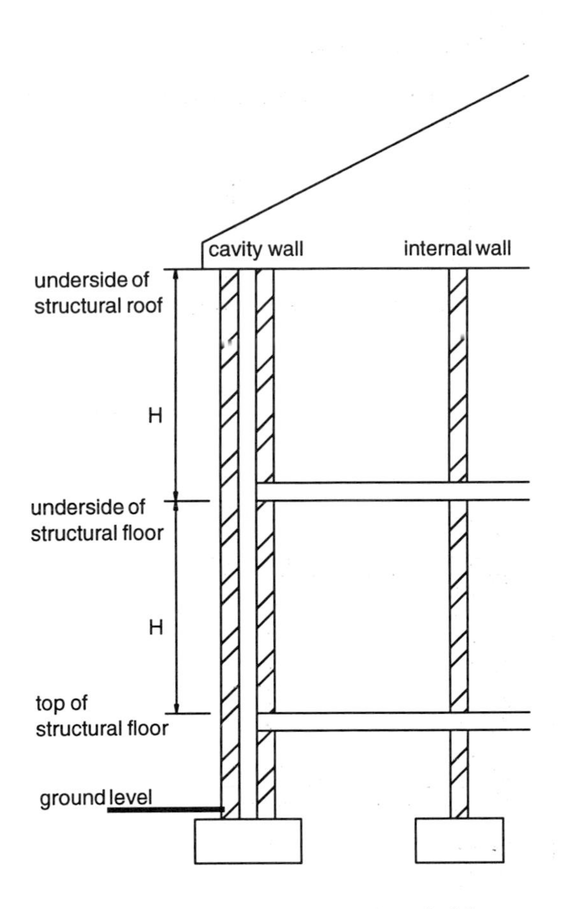

one or two storey building

1. If H is not greater than 2.7 m, the compressive strength of bricks or blocks must be used in walls as indicated in the key.

2. If H is greater than 2.7 m, the compressive strength of bricks or blocks used in the wall must either be 7 N/mm^2 or as indicated by the key, whichever is the greater.

3. If the external wall is solid construction, the bricks or blocks must have a compressive strength of at least that shown for the internal leaf of a cavity wall in the same position.

4. This diagram shall only be used to determine the compressive strength of brick and block units for walls of two and three storey buildings where the roof construction is of timber.

Mortar

Mortar must be:

(1) to the proportions given in BS 5628: Part 1: 1992 for mortar designation (iii);
(2) of equivalent or, where appropriate, of greater strength, compatible with the masonry units and position of use.

Cavity wall ties

The maximum cavity width and spacing of cavity wall ties is given in Table C7.

Table C7 Maximum spacing of cavity ties

Width of cavity	Horizontal spacing	Vertical spacing	Other comment
(mm)	(mm)	(mm)	
50-75	900	450	see notes 1&2
76-100	750	450	see notes 1,2&3

Notes:

1. The horizontal and vertical spacing of wall ties may be varied if necessary to suit the construction provided that the number of wall ties per unit area is maintained.
2. Wall ties spaced not more than 300 mm apart vertically shall be provided within 225 mm from the sides of all openings with unbonded jambs.
3. Vertical twist type ties, or ties of equivalent performance, shall be used in cavities wider than 75 mm.

Parapet walls

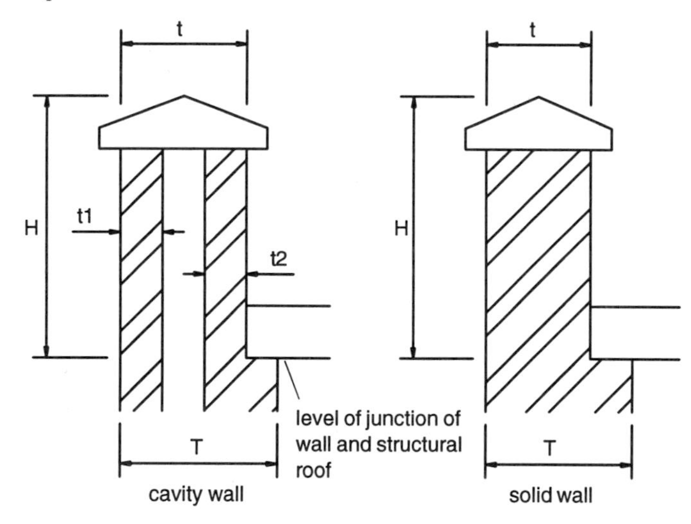

Table C8 Parapet wall thickness

Wall type	Thickness	Parapet height H to be not more than (mm)
Cavity wall	t1 + t2 equal to or less than 200	600
	t1 + t2 greater than 200, equal to or less than 250	860
Solid wall	t = 150	600
	t = 190	760
	t = 215	860

Note: **t** must not be greater than **T**

End restraint of walls and lateral support

The stability of buildings designed in compliance with the guide is dependent on the restraint provided by buttressing walls, piers and chimneys and floors and roof structures. The main requirements are described in this section. Interruption of lateral support should be avoided but if this is not possible the size of openings should be limited.

End restraint of walls

Walls must be bonded or securely tied at each end and throughout their height to a buttressing wall, pier or chimney. For long walls, restraint must also be provided at intermediate points.

Design requirements for buttressing walls

For buttressing walls which are not supported, the thickness must be not less than:

(a) half thickness from Table C6 for an external wall less than 5 mm; or
(b) 75 mm if it is part of a dwelling less than 6 m high and 10 m in length; or
(c) 90 mm in all other cases.

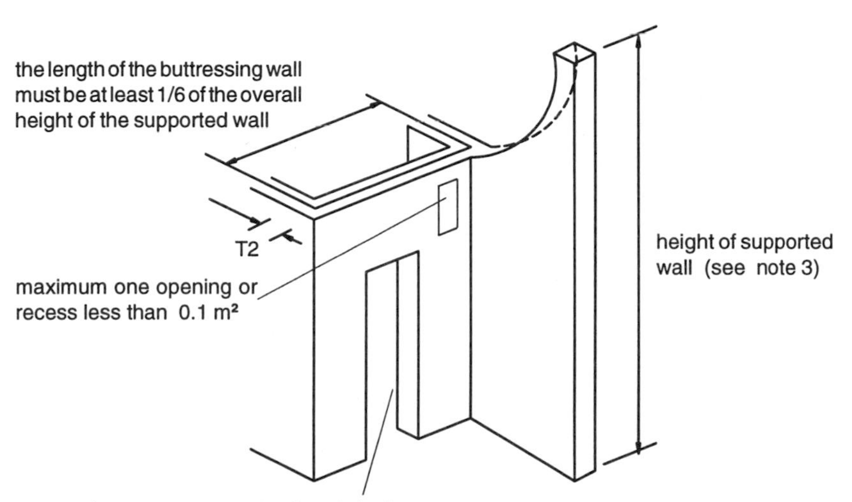

Notes:

1. The buttressing wall must be bonded or securely tied to the supporting wall and at the other end to the buttressing wall, pier or chimney.

2. Openings or recesses in the buttressing wall must be as shown. The position and shape of the openings must not impair the lateral support to be given by the buttressing wall.

3. Refer to pages C13 and C14 for the rules for measuring the height of the supported wall.

Design requirements for buttressing piers and chimneys

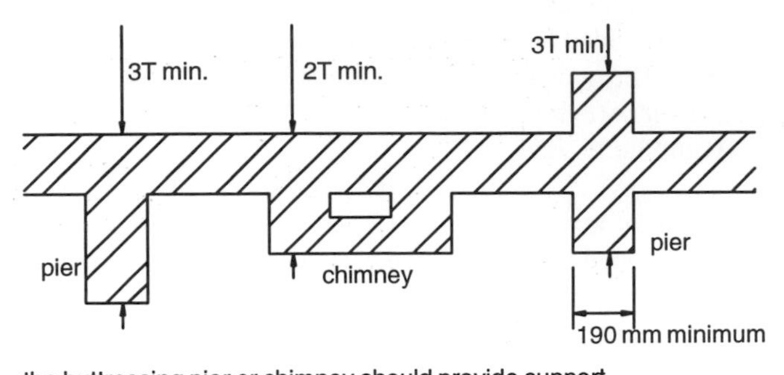

the buttressing pier or chimney should provide support to the full height of the wall from the base to top of wall

Minimum area buttressing support and walls is (570 x T) m²

Minimum requirements for lateral support

Table C9 Lateral support for walls

Wall type	Wall length	Lateral support required
Solid or cavity: external compartment separating	any length	lateral support by roof forming a junction with the supported wall
	greater than 3 m	floor lateral support by every floor forming a junction with the supported floor
Internal load-bearing wall (not being a compartment or separating wall)	any length	roof or floor lateral support at the top of each storey

Lateral support by floors

Walls must be strapped to floors above ground level, at intervals not exceeding 2 m, by galvanized mild steel or other durable metal straps as shown, which have a minimum cross-section of 30 m x 5 mm.

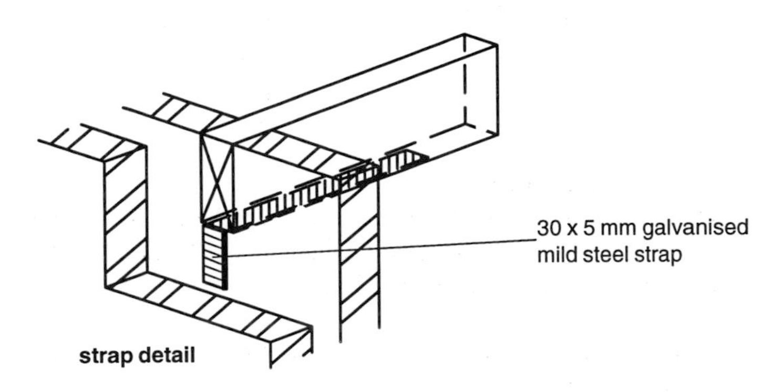

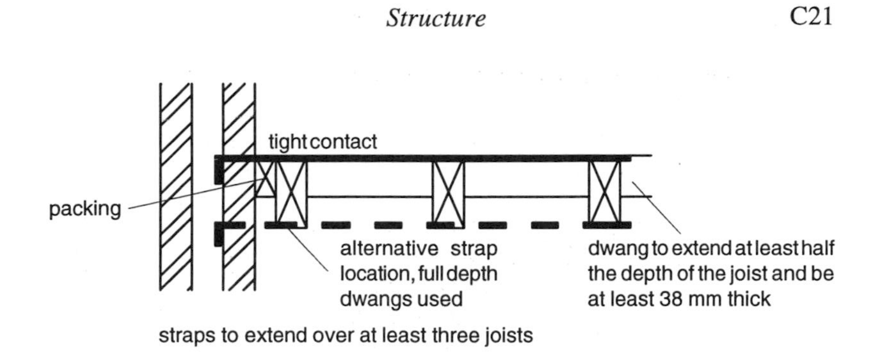

Straps need not be provided:

(1) in the longitudinal direction of joists in houses of not more than two storeys, if the joists are not more than 1.2 m centres and have at least 90 mm bearing on the supported walls or 75 mm bearing on a timber wall-plate at each end;

(2) in the longitudinal direction of joists in houses of not more than two storeys, if the joists are carried on the supported walls by joist hangers of restraint type described in BS 5628: Part 1: 1978 (1985) and are incorporated at not more than 2 m centres;

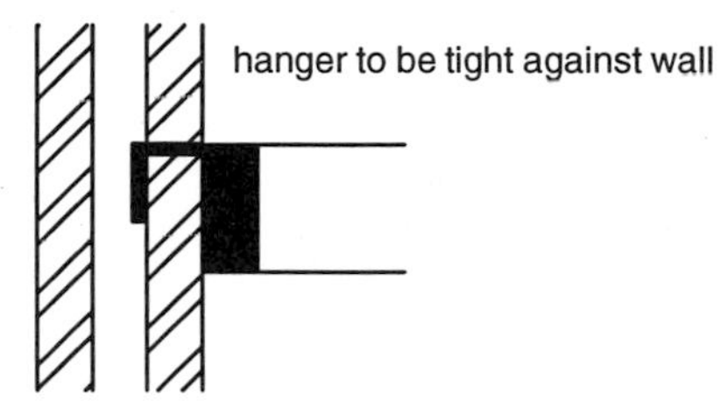

(3) where a concrete floor has at least 90 mm bearing on the supported wall.

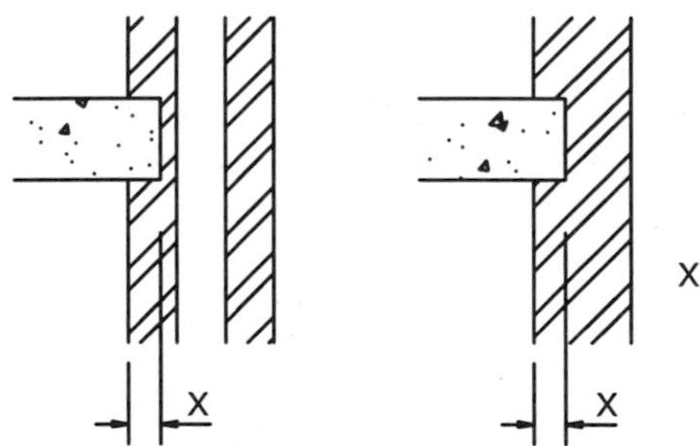

Lateral support at roof level

Walls must be strapped to roofs by galvanized mild steel or other durable metal straps which have a minimum cross-section of 30 mm x 5 mm and extend over at least three rafters or three ceiling ties.

Vertical strapping must be provided at eaves levels at intervals not exceeding 2 m if the roof:

(1) has not got a pitch of 15° or less;
(2) is not tiled or slated;
(3) is not a type known by local experience to be resistant to wind gusts;
(4) has not got main timber members spanning onto the supported wall at not more than 1.2 m centres.

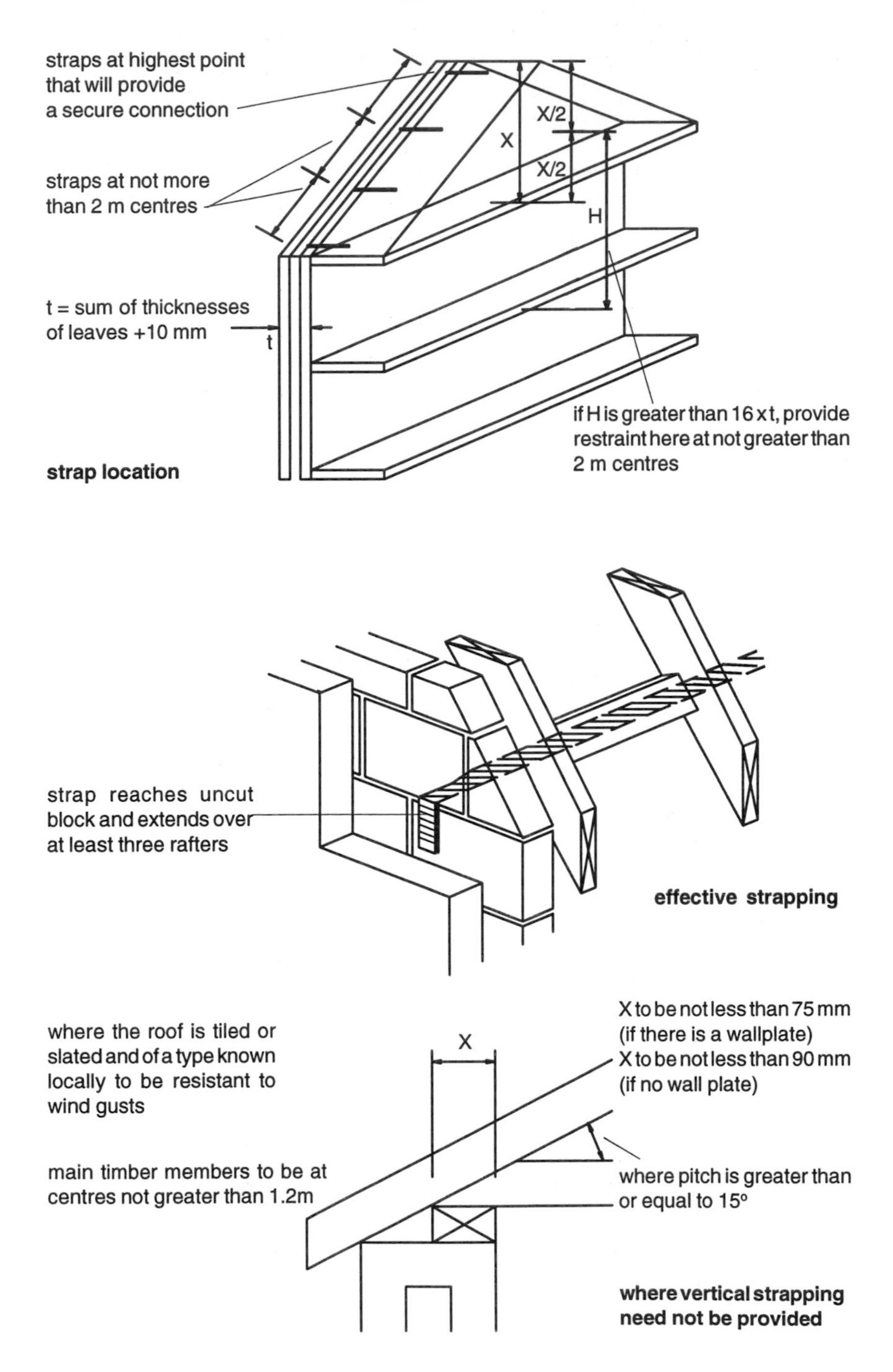

Openings and recesses in walls

Openings and recesses in a wall must not impair its stability, and construction over openings must be adequately supported.

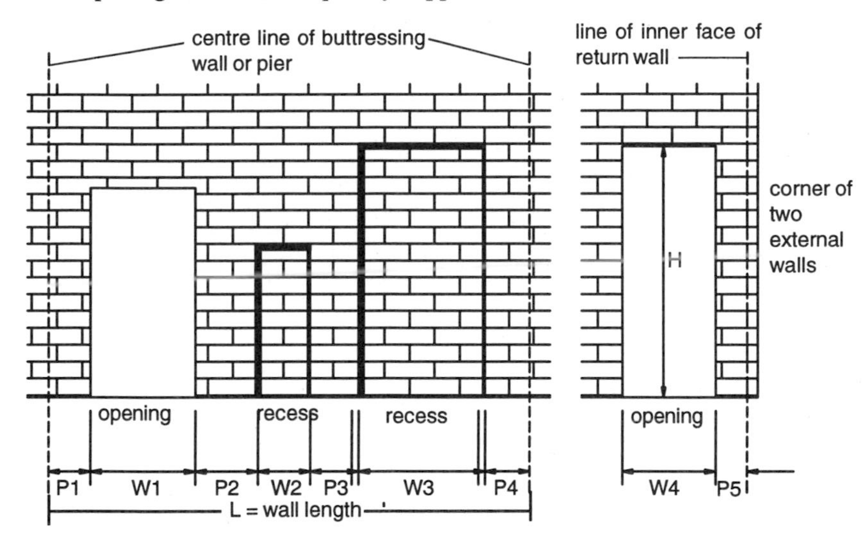

Notes:

Requirements (refer to table below for values of factor X)

1. W1 +W2 +W3 should not exceed 2L divided by 3
2. W1, W2 or W3 should not exceed 3 m
3. P1 should be greater than or equal to W1 divided by X
4. P2 should be greater than or equal to W1 + W2 divided by X
5. P3 should be greater than or equal to W2 + W3 divided by X
6. P4 should be greater than or equal to W3 divided by X
7. P5 should be greater than or equal to W4 divided by X but should not be less than 385 mm. In this case the height H of the opening should not exceed 2.1 m
8. The value of the factor 'X' shall be taken from the table, or it can be given the value 6, provided the compressive strength of the bricks or blocks (in the case of a cavity wall, in the loaded leaf) is not less than 7 N/mm^2

Table C10 Value of factor 'X'

Nature of roof span	Max roof span	Min thickness of wall inner leaf (mm)	Span of floor is parallel to wall	Span of timber floor into wall max 4.5 m	Span of timber floor into wall max 6.0 m	Span of concrete floor into wall max 4.5 m	Span of concrete floor into wall max 6.0 m
roof parallel to wall	not applicable	100	6	6	6	6	6
		90	6	6	6	6	5
timber roof spans into wall	9	100	6	6	5	4	3
		90	6	4	4	3	3

Section 4 Proportions for masonry chimneys

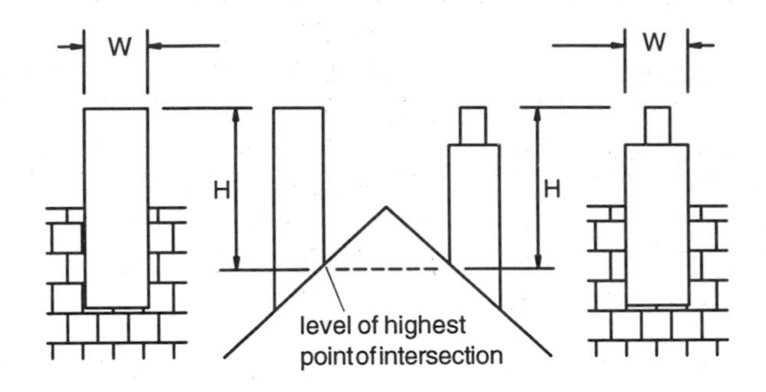

H must be less than 4.5 x W
(this should be reduced for masonry chimneys with a density less than 1500 kg/m³)

Section 5 Sizes of certain timber floor, ceiling and roof members

The tables in this section allow the determination, for given loadings and spans, of suitable section sizes and joist spacings for timber of strength class 3 or 4. The tables are only applicable for dwellings of purpose sub-group 1B and 1C of not more than three storeys and for three types of construction: floors, flat roofs with access for maintenance only and flat roofs with full access. The timber species/grade combinations which meet the requirements for strength classes 3 and 4 upon which the tables are based are set out in BS 5628 : Part 2: 1989. Notches and holes in floor and roof joists should be within the following limits:

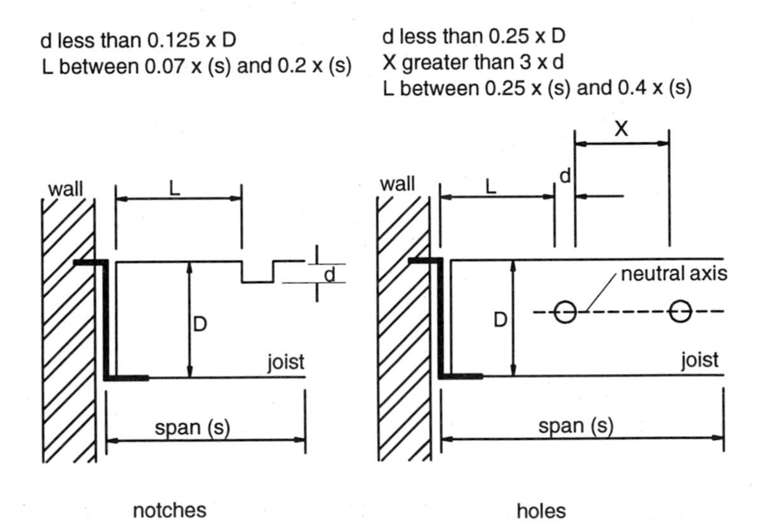

Table C11 Extract from Table 5.10 of the Small Buildings Guide
Floor joists (regularised sizes): timber of strength class SC3

	Dead load (kN/m^2)								
	≤0.25			>0.25 & ≤0.50			>0.50 & ≤1.25		
	Spacing of joists (mm)								
	400	450	600	400	450	600	400	450	600
Size of joist (mm x mm)	**Maximum clear span of joist (m)**								
38 x 97	1.83	1.70	1.31	1.72	1.56	1.22	1.43	1.31	1.04
38 x 122	2.49	2.39	1.93	2.37	2.22	1.76	1.95	1.80	1.45
47 x 147	3.20	3.08	2.79	3.06	2.95	2.61	2.72	2.57	2.17
47 x 170	3.69	3.55	3.19	3.53	3.40	2.99	3.12	2.94	2.55
47 x 195	4.22	4.06	3.62	4.04	3.89	3.39	3.54	3.34	2.90
47 x 220	4.72	4.57	4.04	4.55	4.35	3.79	3.95	3.74	3.24
50 x 97	2.08	1.97	1.67	1.98	1.87	1.54	1.74	1.60	1.29
50 x 147	3.27	3.14	2.86	3.13	3.01	2.69	2.81	2.65	2.27
75 x 195	4.83	4.70	4.31	4.68	4.52	4.13	4.24	4.08	3.65
75 x 220	5.27	5.13	4.79	5.11	4.97	4.64	4.74	4.60	4.07

Note: Dead load is the load supported by the joist, excluding the mass of the joist.

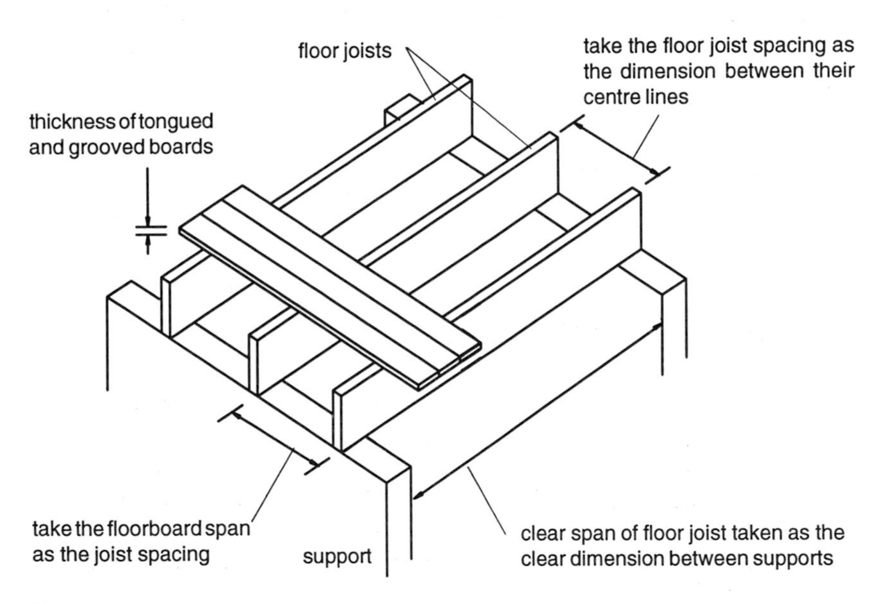

Notes:

1. The table gives sizes, spacings and spans for floor joists. The sizes, spacing and spans given will support the dead loads given in the tables and an imposed load not exceeding 1.5 N/m^2
2. Partition loads have not been allowed for in 5.10 and 5.11
3. Softwood tongued and grooved floorboards if supported at a joist spacing of up to 450 mm shall be at least 16 mm thick; and if supported at wider spacings up to 600 mm should be at least 19 mm thick.

Table C12 Extract from Table 5.12 of the Small Buildings Guide Joists for flat roofs with access only for the purpose of maintenance or repair: timber of strength class SC3

	Dead load (kN/m²)								
	≤0.50			>0.50 & ≤0.75			>0.75 & ≤1.00		
	Spacing of joists (mm)								
	400	450	600	400	450	600	400	450	600
Size of joist (mm x mm)	**Maximum clear span of joist (m)**								
38 x 97	1.74	1.72	1.67	1.67	1.64	1.58	1.61	1.58	1.51
38 x 122	2.37	2.34	2.25	2.25	2.21	2.11	2.16	2.11	2.01
38 x 147	3.02	2.97	2.85	2.85	2.80	2.66	2.72	2.66	2.51
47 x 97	1.92	1.90	1.84	1.84	1.81	1.74	1.77	1.74	1.65
47 x 147	3.30	3.25	3.12	3.12	3.06	2.90	2.96	2.90	2.74
47 x 195	4.68	4.53	4.13	4.37	4.28	3.89	4.14	4.04	3.70
50 x 147	3.39	3.34	3.19	3.19	3.13	2.97	3.04	2.97	2.80
50 x 195	4.79	4.62	4.22	4.48	4.36	3.97	4.23	4.13	3.78
63 x 97	2.19	2.16	2.09	2.09	2.06	1.97	2.01	1.97	1.87
63 x 147	3.72	3.66	3.44	3.50	3.43	3.25	3.33	3.26	3.07
63 x 195	5.14	4.96	4.54	4.86	4.69	4.28	4.61	4.47	4.07
75 x 122	3.17	3.12	3.00	3.00	2.94	2.80	2.86	2.80	2.65
75 x 147	3.98	3.92	3.64	3.75	3.67	3.44	3.56	3.48	3.27
75 x 195	5.42	5.23	4.79	5.13	4.95	4.53	4.89	4.72	4.31
75 x 220	6.07	5.87	5.38	5.76	5.56	5.09	5.50	5.30	4.85

Note: Dead load is the load supported by the joist, excluding the mass of the joist.

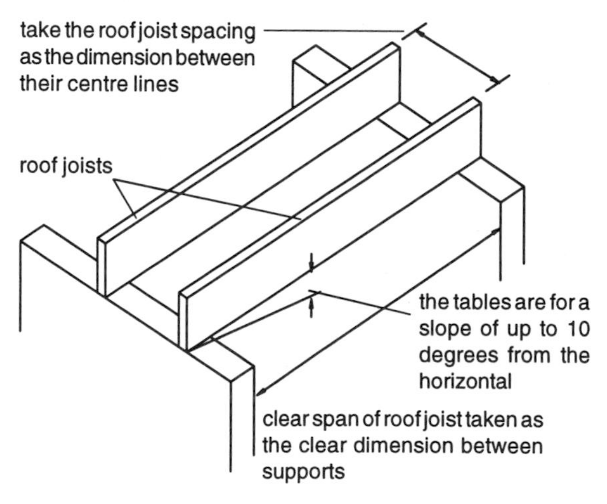

Notes:

1. The table gives sizes, spacings and spans for flat roof joists designed for access only for maintenance. The member sizes, spacings and spans will support the dead loads given in the tables and maximum imposed load of 0.75 kN/m² or an imposed concentrated load of 0.9 kN

Table C13 Extract from Table 5.15 of the Small Buildings Guide Joists for flat roofs with access not limited to the purpose of maintenance or repair: timber of strength class SC4

	Dead load (kN/m^2)								
	≤0.50			>0.50 & ≤0.75			>0.75 & ≤1.00		
	Spacing of joists (mm)								
	400	450	600	400	450	600	400	450	600
Size of joist (mm x mm)	Maximum clear span of joist (m)								
47 x 147	2.75	2.73	2.66	2.66	2.62	2.54	2.57	2.54	2.44
47 x 195	4.03	3.98	3.68	3.85	3.80	3.54	3.71	3.64	3.42
47 x 220	4.71	4.56	4.15	4.49	4.39	3.99	4.31	4.23	3.85
50 x 122	2.19	2.17	2.12	2.12	2.10	2.04	2.06	2.04	1.97
50 x 147	2.83	2.81	2.73	2.73	2.70	2.61	2.65	2.61	2.51
50 x 195	4.14	4.09	3.76	3.96	3.90	3.61	3.81	3.74	3.49
63 x 97	1.77	1.75	1.72	1.72	1.71	1.66	1.68	1.66	1.61
63 x 147	3.15	3.12	3.03	3.03	2.99	2.89	2.93	2.89	2.77
63 x 195	4.56	4.45	4.06	4.36	4.29	3.90	4.19	4.11	3.77
75 x 122	2.64	2.62	2.56	2.56	2.53	2.45	2.48	2.45	2.36
75 x 147	3.40	3.36	3.25	3.27	3.23	3.11	3.16	3.11	2.98
75 x 195	4.79	4.70	4.29	4.67	4.53	4.13	4.49	4.38	3.99

Note: Dead load is the load supported by the joist, excluding the mass of the joist.

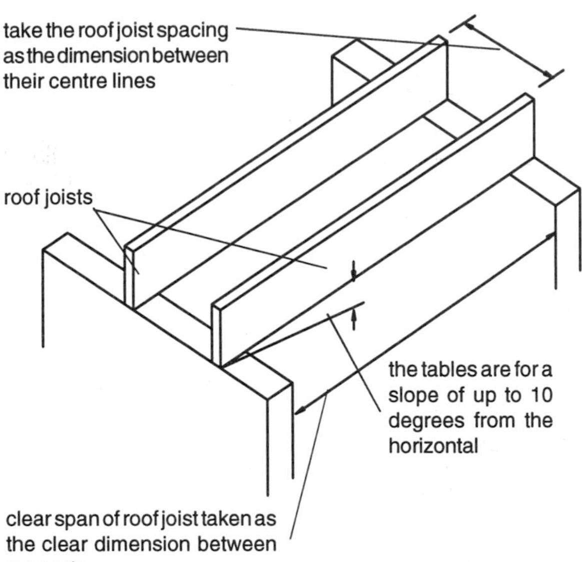

Notes:
1. The table gives sizes, spacings and spans for flat roof joists designed for access generally. The member sizes, spacings and spans will support the dead loads given in the tables and maximum imposed load of 1.5 kN/m^2 or an imposed concentrated load of 1.8 kN

STRUCTURAL FIRE PRECAUTIONS

PART D — Regulation 12

BUILDING STANDARDS

Regulation

12 (1) Every building shall be so constructed that, for a reasonable period, in the event of fire -

a) its stability is maintained;

b) its spread of smoke and fire within the building is inhibited; and

c) the spread of fire to and from other buildings is inhibited.

(2) This regulation shall not be subject to specification in a notice served under Section 11 of the Act.

Aim

The organisation of space within buildings and the design and construction of elements of structure should:

1) limit the spread of fire within and between buildings;
2) ensure the stability of the structure and the safety of escape routes and access routes for fire fighting for a specified minimum period of time.

TECHNICAL STANDARDS

Scope

The requirements apply to all buildings and should be interpreted with reference to Part E of the Technical Standards.

Requirements

Structural fire precautions

Controlling the spread of fire and smoke within the building

To inhibit the spread of fire within a building it must be divided into 'fire tight cells', the walls and floors providing a complete barrier to the passage of fire **D2.1** and smoke. These 'fire tight cells' are called compartments and must remain effective for a specified period of time which is referred to as the fire resistance **D2.2** (Figure D1).

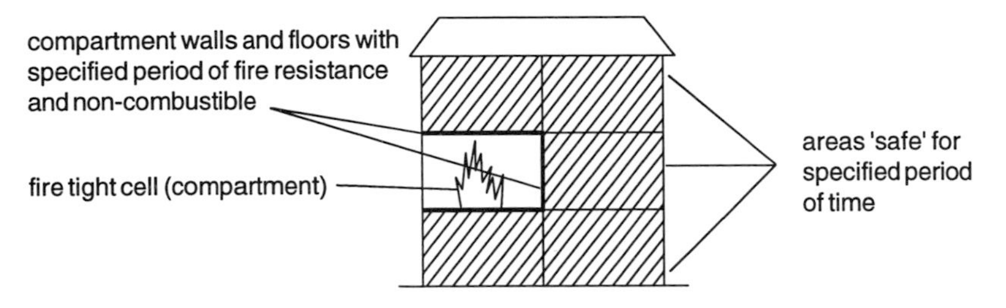

Figure D1 Fire tight cells.

D2.7 Where different parts of a building are under different ownership they must be separated from each other by barriers to the spread of fire and smoke. These are called separating walls and floors. The performance requirements are determined as for compartment walls and floors.

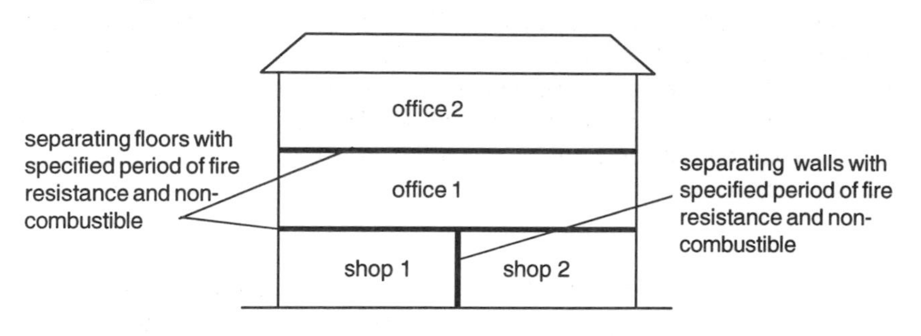

Figure D2 Section of building with separating walls and floors.

D2.9
D2.7-D2.10 Parts of a building where there is a high risk of fire or a high fire load are designated as places of special fire risk. Compartment walls and floors must separate places of special fire risk, stairwells and lift shafts from the rest of the building.

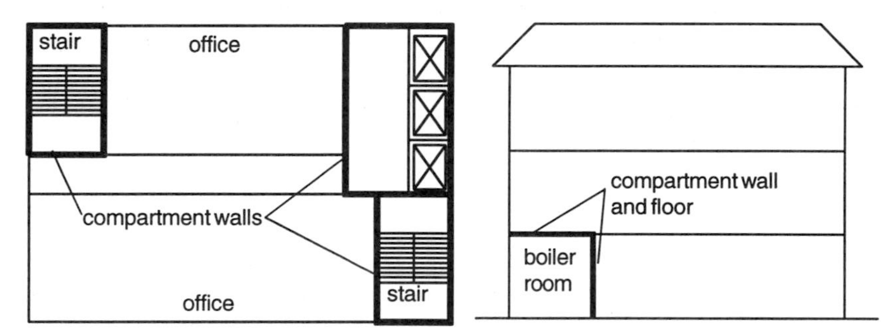

Figure D3 Plan and section of buildings.

D2.11-D2.13 Openings that have to be made in compartment/separating walls and floors for services or for circulation through the building must not reduce the overall effectiveness of the element as a barrier to fire and smoke (Figure D4). Other **D2.19-D2.21** spaces within a building such as cavities in walls, roof spaces and voids above suspended ceilings must also be provided with barriers to prevent the passage of fire and smoke.

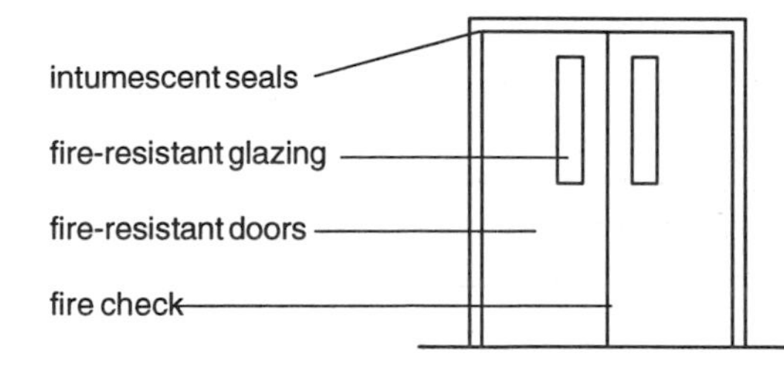

Figure D4 Door within compartment wall.

The requirements for escape and fire access routes and for fire doors are given in Part E of the Technical Standards.

Maintaining the structure of a building in a fire

In a fire the load bearing structure of a building must be able to carry its full load for a specified period of time without loss of stability or collapse. Elements of structure which are not load bearing must also retain their stability and integrity for an adequate period of time if the escape and fire-fighting access routes are to remain safe and compartmentation is to be effective. The term element of structure includes separating/compartment walls and floors described previously. **D2.2**

Fire performance requirements for materials and elements of structure

The fire performance requirements for materials and elements of structure are specified in terms of fire resistance, non-combustibility, surface spread of flame and external roof exposure designations. These measures of performance are referred to in both Parts D and E of the Technical Standards. **D2.2** **D2.3** **DTS 2.2** **App.to Part D**

Controlling the spread of fire between buildings

The spread of fire between buildings by radiant heat, flame or flying brands is prevented by ensuring adequate separation between them. The separation can be provided by structure (e.g. a solid wall and roof of adequate fire resistance, Figure D5) or by distance, from the side of the building to the boundary (Figure D6). The boundary is the edge of the property or the centre line of a facing road, canal or public open space. **D2.5** **D2.6**

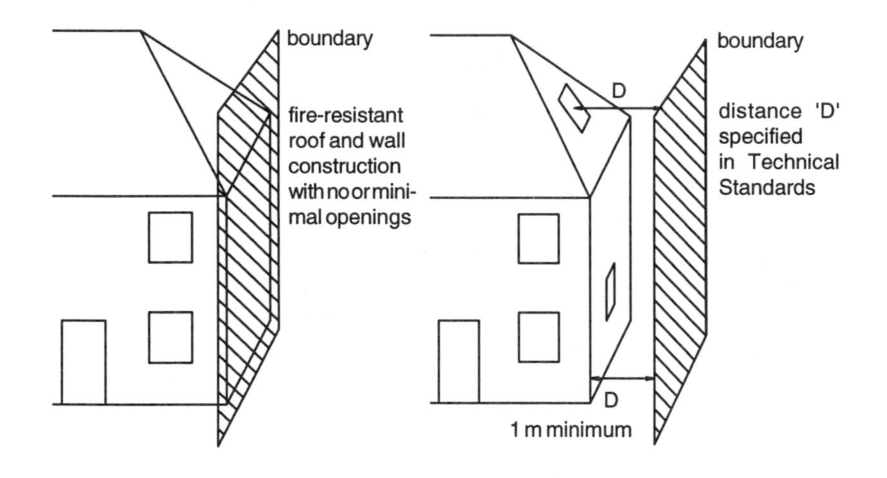

Figure D5 Separation by structure. *Figure D6* Separation by distance.

Buildings ancillary detached/attached to dwellings

Alternative, and generally less onerous standards, can be applied to buildings such as garages, conservatories, porches and carports which are adjacent or attached to dwellings. **D3**

Background

Fire safety

The main aim of the statutory requirements for fire safety is to prevent injury and loss of life, firstly of the occupants of a building in the event of a fire and secondly of the occupants of adjacent buildings by preventing the spread of the fire. The protection of property, either the contents of a building or its structure and fabric, is not a statutory requirement and is therefore not incorporated explicitly in the minimum performance requirements given in the Technical Standards. Inevitably designing a building to protect life also provides some protection for property particularly through the containment of fire within compartments, the provision of fire fighting facilities and the prevention of spread between buildings. However, the financial cost of fires arising from the loss of, or damage to, property has increased to the level where building insurers either require additional fire protection for certain types of building and their contents, or alternatively offer incentives to adopt additional fire protection.

The additional protection ranges from increased fire resistance of walls and floors to heat and smoke detectors and automatic sprinkler systems. The recommendations of the building insurers for the limitation of fire damage and associated losses are described in the Code of Practice for the Construction of Buildings published by the Loss Prevention Council. The recommendations of the LPC code are intended to '...have effect at a later stage (than requirements for the protection of life) in fire development by controlling the spread of fire through, and between, buildings and preventing building collapse...'.

The development of two sets of fire standards for buildings is an understandable but short term solution to the problem of meeting the objectives of life and property safety separately. As both sets of standards are potentially enforceable, given the need of building owners and users to carry insurance, and have a considerable area of overlap, the development of a more unified approach would clearly be useful. Perceived problems of a unified approach include the effect on the cost of building projects of more onerous fire protection systems and importantly, in what is already a litigious industry, issues of liability. In Scotland a further problem is that the LPC code was produced to complement the Approved Document B supporting the Building Regulations of England and Wales (1991) and this makes parallel reading of the code with the Technical Standards difficult.

The development of fire engineering through the use of quantitative techniques for risk and safety evaluation and computer modelling of fire growth underlies current initiatives to assess the overall fire safety of a building. The aim is to optimise the balance between risk, safety (life and property) and cost by assessing alternative safety strategies against different fire situations. This approach allows more flexibility than the prescriptive requirements in the Regulations and can integrate specific statutory or insurance requirements with other good practice guidance such as British Standard 5588 'Fire Precautions in the Design, Construction and Use of Buildings'. Whilst not necessary, or possibly desirable, for routine building types this approach is often essential for innovative building designs and refurbishment projects where spatial organisation, materials and construction mean that strict compliance with statutory requirements cannot be achieved (e.g. timber floors in a listed building cannot meet compartmentation standards as they are combustible). Alternatively there are some cases where statutory requirements may not be relevant to a particular situation in a building (e.g. travel distance to a protected zone for hospital patients that cannot be moved).

Active and passive fire protection

In practice the overall fire safety of a building is affected by the active and the passive fire protection which is incorporated into its design and construction. Passive fire precautions include the use of fire resistant materials and forms of construction, the separation by fire resistant construction of spaces within a building and the provision of adequate separation between buildings. Active fire precautions include the use of heat and smoke detection systems, alarms, and the provision of fire fighting facilities such as wet and dry risers, automatic sprinkler systems and fire lifts.

Part D

Part E DTS D2.1

The Technical Standards describe the main requirements for passive protection in Part D and for active protection in Part E. The requirements for escape routes, the fire performance of their construction, fire doors and emergency lighting are described in Part E of the Technical Standards. These elements are in practice part of the passive, or in-built, fire protection of the building as they neither play a part in detecting fires nor in actively controlling them; however like active systems their effectiveness is dependent on adequate maintenance. Parts D and E of the Technical Standards are complementary in setting the minimum statutory requirements for fire safety and should therefore be considered together when designing a building or applying the regulations.

The requirements of Part S, in relation to escape routes, and Part F, with respect to openings in compartment/separating walls and floors are also affected by the requirements for fire safety.

Fire performance requirements for materials and elements of structure

The fire performance requirements for materials and elements of structure are specified in terms of fire resistance, non-combustibility, surface spread of flame and external roof exposure designations. These measures of performance are referred to in both Parts D and E of the Technical Standards.

Fire resistance

The fire resistance of an element of structure is a measure of the length of time it can continue to function during a fire and is expressed in minutes. Elements of structure such as upper floors and load bearing walls must remain effective long enough for the Fire Service to bring a fire under control without it spreading through the building or to adjacent buildings. For a compartment wall or floor the fire resistance describes the length of time it can resist the passage of heat, smoke and flame. For a structural column it describes the length of time that it can continue to carry its full load without loss of stability or collapse.

BS 476

D2.2 Table1 Table2

BS 476 Fire resistance is determined with respect to three criteria: stability, integrity and insulation (Figure D7). Stability applies to all load bearing elements and assesses the length of time the element can carry its design load during a fire. Integrity applies to all elements which act as fire barriers such as walls, floors and fire doors. It assesses the length of time the element can resist the passage of fire and smoke. Insulation applies to some elements which act as fire barriers such as walls and cavity barriers. It assesses the length of time the element can effectively resist the passage of heat during a fire.

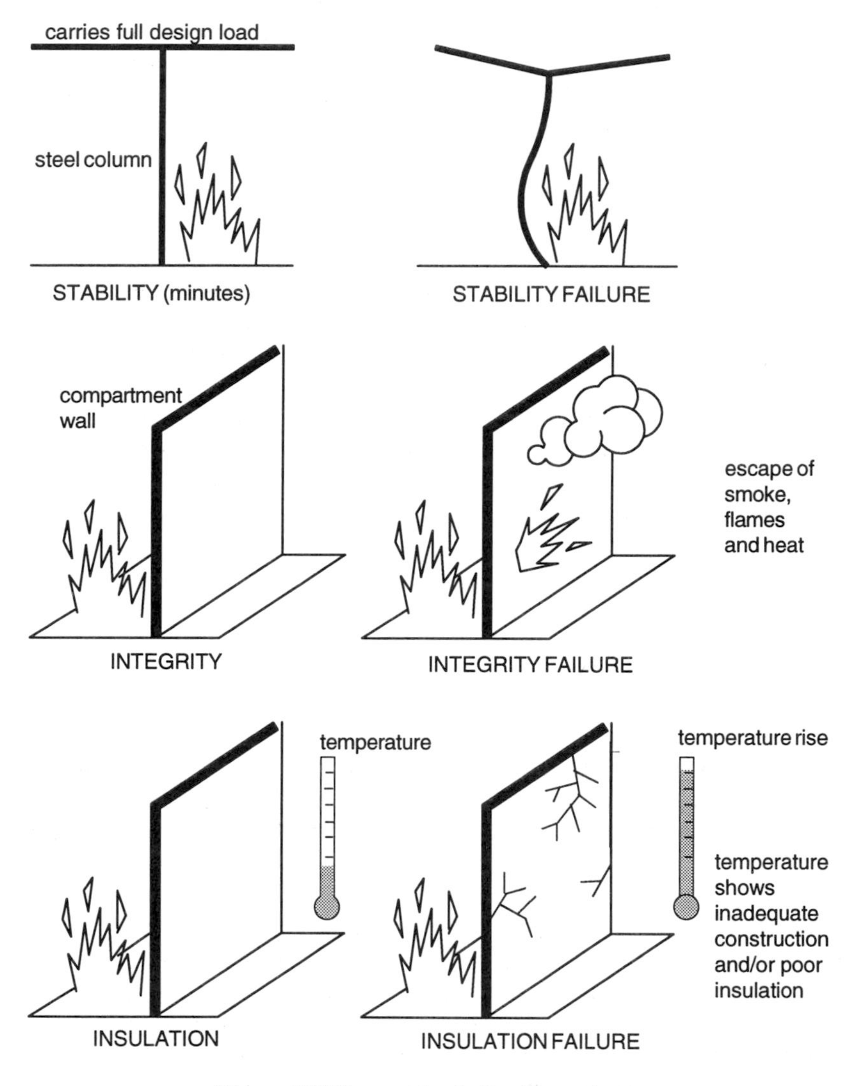

Figure D7 Three criteria for fire resistance.

When assessing the fire resistance of an element it is the performance *in situ* which is important; this includes all the joints and the actual construction details. It is therefore essential that the materials, components and construction in the finished building are those approved in the design drawings and specification.

Verification of fire resistance can be by test, by calculation or by assessment. For components or constructions tested since January 1989 the test should be in accordance with the procedures in the appropriate part of BS 476: Parts 20-24: 1987. The appropriate standard prior to January 1989 was BS 476: Part 8: 1972. Assessment must be carried out by a suitable laboratory, the Building Research Establishment, the British Board of Agrément or a consultant with 'appropriate expertise'. A suitable independent laboratory is defined as one that has NAMAS accreditation; however no guidance is given on how to determine an independent consultant's 'appropriate expertise'. It is unlikely that any assessment carried out by a NAMAS laboratory, BRE, BBA or a fire consultant would not include reference to the appropriate test in BS 476. **App. Part D**

The results of tests on proprietary materials and elements of structure are given in:

(1) BRE Report Volumes 1 and 2 *Results of Fire Resistance Tests on Elements of Building Construction* (HMSO 1975/76);
(2) FPA Report *Fire Test Results on Elements of Building Construction* (FPA 1983);
(3) Association of Structural Fire Protection Contractors and Manufacturers Report 1983 *Fire Protection for Structural Steel in Buildings*;
(4) LPC Report *Rules for the Construction and Installation of Firebreak Doors and Shutters* (LPC 1985).

Prescriptive guidance on the fire resistance of elements is available from many sources; reference is made in Part D to: **App. Part D**

(1) BS 5950: Part 8:1990 for structural steelwork;
(2) BRE Report BR 128.

The need to avoid creating barriers to trade within the European community, means that equivalent tests and 'assessments' carried out by organisations and consultants from other member states can be used to demonstrate compliance with the fire resistance requirements. **pp.B2**

Non-combustibility

In addition to having a minimum fire resistance, those parts of a building which provide a barrier to the passage of heat, smoke and flame (e.g. separating walls and floors) or form part of an escape route must be constructed of non-combustible materials. Floor finishes, wall and ceiling linings and external claddings which do not contribute to the fire resistance requirement of the element are not required to be non-combustible. The test for non-combustibility involves noting flaming and measuring temperature rise when a small sample of material is placed in a pre-heated furnace. A material which burns can be classified as non-combustible if the flaming is brief and the temperature rise in **D2.3 Table** **BS 476** the furnace is below a specified limit. Materials must be tested in accordance with BS 476: Part 4:1970 (1984), although plasterboard conforming with BS 1230: Part 1: 1985 with a class 0 surface can be regarded as non-combustible. The need to avoid creating barriers to trade within the European Community means that results of equivalent tests to BS 476: Part 4:1970 (1984) used in other member states can be used to demonstrate compliance with the non-combustibility requirement.

pp.B2

Surface spread of flame

BS 476

The surface spread of flame test measures the rate at which a flame front can move across the surface of a material or product used in walls or ceilings. The standards for surface spread of flame for internal walls and ceilings are given in Part E and for external wall claddings in Part D. Materials and products are classified as either Class 0, 1, 2 or 3 in order of decreasing performance. Classes 1, 2 and 3 are classifications used by BS 476: Part 7: 1987 (or if tested prior to January 1989, BS 476: Part 7: 1971).

D2.4

App. Part D

To be rated Class 0 a material or product must meet the following criteria:

(1) non-combustible;
(2) Class 1 surface spread of flame;
(3) performance index less than 12 and sub-index less than 6 (when tested in accordance with BS 476: Part 6: 1981 or 1968).

Walls and ceilings of all protected zones must have a Class 0 rating.

Typical classifications for common materials and products are given in Table D1. The need to avoid barriers to trade within the European Community means that results of equivalent tests to BS 476: Part 7: 1987 or classification equivalent to Class 0 can be used to demonstrate compliance with the requirements for surface spread of flame.

pp.B2

Table D1 Classification of common materials

Classification		Material
Class 0 / Class 1	1	brickwork, blockwork, concrete and ceramic tiles.
	2	plasterboard (painted or not) with or without an airgap or fibrous or cellular insulating material behind.
	3	mineral fibre tiles or sheets with cement or resin binding.
Class 3*	4	timber or plywood with density > 400 kg/m^3 painted or unpainted.
	5	wood particle board or hardboard, either treated or painted.

*Timber products listed under Class 3 can be brought up to Class 1 with appropriate proprietary treatments.

External roof exposure designations
The external fire exposure test assesses roofs with respect to two criteria:

(1) resistance to penetration by fire;
(2) resistance to surface ignition and spread of flame.

BS 476

For both of these performance is classified, in decreasing order, by the letters A, B, C or D. The test assesses the risk of heat and flying brands spreading a fire from an adjacent building either into the building through the roof, or across the roof. The designation is used to determine the minimum distance permissible from the building to the boundary and hence the minimum separation between adjacent buildings. Changes included in the most recent version of the BS 476 test have not been incorporated into the Technical Standards so the relevant test procedures are those described in BS 476: Part 3: 1958. Typical roof coverings and their designations are given in Table D2.

Table D2 Designation of typical roof coverings

Roof covering	Roof structure	Designation
Pitched roofs		
1. Natural slate, clay and concrete tiles	timber rafters and most constructions	AA
2. bitumen felt strips type and appropriate underlay	as above	BB
3. corrugated metal sheet (most underlays and infill material)	timber, steel or concrete	AA
4. metal sheet and mastic asphalt	timber, steel joists and T&G board, chipboard, fibre board, plywood	AA
5. Designation of bitumen felt roofs varies with both types of upper layer and substrate		
Flat roofs		
6. felt with surface finish: chippings (12.5 mm), or sand-cement screed, or non-combustible tiles	any	AA

For plastics materials BS 476: Part 3 is not appropriate and roof coverings and rooflights made of plastics materials must satisfy the following criteria:

DTS D2.5

(1) minimum softening point
(tested in accordance with BS 2782: Part 1: 1976: Method 120A);
(2) limited flaming and spread of flame
(tested in accordance with either BS 2782: 1970: Method 508A, 508C or 508D or BS 2782: 1982: Part 1: Method 140E);

(3) limited flaming drops of molten plastic
(tests as for 'limited flaming and spread of flame' including Method 508C).

The need to avoid creating barriers to trade within the European Community means that results of equivalent tests to BS 476: Part 3: 1958 (BS 2782 for plastics materials) used in other member states can be used to demonstrate compliance with the requirements for separation between buildings.

COMPLIANCE

Introduction

The requirements of Part D are summarised in the preceding section. The link between the requirements and the three main aims of structural fire precautions are shown in Table D3. To control the spread of fire within buildings and to maintain the structure of the building requires:

(1) the provision of barriers to smoke and fire, based on floor areas and volumes within the building; and
(2) the use of materials and construction forms which are sufficiently fire resistant and in some cases non-combustible.

To prevent the spread of fire between buildings requires a minimum separation to avoid ignition by radiant heat and flying brands. The separation is achieved in practice by specifying a minimum distance from the building to the boundary based on the number of openings and the construction of the external wall. The methods for calculating the minimum distance to the boundary are described in the provisions deemed to satisfy the requirements in Part D.

Table D3 Contents of Part D

Aim	Requirements in Technical Standards
Controlling the spread of fire within buildings	Compartmentation/ separation
	D2.1 and Tables
	D2.2 and Tables
	D2.3 and Tables
	D2.7 - D2.10
	Junctions
	D2.14 - D2.14
	Openings
	D2.11 - D2.13
	D2.16 - D2.18
	Cavities
	D2.19 - D2.21 and Table
Maintaining the structure of a building in a fire	Fire performance of materials and elements
	D2.2 and Tables
	D2.3 and Tables

Controlling the spread of fire between buildings	Distance from boundary/ separation D2.4 and Table D2.5 D2.6
Alternative standards for buildings ancillary to dwellings	D3

Controlling the spread of fire within buildings

Compartmentation

The maximum size of compartments (based on storey area and volume) and the fire resistance requirements for compartment walls and floors are determined by:

(1) the purpose group – an estimate of the risk of fire and the fire load;
(2) the height of the topmost storey of the compartment (or building) which affects ease of escape and fire fighting.

The limits on storey area and cubic capacity of a compartment for different purpose groups and building forms are given in Table D4. The figures can be doubled (except for purpose groups 1 and 2) if an automatic fire control system is installed which meets either the requirements of BS 5306: 1986-89, or the 1990 Loss Prevention Council recommendations for automatic sprinkler installations. Compartment walls must separate lift shafts from the rest of the building but are not required between lift shafts and adjoining stair enclosures or protected lobbies. Places of special fire risk must also be separated from the rest of the building by compartmentation. The minimum period of fire resistance for a place of special fire risk is 60 minutes and if it contains any hazardous liquid the access must be constructed so that all the liquid can be contained in the event of leakage or spillage.

D2.1 Table DTS D2.1 D2.8 D2.9 D2.10

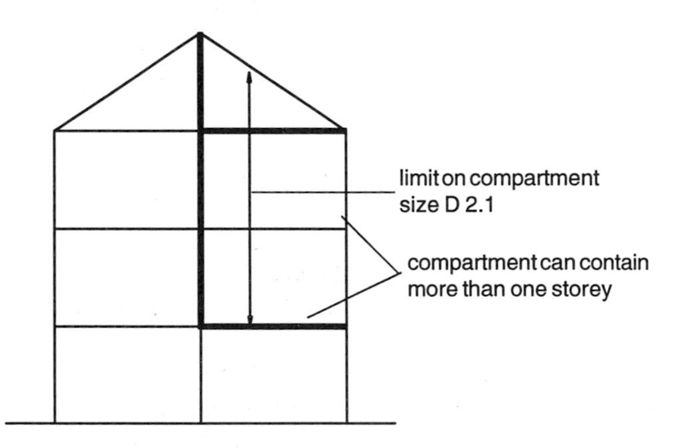

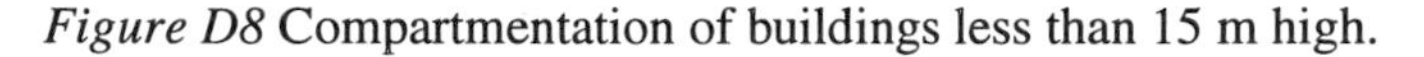
Figure D8 Compartmentation of buildings less than 15 m high.

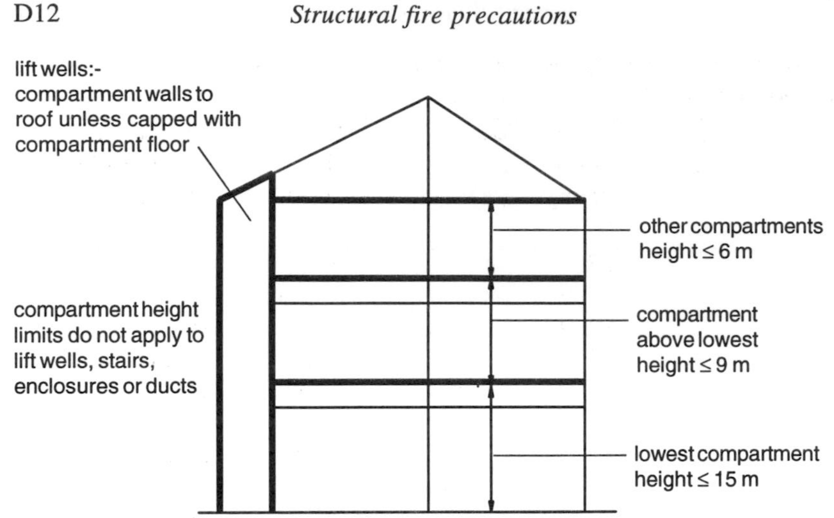

Figure D9 Compartmentation of multi-storey buildings greater than 15 m high (except buildings of purpose group 7C and roof spaces in buildings of purpose group 5).

Table D4 Maximum storey area and cubic capacity of a building or compartment

Purpose group or sub-group	Building form	Maximum storey area (m^2)	Maximum cubic capacity (m^3)
1A (flats and maisonettes)	any	no limit	no limit
1B (houses)	any	no limit	no limit
1C (houses)	any	no limit	no limit
2A (institutional)	any	1500	8500
2B (other residential)	any	2000	14000
3 (offices)	any	4600	28000
4 (shops and commercial)	any	2800	7100
5A (high hazard assembly)	any	1900	no limit
5B (low hazard assembly)	any	no limit	21000
6A (high hazard industrial)	single storey	33000	no limit
	> one storey	2800	28000
6B (low hazard assembly)	single storey	93000	no limit
	> one storey	7500	no limit
7A (high hazard storage)	single storey	1000	no limit
	> one storey	500	4200
7B (low hazard storage)	single storey	14000	no limit
	> one storey	2800	21000
7C (open-sided car parks)	any	no limit	no limit

Table D5 Periods of fire resistance for building/compartment according to height, area and cubic capacity

Purpose group	Building form	Height of top storey (m)	Storey area (m^2)	Maximum cubic capacity (m^3)	Minimum period of fire resistance (mins)
1A	any	≤ 7.5	no limit	no limit	30(2)
		≤18.0	no limit	no limit	60
		>18.0	no limit	no limit	90(3)
1B	any	≤ 7.5	no limit	no limit	30(2)
		> 7.5	no limit	no limit	60(4)
1C	any	≤4.5	no limit	no limit	30(2)
2A	any	≤ 7.5	≤ 1500	≤ 3000	30(2)
		≤18.0	≤ 1500	≤ 5700	60
		>18.0	≤ 1500	≤ 8500	90
2B	any	≤ 7.5	≤ 2000	≤ 4200	30(2)
		≤18.0	≤ 2000	≤ 8500	60
		>18.0	≤ 2000	≤ 14000	90
3	any	≤ 7.5	≤ 4600	≤ 4200	30(2)
		≤18.0	≤ 4600	≤ 14000	60
		>18.0	≤ 4600	≤ 28000	90
4	any	≤ 7.5	≤ 2800	≤ 2200	30(2)
		≤18.0	≤ 2800	≤ 4200	60
		>18.0	≤ 2800	≤ 7100	180
5A	any	≤ 7.5	≤ 1900	≤ 4900	30(2)
		≤18.0	≤ 1900	≤ 14000	60
		>18.0	≤ 1900	no limit	120
5B	any	≤ 7.5	no limit	5600	30(2)
		≤18.0	no limit	≤ 14000	60
		>18.0	no limit	≤ 21000	90
7C	any	no limit	no limit	no limit	15

Notes:
(1) Except in the case of purpose groups 1 and 2, the storey area or cubic capacity, or both, may be doubled for a stated period of fire resistance if the building or compartment is provided with an appropriate automatic fire control system.
(2) 60 minutes for a separating wall or separating floor or part of the enclosing structure of a protected zone.
(3) Reduced to 60 minutes for any non-loadbearing separating wall between flats or maisonettes.
(4) Reduced to 30 minutes for a floor wholly within a dwelling which does not contribute to the support of the building as a whole.

D2.2 D2.3 D2.2 Table1 DTS D2.2

Compartment floors and walls must be non-combustible and comply with the minimum period of fire resistance required by Table D5. The period of fire resistance required applies to stability, integrity and insulation tested or assessed for exposure to fire from either side for a compartment wall but only from the underside for a compartment floor. A suspended ceiling can contribute to the fire resistance of a compartment floor if it is either of jointless construction with no openings, or it is built as shown in Figure D10.

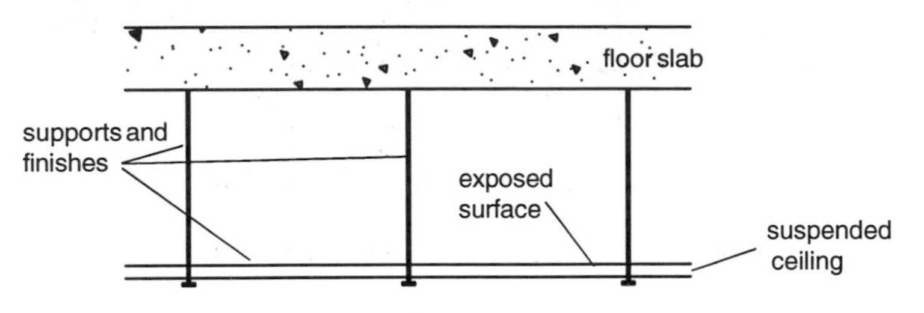

Figure D10 Suspended ceiling.

Height of topmost compartment (m)	Required fire resistance	Exposed surface	Supports and fixings	Joints	Ceiling construction
≤ 11	≤ 30	Class 0/1	—	—	—
	≤ 60	Class 0	Non-combustible	—	—
> 11	≤ 60	Class 0	Non-combustible	—	—
	> 60	Class 0	Non-combustible	Jointless	Non-combustible

Separation

D2.7

Separating walls and/or floors must be provided where adjoining buildings or parts of a building are in either different occupation (owners/tenants/users) or different purpose groups (use). The performance requirements are determined as for compartment walls and floors but there is a minimum period of fire resistance of 60 minutes required for separating walls and floors. Where separation is between different purpose groups the more onerous requirement applies. A suspended ceiling can contribute to the fire resistance of a separating floor in the same way as for a compartment floor. Generally separating walls and floors should be constructed of non-combustible materials. A few exceptions relate to buildings of purpose group 1 which allow the use, in certain situations, of a timber structure if the overall construction meets the requirements for fire resistance.

D2.3 Table

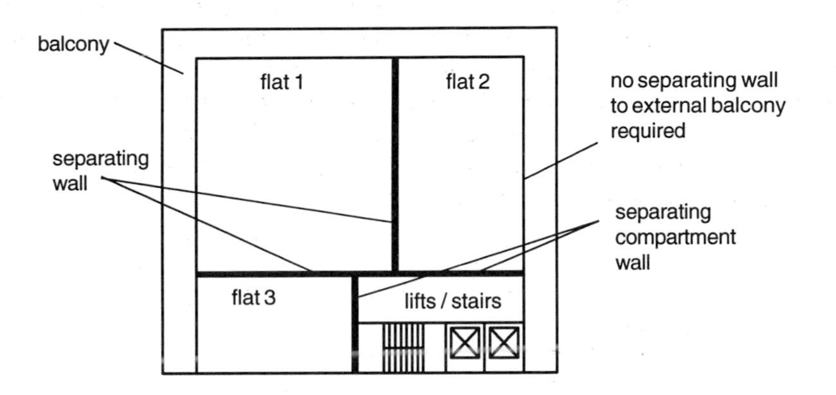

Figure D11 Flats with communal external balcony.

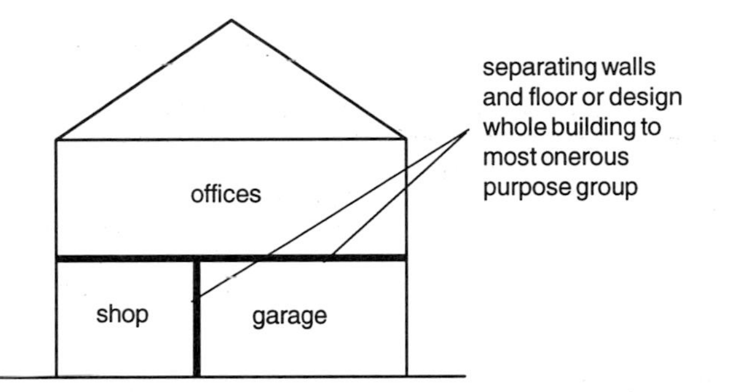

Figure D12 Building with different purpose groups.

Junctions

The detailing of the junctions between compartment/separating walls and floors with external walls and the roof of a building must ensure that it is not possible for fire or smoke to pass through or over them. Junctions between compartments/separating/external walls and compartment/separating floors must be suitably fire-stopped in the same way as service openings. Suitable details for junctions between compartment/separating walls and roofs are illustrated below. **D2.14 D2.15 DTS D2.15**

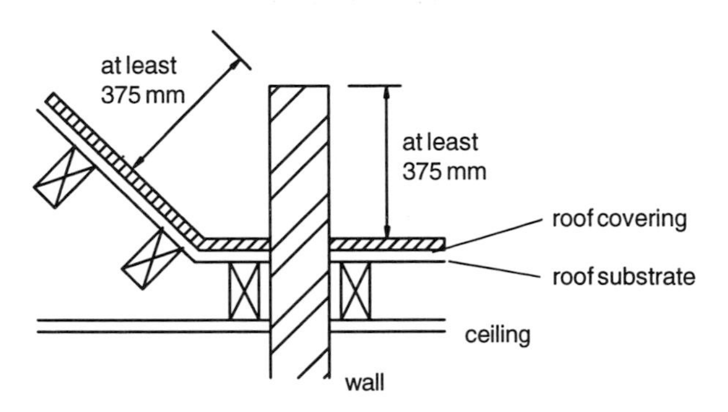

Figure D13 Wall carried through roof.

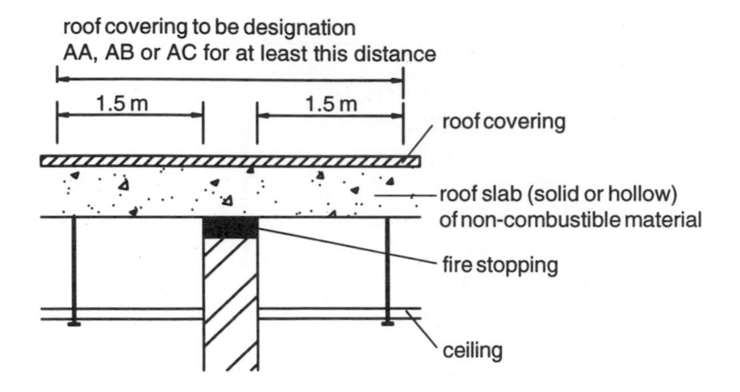

Figure D14 Roof carried over wall (non-combustible substrate).

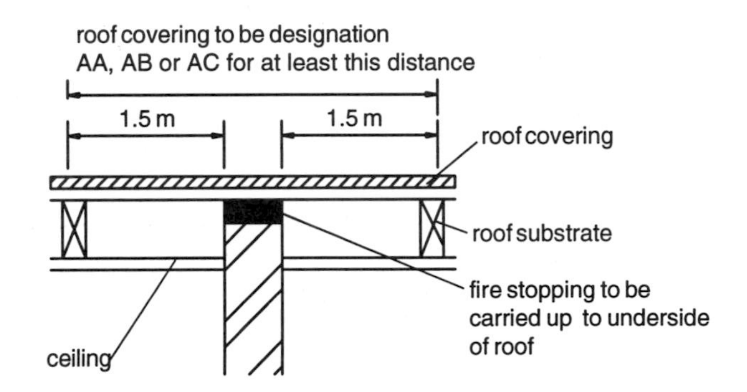

Figure D15 Roof carried over wall (combustible substrate).

Openings in compartments/separating walls and floors

Openings must not reduce the overall effectiveness of the floors or walls. The only openings that can be made in compartment/separating walls are for fire doors or services. In compartment/separating floors, openings can be made for a lift well, a flue or a stairwell in a protected zone. If appropriate alternative protection is provided, an accommodation stair or an escalator is allowed to pass through a compartment floor.

D2.12
D2.13
D2.16

(1) *Fire doors*

D2.2
Table 1

The integrity of fire doors (tested fitted in a frame from both sides) must be at least that of the separating wall or the compartment floor or wall in which the opening is made. One exception is for a fire door between a flat or maisonette and a space in communal occupation where the requirement is reduced to 30 minutes. The requirements for fire doors in protected zones or cavity barriers are discussed in the section on Part E.

(2) *Service openings*

D2.16 -
D2.18

Services (e.g. chimney, flue-pipe, pipe, duct) should either meet the fire resistance requirement for the wall or floor, or be enclosed within a duct which meets the requirement. All service openings must be fire-stopped using materials that can maintain the fire resistance of the element. Suitable materials include:

DTS
D2.16

(1) cement mortars;
(2) gypsum based plasters;
(3) glass or ceramic based products;
(4) intumescent mastics;
(5) proprietory sealing systems.

Viscoelastic fire stopping materials should always be reinforced or supported by a non-combustible material to avoid the risk of movement. Where the unsupported span in the opening is greater than 100 mm, all fire-stopping materials should be supported or reinforced.

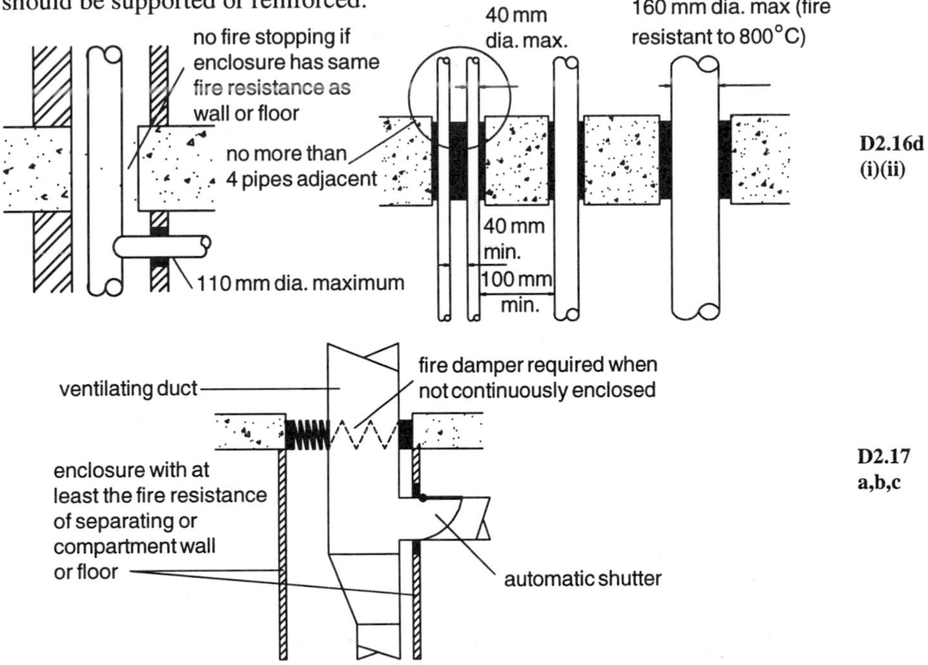

D2.16d (i)(ii)

D2.17 a,b,c

Figure D16 Main deemed-to-satisfy provision for service openings.

Cavities

Cavities within buildings range from cable ducts and wall cavities to voids above suspended ceilings and loft spaces. All cavities provide a possible route for smoke and flame through a building and therefore where the limits on area given in Table D6 are exceeded cavity barriers must be provided. As shown in the table the maximum permissible area of the cavity is linked to the purpose group and the surface spread of flame classification for the surfaces in the cavity. Where a cavity barrier is required it must:

DTS D2.16 D2.19

DTS D2.19

(1) be imperforate (fire doors and service openings meeting the requirements for compartment walls are allowed);
(2) provide 30 minutes fire resistance with respect to integrity and 15 minutes with respect to insulation (for cross-sections of less than 1 m x 1 m the requirements are less onerous);
(3) be tightly fitted to the surrounding construction, or at least the junction should be fire stopped;
(4) be suitably fixed so that loss of performance is unlikely as a result of movement of the building in a fire.

Table D6 Maximum dimensions of cavities

Purpose group or sub-group	Location of cavity	Class of surface exposed in cavity	Maximum distance between cavity barriers in any direction (m)	Maximum area of cavity (m^2)
1	between a roof and ceiling	any	any	no limit
	any other	0	20	no limit
		any other	15	no limit
2	between a roof and a ceiling above an undivided space not used for sleeping	0	no limit	no limit
		any other	15	199
	between a roof and a ceiling above an undivided space used for sleeping	any	15	100
	any other	0	20	no limit
		any other	15	no limit
3,4,5,6 and 7	between a roof or floor and a ceiling above an undivided space	0	no limit	no limit
	between a roof or floor and a ceiling other than above an undivided space	any	20	no limit
	any other	0	20	no limit
		any other	15	no limit

Maintaining the structure of a building in a fire

D2.2
Table 1
Table 2

The fire resistance of structural elements is determined by the purpose group of the building, the number of storeys, the height of the topmost storey/compartment, the storey area, compartment/building volume and the nature of the element of structure (e.g. structural beam, loadbearing wall, floor, external wall). The load bearing structure of a building must have the minimum period of fire resistance with respect to stability specified in Table D5. Floors and external walls less than one metre from the boundary must meet the fire resistance requirement with respect to stability, integrity and insulation. For the upper storeys of a house of purpose group 1C the requirement for floors is reduced to 30 minutes for stability and 15 minutes for integrity and insulation. Where support is provided to other elements of structure or where a structural element is built into, or forms part of, another element then it should attract the most onerous requirement for fire resistance and may in addition have to be non-combustible (e.g. a column forming part of a compartment wall). Non-combustibility is normally required for the following elements of structure:

D2.3 Table

(1) external walls on or within 1 m of the boundary;
(2) enclosing structure of protected zones;
(3) stairs, ramps and landings or passages within stair enclosures.

Example D1

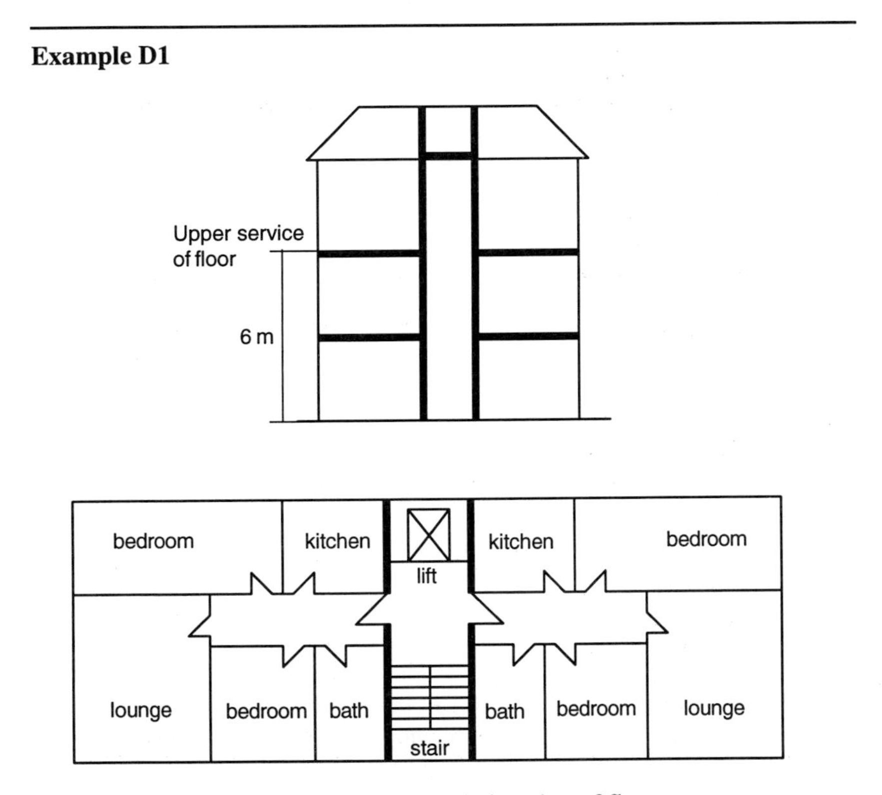

Figure D17 Plan and elevation of flats.

The main requirements for structural fire protection for the flats shown in Figure D17 are as follows:

1. Compartmentation/separation within building

(1) separating floors must be provided between flats but there is no compartmentation requirement for a building of purpose sub-group 1A based on storey area or cubic capacity; **D2.7** **D2.1 Table**
(2) compartment walls must separate the lift well from the flat, but are not required to the stair enclosure; **D2.8**

D2.2 D2.3 (3) the stair enclosure is a protected zone (i.e. part of the escape route leading directly to a place of safety) and therefore the fire resistance and non-combustibility requirements must be checked.

2. Fire resistance and non-combustibility

Reference	Element	Requirement
D2.2 Table 2	(1) separating floors	- as the height of the topmost storey is less than 7.5 m the minimum required fire resistance is 60 minutes;
D2.3 Table		- as the building is of purpose sub-group 1A and has no storey above 11 m the floors do not have to be non-combustible.
D2.2 Table 1	(2) compartment walls to lift well	- the minimum required fire resistance is 30 minutes in terms of stability, integrity and insulation;
D2.3 Table		- the construction must be of non-combustible materials;
D2.8		- the compartment walls must continue for the full height of the building.
	(3) stair enclosure (protected zone)	- as for the separating floors the minimum required fire resistance is 60 minutes; - the construction must be of non-combustible materials.
D2.3 Table	(4) stairs and landing within stair enclosure	- the construction must be of non-combustible materials.
D2.2	(5) external walls (load bearing elements of structure)	- the requirement is only 30 minutes but as the walls support a floor with a requirement of 60 minutes they must also have a fire resistance of at least 60 minutes;
D2.3 Table		- if the external walls are on or within 1m of the boundary they must be non-combustible.
	(6) internal walls (load bearing elements of structure)	- the internal load bearing walls of the ground floor flat will need to meet the fire resistance requirement of the floors they support (60 minutes as for the external walls) but do not have to be non-combustible.

The following points should also be noted:

(1) For this particular building it would be appropriate to use the same construction for all the walls which have a fire resistance requirement.

(2) The ground floor is neither a separating floor nor an element of structure and does not therefore have to meet fire resistance of non-combustibility requirements.

3. Openings, junctions and cavities

(1) Openings - Access from the flats to the stair enclosure must meet the requirement of Part E;

- the detailing of service openings through the separating floor must not adversely affect their performance. **D2.16 - D2.18**

(2) Junctions - junctions between the separating floor and the external walls and stair enclosure walls must be fire stopped. **D2.14**

(3) Cavities - cavity barriers are not required between the roof and the ceiling of the top flat but will be required below the ground floor unless the cavity is inaccessible or not more than 1m high. **D2.19**

Controlling the spread of fire between buildings

The methods for ensuring adequate separation between buildings are described in the provisions deemed to satisfy Part D. The minimum permissible distance from the boundary is determined from:

(1) the materials and construction of the roof taking into account any rooflights (the primary objective is not to contain a fire within the building but to avoid fire spreading from an adjacent building by penetrating through the roof, or across the roof to another building); **D2.5**

(2) the materials, construction and number of openings in the side of the building (the primary objective is to control the radiant heat at the boundary during a fire by taking into account the fire load, the fire resistance of the external wall, the number of openings in the wall and the amount of combustible cladding on the building). **D2.6**

Roofs

The minimum distance that a roof, or rooflight, must be from a boundary is determined by its fire performance designation. A description of how the designation is determined and examples of designations for typical roof types are given in Table D2. The table specifying the minimum boundary distances for different roof coverings and rooflights is shown below. Where the roof covering or a rooflight is within the building it is also subject to the requirements of Part E. **DTS D2.5**

Table D7 Minimum boundary distances for roof coverings and rooflights

Designation or description of roof covering (1)	Minimum distance from a boundary (m) less than 6	at least 6	~~less than~~ at least. 12	at least 20
AA,AB or AC	Yes	Yes	Yes	Yes
BA,BB or BC	No	Yes	Yes	Yes
CA,CB or CC	No	Yes(2)	Yes(3)	Yes
AD,BD or CD	No	Yes(2)	Yes(3)	Yes(3)
DA,DB,DC or DD	No	No	No	Yes(2)

Thatch or wood shingles	No	Yes(2)	Yes(3)	Yes(3)

'Yes' means acceptable; 'No' means not acceptable

Notes:

(1) For this purpose glass (wired or unwired) not less than 4 mm thick can be regarded as having an AA designation.

(2) Acceptable only in the buildings listed in Note (3), in a part of the roof being of an acceptable designation related to its boundary distance.

(3) Acceptable only in -

(a) a detached house; or

(b) a building of purpose group 2, above a compartment of not more than 1500 m^3.

If plastics materials are used they must either comply with the boundary distance requirements given above, or they must be at least 6 m from the boundary and meet the test criteria described on page D9. This requirement is relaxed where the plastics materials are used to cover open areas such as loading bays. In these cases there is no minimum boundary distance although the plastics must still meet the test criteria.

DTS D2.5

Sides of a building

The minimum distance to the boundary is normally determined using one of the three methods described in the provisions deemed-to-satisfy. Which method is used depends on the purpose group of the building and how critical precise calculation is. The methods are ranked in terms of the complexity of the procedure, with use of the simplest method producing the most onerous requirement. Method 1, the most restrictive, can only be used for certain purpose groups such as dwellings, institutional and residential buildings and offices (purpose groups and sub-groups 1,2,3,5B,6B). Methods 2 and 3 can be applied to buildings of any purpose group. Each of the three methods is described below. All of the methods require calculation of the total unprotected area of external wall. Unprotected areas are defined as openings, areas covered with combustible material more than 1 mm thick and sections with a fire resistance less than that prescribed for the external wall.

DTS D2.6

An alternative method for determining the minimum distance from the boundary is based on the use of the total unprotected area and a specified value of radiation intensity (dependent on the building type) to calculate the distance at which the radiation intensity would be dissipated to 12.6 kW/m^2. This calculation is only carried out in special circumstances. Additional requirements which are applicable to all buildings are summarised in Figures D18 to D21.

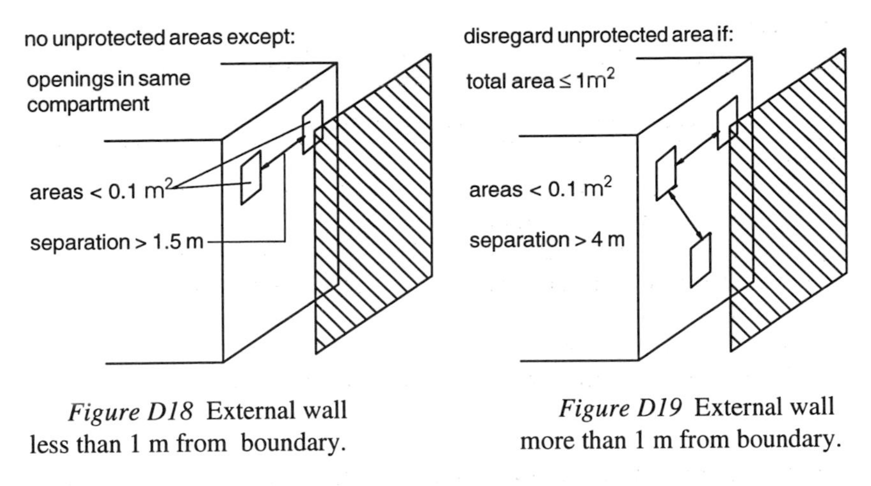

Figure D18 External wall less than 1 m from boundary.

Figure D19 External wall more than 1 m from boundary.

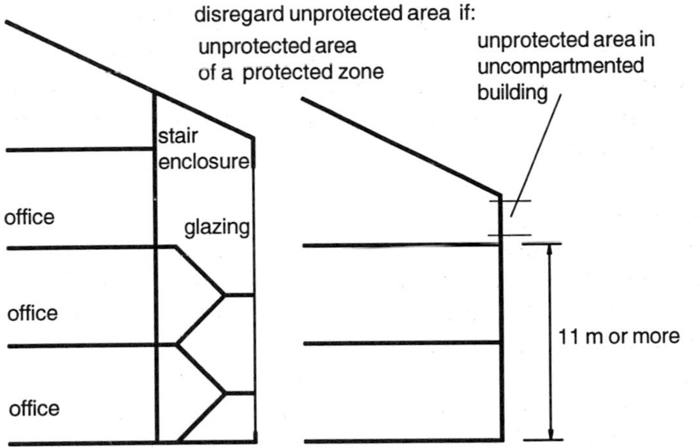

Figure D20 External wall 1 m or more from boundary.

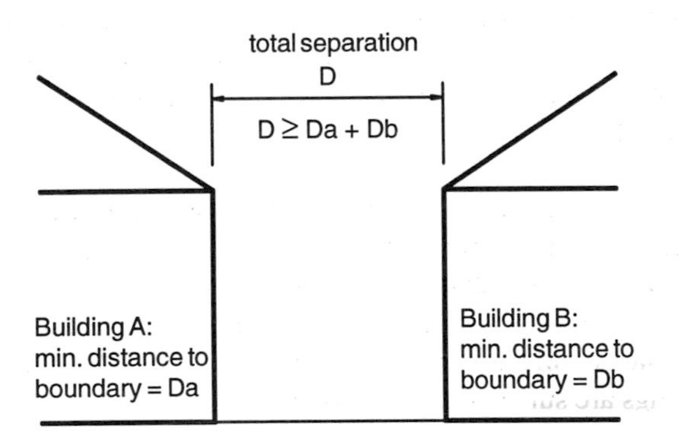

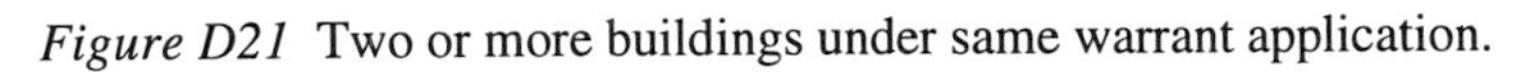

Figure D21 Two or more buildings under same warrant application.

Method 1 – Determining minimum distance to the boundary by simple geometry
The total unprotected area of the elevation and the length of the side of the building are used in Table D8 to find the minimum permissible distance to the **DTS D2.6**

boundary. This method is restricted to buildings of purpose groups and sub-groups 1,2,3,5B and 6B which are not more than 9 m in height and are at least 1 m from the boundary.

Table D8 Separation of sides of buildings from boundaries by simple geometry

Maximum total unprotected area (m^2)	Maximum length of side (m)	Minimum distance between side of building and boundary (m)
5.6 (2)	24	1.0 (1)
15.0 (2)	24	2.5 (1)
no limit	12	5.0
no limit	24	6.0

Note:

(1) The minimum distance from the boundary may be arrived at by interpolation between the figures shown for the maximum unprotected area.

(2) For those parts of the external wall clad in combustible material more than 1 mm thick, the whole area if the wall behind does not have the required fire resistance, or half the area where it does.

Example D2

Determination of minimum boundary distance by simple geometry for the building in Figure D22.

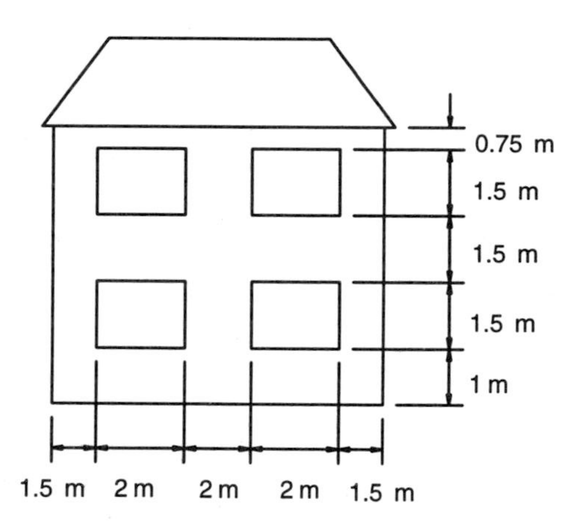

Figure D22 Elevation of building.

Solution.

Unprotected area $= (4 \times 1.5 \times 2)\ m^2 = 12\ m^2$

Length of side of building $= 9$ m

12 m^2 lies between 5.6 and 15.0 in table D8 and by interpolation results in a minimum distance between side of building and boundary of 2.0 m

Method 2 – Determining minimum distance to the boundary by enclosing rectangles

In this method the unprotected area is expressed as a percentage of the area of a rectangle taken from the tables to method 2 which projected onto the elevation encloses all of the unprotected area(s). The percentage is then used in the tables to determine the minimum distance to the boundary. The method can be applied to all buildings 1 m or more from the boundary. Table D9 gives extracts from Table A to method 2.

Step 1. Establish projected rectangle – Project a rectangle onto the plane of reference which encloses all of the unprotected areas. In the examples below the projected rectangle is shown by shading.

Step 2. Determine enclosing rectangle – From the tables to method 2 (extracts in Table D9) select a height and width for an enclosing rectangle equal to or greater than that of the projected rectangle. The heights for enclosing rectangles given in the tables range from 3 m to 15 m in 3 m increments and the widths from 3 m to no limit.

Example D3

Determination of the minimum boundary distance by enclosing rectangle for the buildings shown in Figures D23-D25.

D23-D25

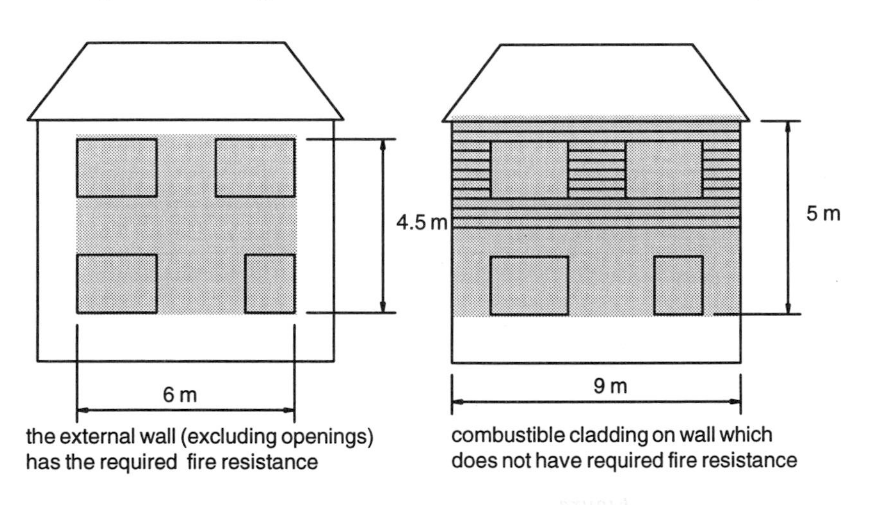

Figure D23 Building A.

Figure D24 Building B.

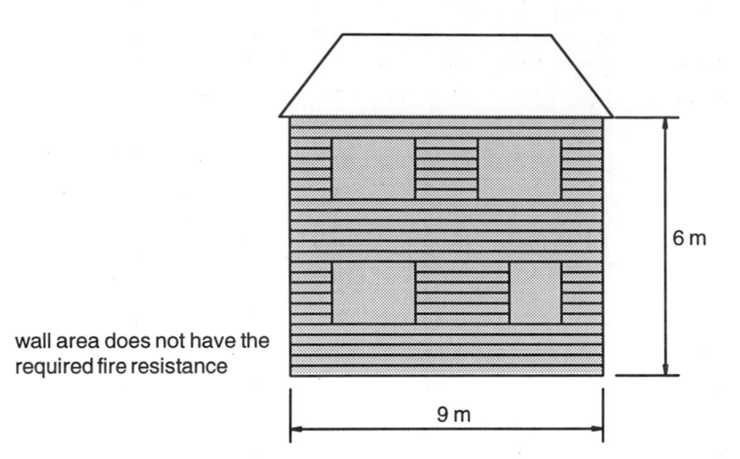

Figure D25 Building C.

In these examples it has been assumed that the facade is parallel to the 'plane of reference'. The method for projecting the facade onto the plane of reference where the two are not parallel is explained later.

Table D9 Assuming buildings (A),(B) and (C) are of purpose group 2 the dimensions of the enclosing rectangles are:

Building (A):	Height: 6 m	Width: 6 m
Building (B):	Height: 6 m	Width: 9 m
Building (C):	Height: 9 m	Width: 9 m

Table D9 Permitted unprotected percentages in relation to enclosing rectangles for purpose groups 1,2,3 or purpose sub groups 5B,6B or 7C

Height of enclosing rectangle	**Width of enclosing rectangle**	**Minimum distance from relevant boundary (m) for unprotected percentage not exceeding**								
(m)	(m)	20%	30%	40%	50%	60%	70%	80%	90%	100%
3	3	1.0	1.0	1.0	1.5	1.5	1.5	2.0	2.0	2.0
	6	1.0	1.0	1.5	2.0	2.0	2.0	2.5	2.5	3.0
	9	1.0	1.0	1.5	2.0	2.5	2.5	3.0	3.0	3.5
	30	1.0	1.5	2.0	2.5	3.0	3.5	4.0	4.0	4.5
	no limit	1.0	1.5	2.0	2.5	3.0	3.5	4.0	4.0	5.0
6	3	1.0	1.0	1.5	2.0	2.0	2.0	2.5	2.5	3.0
	6	1.0	1.5	2.0	2.5	3.0	3.0	3.5	4.0	4.0
	9	1.0	2.0	2.5	3.0	3.5	4.0	4.5	4.5	5.0
	12	1.5	2.5	3.0	3.5	4.0	4.5	5.0	5.0	5.5

	15	1.5	2.5	3.0	4.0	4.5	5.0	5.5	5.5	6.0
	18	1.5	2.5	3.5	4.0	4.5	5.0	5.5	6.0	6.5
	80	1.5	2.5	3.5	5.0	6.0	7.0	7.5	8.5	9.5
	no limit	1.5	2.5	3.5	5.0	6.0	7.0	8.0	8.5	10.0
9	3	1.0	1.0	1.5	2.0	2.5	2.5	3.0	3.0	3.5
	6	1.0	2.0	2.5	3.0	3.5	4.0	4.5	4.5	5.0
	9	1.5	2.5	3.5	4.0	4.5	5.0	5.5	5.5	6.0
	12	1.5	3.0	3.5	4.5	5.0	5.5	6.0	6.5	7.0
	15	2.0	3.0	4.0	5.0	5.5	6.0	6.5	7.0	7.5
	60	2.0	4.0	5.5	7.0	8.0	9.5	11.0	11.5	13.0
	80	2.0	4.0	5.5	7.0	8.5	10.0	11.5	12.5	13.5
	no limit	2.0	4.0	5.5	7.0	8.5	10.5	12.0	12.5	15.0
12	3	1.0	1.5	2.0	2.0	2.5	3.0	3.0	3.5	3.5
	18	2.5	4.0	5.0	6.0	7.0	7.5	8.5	9.0	10.0
	30	2.5	4.5	6.5	7.5	8.5	9.5	10.5	11.5	12.5
	40	2.5	5.0	6.5	8.0	9.5	10.5	12.0	13.0	14.0
	60	2.5	5.0	7.0	9.0	10.5	12.0	13.5	14.5	16.0
	100	2.5	5.0	7.5	9.5	11.5	13.5	15.0	16.5	18.0
	no limit	2.5	5.0	7.5	9.5	12.0	14.0	15.5	17.0	19.0
15	3	1.0	1.5	2.0	2.5	2.5	3.0	3.5	3.5	4.0
	6	1.5	2.5	3.0	4.0	4.5	5.0	5.5	5.5	6.0
	9	2.0	3.0	4.0	5.0	5.5	6.0	6.5	7.0	7.5
	120	3.5	6.5	9.0	11.5	14.0	16.5	18.5	20.5	22.5
	no limit	3.5	6.5	9.0	12.0	14.5	17.0	19.0	21.0	23.0

Note: The minimum distance from the boundary may be arrived at by interpolation between the percentages shown.

Determining the minimum distance to the boundary

Step 1. Calculate the total unprotected area – Any area covered by combustible cladding is considered as an unprotected area if it is more than 1 mm thick and if the wall behind it does not meet the requirements for fire resistance. If the wall does meet the requirements for fire resistance half the area of cladding is used in the determination of total unprotected area.

Step 2. Determine the area of the enclosing rectangle.

Step 3. Express the unprotected area as a percentage of the area of the enclosing rectangle.

Step 4. From the table find the minimum allowable distance for the unprotected percentage.

Example D3 (cont.)

Assuming that the unprotected areas for Buildings (A) and (B) in Figures D23 and D24 are 11 m^2 and 27 m^2 respectively, the minimum distances to the boundary are calculated as follows:

Building (A)
(1) Total unprotected area = 11 m^2
(2) Area of enclosing rectangle = 6 m x 6 m = 36 m^2
(3) Unprotected percentage = 11/36 x 100 = 31%
(4) Minimum distance to the boundary = 1.550 m
(by interpolation between 30% and 40% values)

Building (B)
(1) Total unprotected area = 27 m^2
(2) Area of enclosing rectangle = 6 m x 9 m = 54 m^2
(3) Unprotected percentage = 27/54 x 100 = 50%
(4) Minimum distance to the boundary = 3.0 m

The effect of compartmentation/separation

DTS D2.6 Where a building is divided into fire resistant cells by compartment/separating walls and floors the minimum distance to the boundary is determined for each compartment separately.

Example D4

Effect of compartmentation on the determination of minimum boundary distance

(a) An uncompartmented building

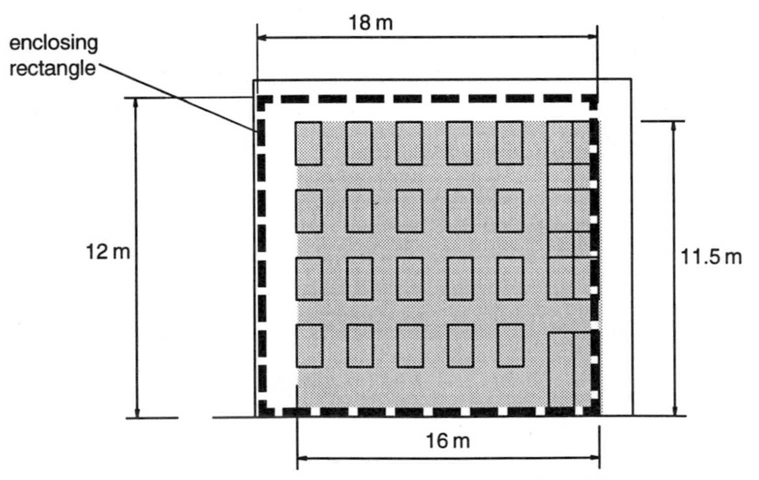

Figure D26 Uncompartmented building.

(1) Projected rectangle (enclosing unprotected areas) = 11.5 m x 16 m
(2) From Table D9 enclosing rectangle = 12 m x 18 m = 216 m^2
(3) Assume unprotected areas = 105 m^2
(4) Unprotected percentage (unprotected areas as percentage of enclosing rectangle) = 105 m^2 as percentage of 216 m^2 in Table D9
= 48.5% use 50% column
(5) From Table D9 minimum distance from boundary = 6 m

(b) Compartmented building

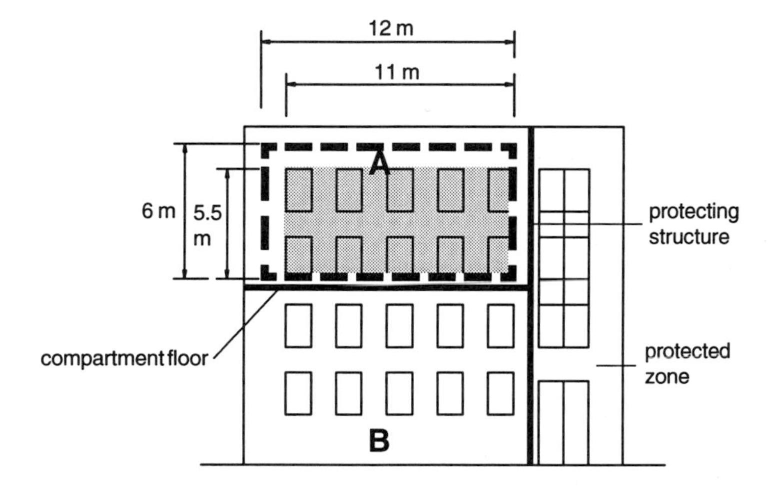

Figure D27 Compartmented building.

As the entrance and stairways are now isolated the area becomes a protected zone and the glazed area does not now count as part of the unprotected area. The remainder of the building is divided by the compartment floor into compartments A and B. In this example the compartments have the same unprotected area. But where there are two (or more) compartments with different unprotected areas take the compartment with the greatest unprotected area.

(1) Projected rectangle = 5.5 m x 11 m
(2) From Table D9 enclosing rectangle = 6 m x 12 m
= 72 m^2
(3) Assume unprotected areas = 26 m^2
(4) Unprotected percentage = 26 m^2 as percentage of 72 m^2
= 36% – use 40% column in Table D9
(5) From Table D9 minimum distance from boundary = 3 m

(c) Compartmented building

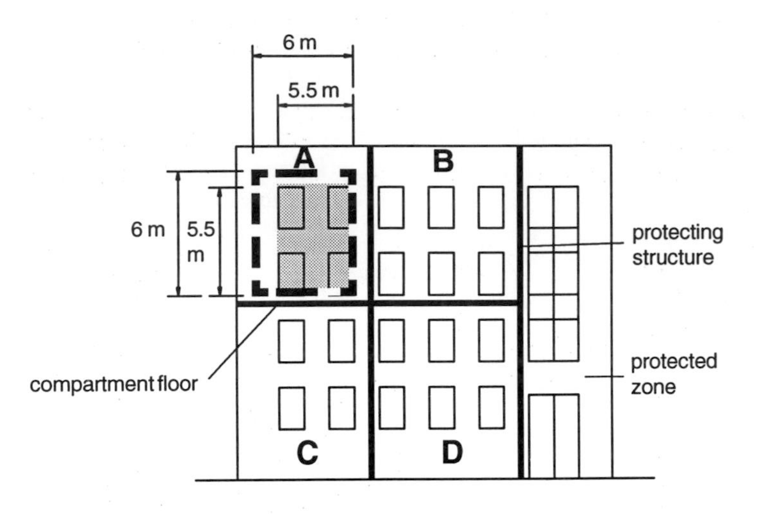

Figure D28 Compartmented building.

With the inclusion of a compartment wall the building is now divided into compartments A,B,C and D, each having the same unprotected area for the purpose of this example.

(1) Projected rectangle	= 5.5 m x 5.5 m
(2) From Table D9 enclosing rectangle	= 6 m x 6 m = 36 m^2
(3) Assume unprotected areas	= 13 m^2
(4) Percentage of unprotected area	= 36% – use 40% column in Table D9
(5) From Table D9 minimum distance from boundary	= 2 m

Notes:

(1) In the above diagrams the relevant boundary is assumed as parallel with the wall face and the plane of reference to coincide with the wall face. But this will not always be so.

(2) In a building of purpose sub-group 7C (open sided car park) unprotected areas of each storey are to be considered separately notwithstanding that floors are not required to be compartment floors.

Establishing the plane of reference

Where the wall face is not flat a plane of reference must be established and the unprotected areas projected onto it. While the plane of reference should normally be approximately parallel to the relevant boundary it must also comply with the following requirements when drawn on a plan of the building:

(1) touches all or part of the building;

(2) however far extended does not pass within the building (but may pass through projections such as balconies and copings); and
(3) does not cross the relevant boundary if this is already established.

When the plane of reference is established, unprotected areas in the wall faces are projected perpendicularly into it as shown. The minimum distance from the boundary is then determined as before using the elevation of the plane of reference.

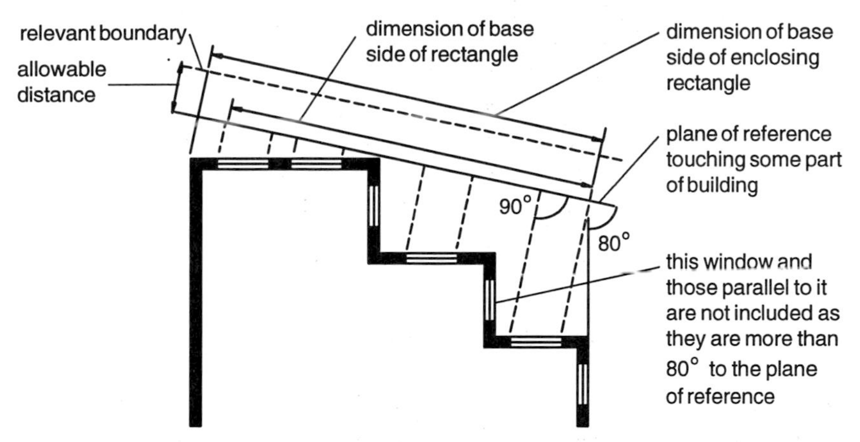

Figure D29 Establishing length of baseline of enclosing rectangle.

Notes:
(1) Project on to the plane of reference the unprotected areas on all floors to find the dimension of the base side of the rectangle.
(2) The relevant boundary could be a notional boundary depending on the use of the building and the circumstances.

Method 3 – Determining minimum distance to the boundary by calculating aggregate notional areas
The basis of this method is to assess the effect of a fire at a series of points on the boundary using the unprotected areas that would actually contribute to radiation at that point. The method is more complicated than method 2 but as it is possible to exclude certain unprotected areas it may be advantageous where:

(1) the unprotected areas are to be maximised, particularly in borderline cases;
(2) a building has setbacks in plan or section;
(3) the boundary is irregular;
(4) the distance determined using method 1 or 2 cannot be complied with.

The unprotected areas which are not excluded are multiplied by a factor which varies inversely with the distance of the unprotected area from the point on the boundary being considered. The multiplication factors are given in Table D10.

Table D10 Multiplication factors for aggregate notional area

Distance of unprotected area from vertical datum		
Not less than (m)	**Less than (m)**	**Multiplication factor**
1.0	1.2	80.0
1.2	1.8	40.0
1.8	2.7	20.0
2.7	4.3	10.0
18.5	27.5	0.25
27.5	50.0	0.1
50.0	no limit	0.0

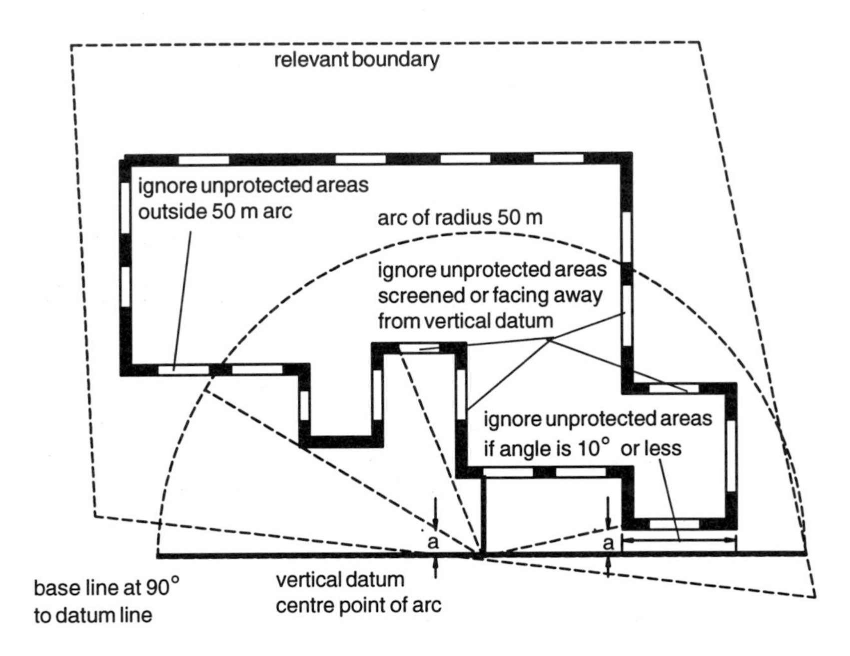

Figure D30 Distance to boundary.

With reference to Figure D30, the procedure for determining whether or not the distance to the boundary meets the requirements of the Technical Standards is as follows:

Step 1. Set a vertical datum of unlimited height at any position on the relevant boundary, and from this point draw a datum line to the nearest point on the building or compartment. (Where the boundary distance has not been set, an assumed relationship with the relevant boundary should be made.)

Step 2. From the vertical datum:

(a) draw a base line set at 90° to the datum line; and

(b) describe an arc with a radius of 50 m to meet the base line.

Step 3. Establish which unprotected areas can be discounted, namely those which:

(a) are outside the 50 m arc; or
(b) are screened from the vertical datum by any part of the external wall which is not an unprotected area; or
(c) are facing away from the vertical datum; or
(d) make an angle not exceeding 10° with a line drawn from the unprotected area to the vertical datum.

Step 4. For those unprotected areas which cannot be discounted under step 3:

(a) measure the distance between each unprotected area and the vertical datum and find the appropriate multiplication factor given in Table D10;
(b) multiply each actual unprotected area by its factor; and
(c) total these to give the aggregate notional area (in m^2) for that vertical datum.

The aggregate notional area should not be more than:

(1) 210 m^2 (in the case of purpose groups 1, 2, 3 or purpose sub-groups 5B, 6B or 7C); or
(2) 90 m^2 (in the case of purpose group 4 or purpose sub-groups 5A, 6A, 7A or 7B).

Step 5. On either side of the vertical datum mark off a series of points not more than 3 m apart along the length of the relevant boundary then, using each of these points, repeat steps 2, 3 and 4. (In practice it is usually possible, by observation, to place the first vertical datum at the worst position thereby obviating the need for further calculations.)

Step 6. If any of the resulting aggregate notional areas is greater than the allowance (given in step 4), the design must be modified accordingly until the proposed unprotected areas are suitable for the boundary distance.

Step 7. Repeat the process for all sides of the building situated not less than 1 m from any point of the relevant boundary.

Alternative standards for buildings ancillary to dwellings

For buildings attached to dwellings of purpose group 1, or on the property of dwellings of purpose group 1, and which are in the same occupation as the dwelling, alternative standards may be applied. These are illustrated in Figures D31-D33. **D3.1**

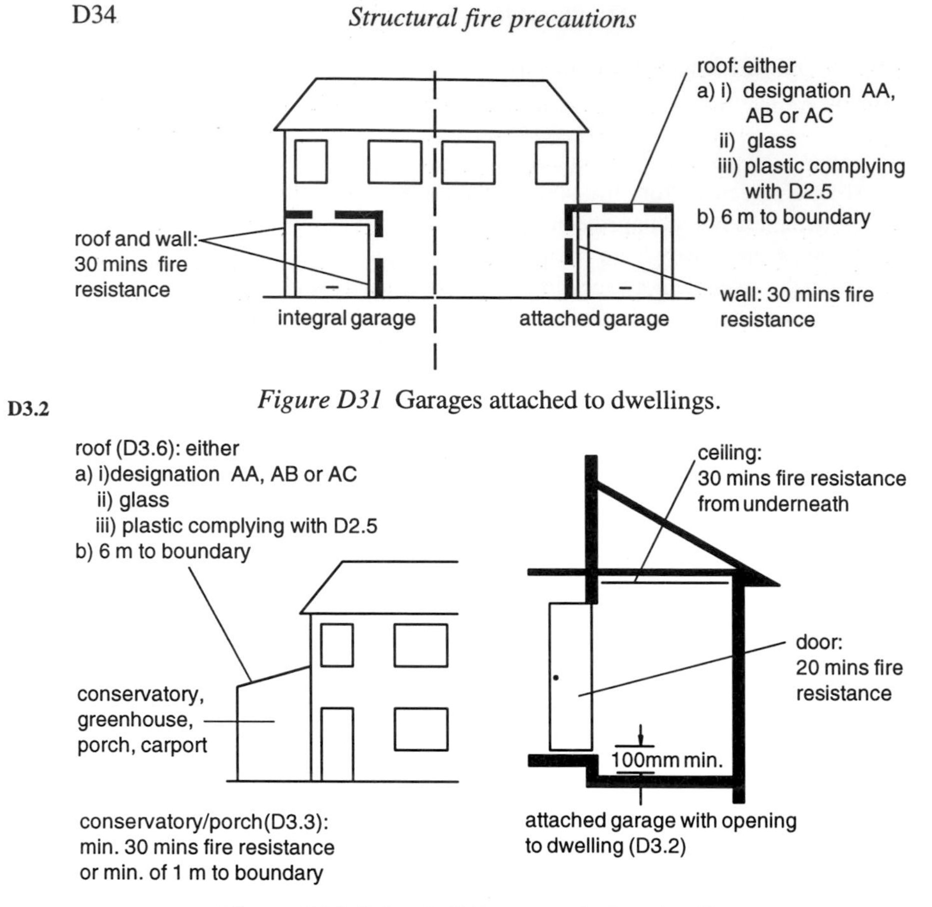

D3.2

Figure D31 Garages attached to dwellings.

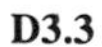

D3.3

Figure D32 Other buildings attached to dwellings.

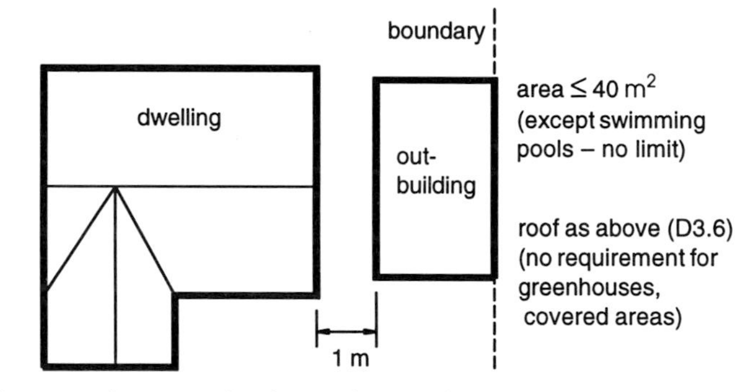

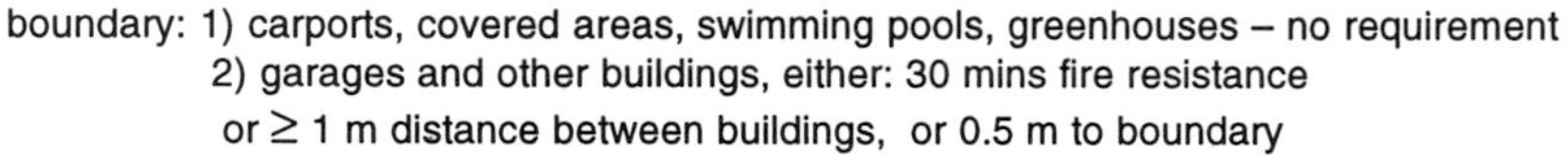

D3.4
D3.5

Figure D33 Detached buildings ancillary to dwellings.

Amendments to Part D

The 1993 amendments to the Technical Standards included a Part D amendment which consisted in the main of tidying up information in tables D4, D5, D6 and D7 presented earlier (in their current amended form).

Part D is presently under further review by the Scottish office (Construction and Building Control Group). The consultation document has been circulated for comment and the Scottish Office is formulating the amendment in the light of these responses.

MEANS OF ESCAPE FROM FIRE AND FACILITIES FOR FIRE FIGHTING

PART E **Regulation 13**

BUILDING STANDARDS

Regulation

13 (1) Every building shall be provided with –
  (a) adequate means of escape in the event of fire; and
  (b) adequate fire-fighting facilities.
(1A) Every dwelling shall be provided with means of warning the occupants of an outbreak of fire.
(2) This regulation shall not be subject to specification in a notice served under section 11 of the Act in respect of –
  (a) buildings of purpose sub-groups 1B and 1C; and
  (b) buildings to which the Fire Certificates (Special Premises) Regulations 1976(a) apply.

Aim

That a building is designed and constructed to provide the safest possible means of escape for all users of the building. In pursuit of this aim regulation (1A) relating to installation of smoke alarms in dwellings was added as a result of the July 1993 amendment. Also that the building is provided with appropriate fire-fighting equipment and safe means of access for the fire brigade.

TECHNICAL STANDARDS

Scope

The provisions for means of escape apply to all buildings, dwellings are specifically targetted for means of warning of outbreak of fire and the provisions for fire-fighting facilities apply to only certain purpose groups.

Requirements

The design of a building must include the provision of effective means of escape. The escape routes from any point within the building can be broken down into three zones:

(1) unprotected zone – it should take no more than 2.5 minutes to reach either a protected doorway or a place of safety; this is ensured by specifying a maximum travel distance which is related to purpose group;
(2) protected zone – stair enclosure, corridor, etc., providing at least one hour fire resistance;
(3) place of safety – a space in the open air at ground level.

It is not a requirement that all three zones form part of the escape route as long as a place of safety is reached.

The number and width of escape routes (including doorways, corridors, stairs) are determined by the occupancy capacity of the locations from which escape must be made. There is a requirement for half an hour fire resistance for corridors/ fire doors which will provide for the safe evacuation of everyone in the building. Protected zones are required to provide a minimum of one hour fire resistance to allow half an hour for the fire brigade to reach the building and pass through the protected zone to the source of the fire.

Where a building has only one normal means of escape then emergency windows should generally be provided.

The provision of means of warning of fire in dwellings is to increase significantly the level of safety of the occupants.

Background

The implementation of this part has always been the responsibility of building control departments in Scotland. The implementation of the Fire Protection Act is the responsibility of the fire service. Building Control Officers and Fire Prevention Officers work together to scrutinise the legislative requirements.

The introductory section in Part D should be read as an introduction to this section. Part E should always be applied in conjunction with Part D.

COMPLIANCE

There is considerable variation in the application of this regulation to buildings of different purpose group or sub group. Therefore care must be taken when applying the provisions of the relevant clauses.

In particular it should be noted (when applying Part E) that:

(1) a flat or maisonette entered only from the open air at ground level and with no storey at more than 4.5 m should be considered as purpose sub-group 1C;

E1.3 (2) a maisonette entered only from the open air at ground level and with a storey at more than 4.5 m should be considered as purpose sub-group 1B;

E1.4 (3) 'smoke alarm' means a device powered by mains electricity, with or without a back-up power source and containing within one housing all the components necessary for detecting fire and thereupon giving an audible alarm.

Means of escape from fire

Provision of escape routes

E2.1 Every part of the building must have easy access to an escape route. The term used in the regulation is from every 'point of origin'. Briefly this can be defined as:

(1) any point within a room or storey; or
(2) the main entrance door to a flat or maisonette.

This provision also includes roof areas that are accessible (e.g. flat roofs and escape routes over roof areas) but excludes roof areas that are accessible for maintenance purposes only.

These escapes routes must be designed to lead to a place of safety; this is defined as:

(1) an unenclosed space in the open air at ground level; or
(2) an enclosed space in the open air at ground level leading to an unenclosed space, via an access not narrower than the total width of the exits leading from the building to that enclosed space.

The number of escape routes is determined by the following:

(1) the purpose group of the building; **E2.2**
(2) the occupancy capacity of the building;
(3) the height of a storey above or below ground level;
(4) the travel distance involved.

Each of these factors influences the appropriate number of escape routes. The minimum requirement in relation to occupancy capacity is set out in Table E1. **E2.3** As an example, for a cinema with a seating capacity of 700 persons, the minimum requirement is three escape routes. However, if the cinema had a basement, the minimum number of escape routes would be as follows:

(1) from the basement – two escape routes;
(2) from the seated area – three escape routes.

Similarly buildings with upper storeys above ground level must have two escape routes from the upper storeys with the following exceptions: **E2.4**

(1) buildings of purpose group 1;
(2) buildings of purpose group 2b, 3, 4, 5, 6 and 7 if the upper storey height is not more than eleven metres and the limit on occupancy capacity is met;
(3) upper storeys containing only plant (including lift machinery) unless the room is a place of special fire risk;
(4) a basement storey where the floor level is not more than three metres below the ground which is used only for storage, or contains only plant, or has sanitary accommodation or a combination of the above.

Table E1 Minimum number of escape routes in relation to occupancy capacity

Occupancy capacity of room or storey	Minimum number of escape routes
Not more than 60	1
61–600	2
over 600	3

Requirements for escape routes

Maximum travel distance

The number of escape routes and the purpose group together determine the

E2.5 maximum distance to be travelled along an escape route to a protected doorway. The distances are set out in Table E2 (which is an extract from Table E25 of the standard).

Table E2 Maximum travel distance

Available directions of travel from any point within storey (m)		one	more than one
2A		9	18
2A	hospitals	15	32
2B		15	32
3		18	45
4		15	32
5	generally	15	32
5B	(a) buildings primarily for use by disabled people	9	18
	(b) swimming pool in air supported structures	9	15
6		18	45

If the travel distance requirement cannot be complied with then extra escape routes will have to be provided.

Measuring travel distance

E2.71 Travel distance is measured along the escape route *from* either:

(1) main entrance of a flat or maisonette; or
(2) any point within a room or storey.

To the nearest protected door which must open on to either:

(1) a place of safety;
(2) a protected zone leading to a place of safety;
(3) an unenclosed external escape stair;
(4) an escape roof across a flat roof to a place of safety;
(5) a doorway in a compartment wall.

A corridor providing escape in only one direction in a building, other than a building of purpose group 1, cannot be considered a protected zone irrespective of its construction.

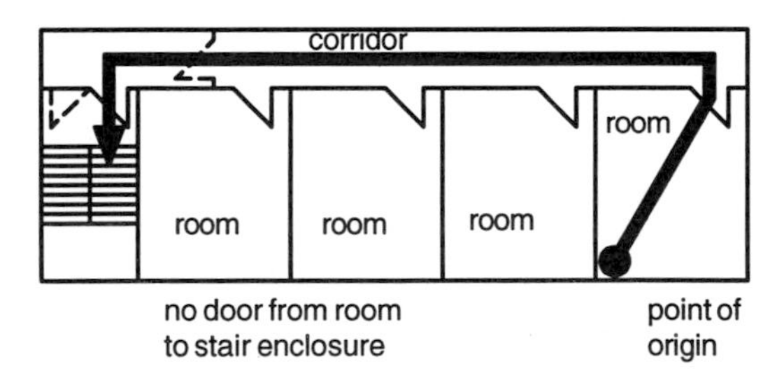

Notes:
1. Travel distance = maximum for one direction (see table E2.5) to protected door of stair enclosure. This is regardless of status (protected or otherwise) of the corridor.
2. Alternative positions of fire doors shown dotted.

Figure E1 Travel distance in one direction – upper storey corridor.

If more than one direction of travel is available on an escape route then if these are to be treated as two escape routes the requirements illustrated should be met. **E2.73**

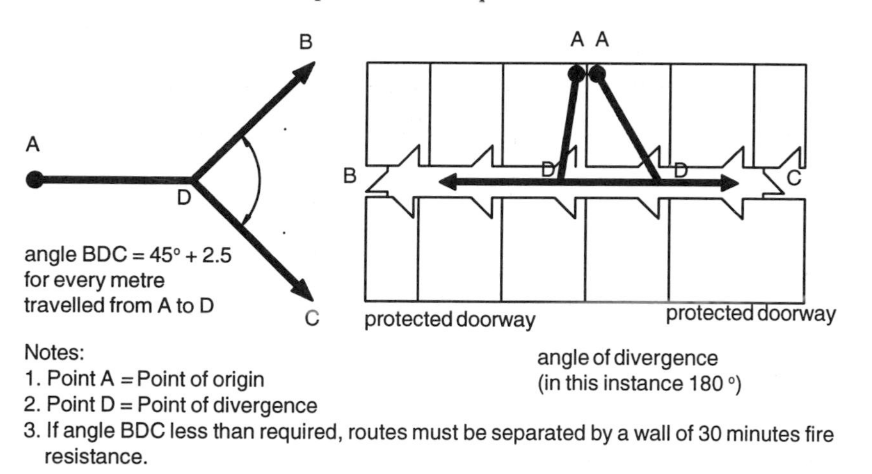

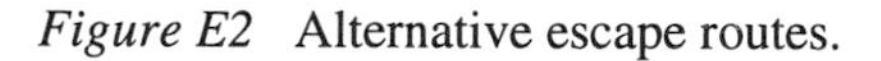
Figure E2 Alternative escape routes.

Width of escape route from a room or storey

Table E3 Minimum unobstructed width of escape routes from a room or storey

Occupancy capacity of room or storey	Minimum width (mm)
Not more than 60	800
61-100	1000
over 100	1100

For a storey with over 100 persons the minimum width of an escape route would be 1100 mm. However, this width can be reduced at door jambs as illustrated in Figure E3. **E2.6** **E2.24**

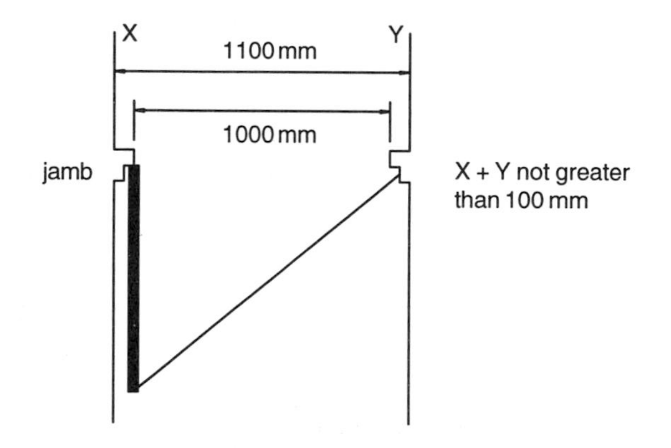

Figure E3 Width of door jambs.

For rooms containing fixed seating or fixed storage, the widths will vary from the requirements given in this table.

E2.7 The total width of all escape routes when added together must be greater than 5.3 x occupancy capacity.

The minimum width of escape stairs is based on this figure as shown in example E1; the exception for protected lobbies is illustrated in example E2.

Example E1

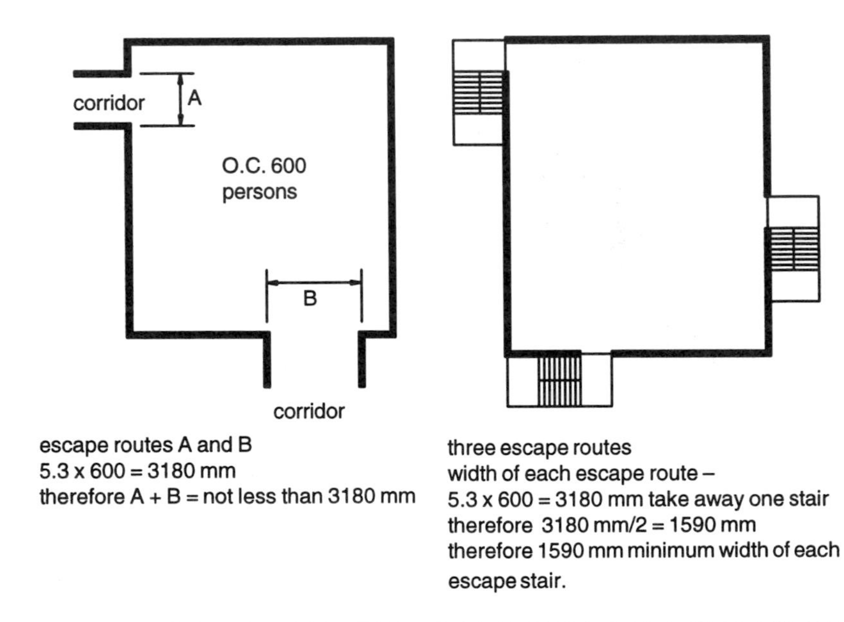

Example E2

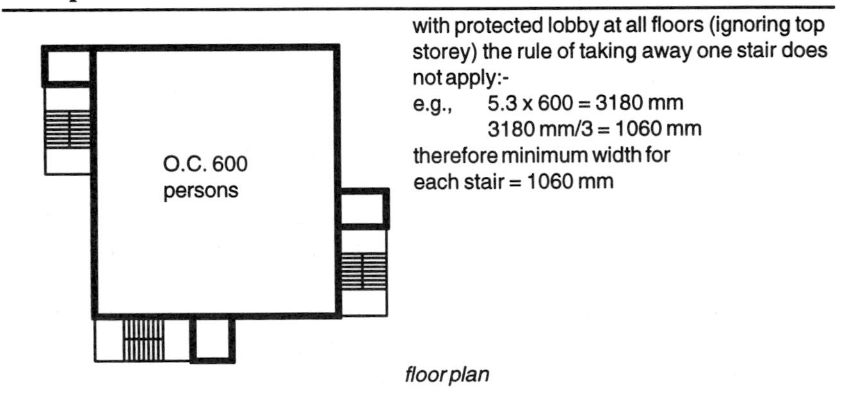

On a ground floor with an escape route passing through an escape stair enclosure, the width of the combined ground storey and escape stair must be sufficient to accommodate the occupancy capacities for the stair and the ground floor served by the escape route. **E2.9**

Example E3

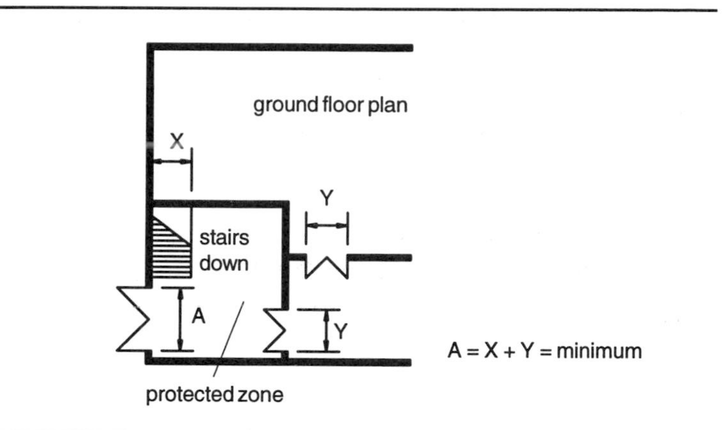

Height of escape routes

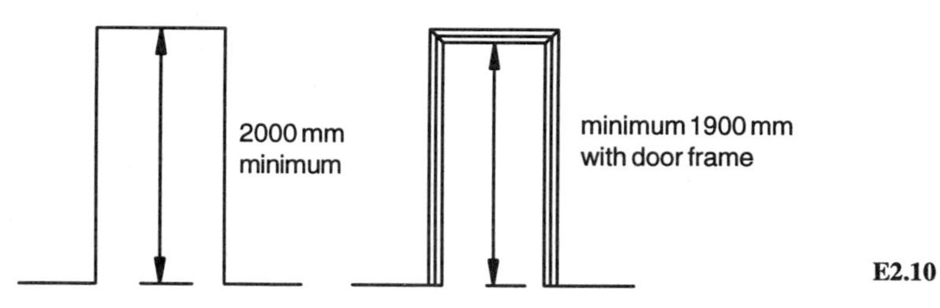

E2.10

Figure E4 Minimum height requirement in escape route corridor.

Capacity of a storey in relation to escape stairs
This can be determined as: **E2.75**

(1) the occupancy capacity of the storey provided it does not serve storeys above;
(2) if it serves two or more storeys which are not compartmented 80% of the sum of the occupancy capacity of each storey;
(3) if it serves two or more storeys each separated by compartment floors 80% of the two immediately adjacent storeys with greatest combined occupancy capacity;
(4) if it serves two or more storeys in a compartmented building 80% of the total occupancy of either the two adjacent storeys or the compartment with the greatest occupancy capacity.

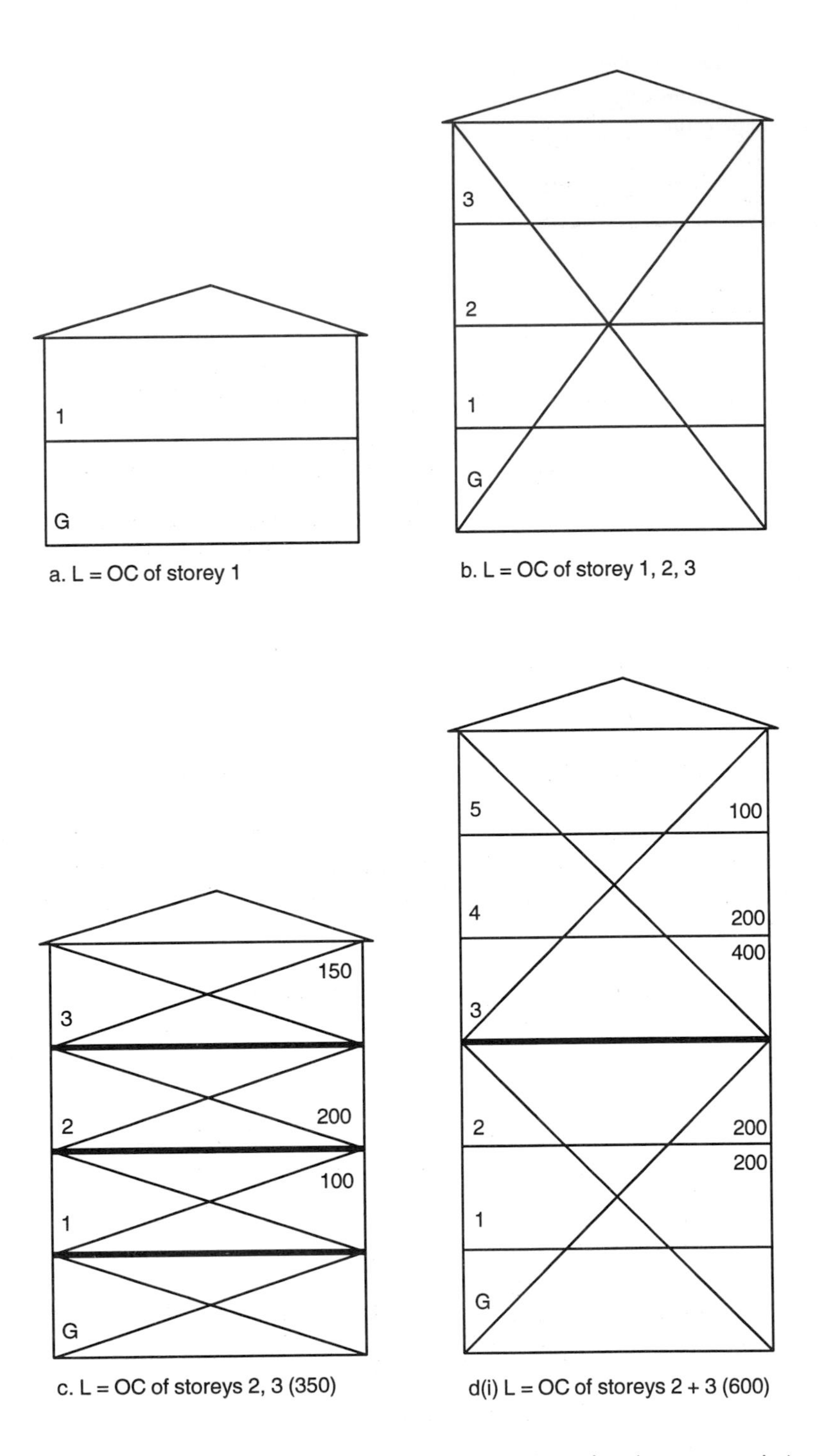

Figure E5 Appropriate capacity of escape stairs, (cont. opposite).

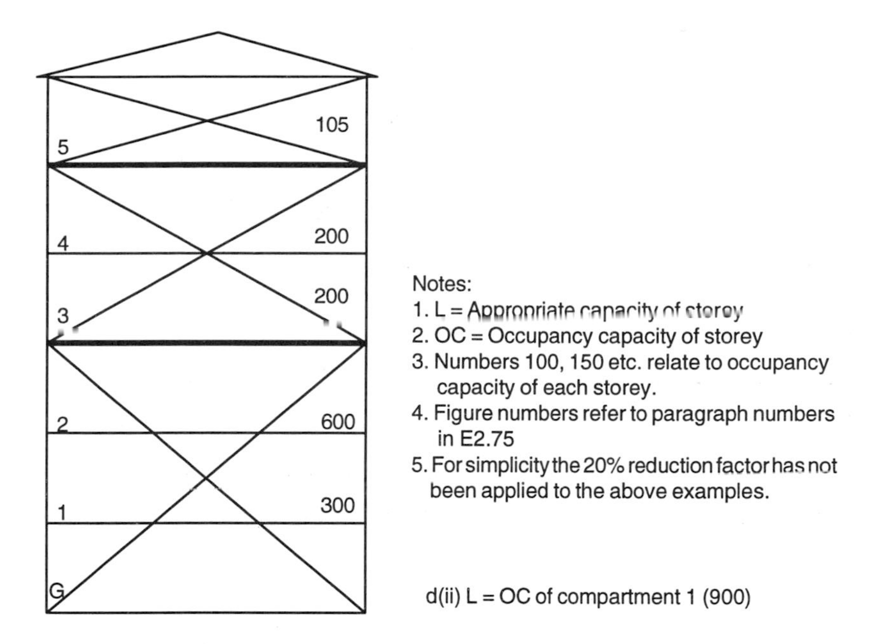

Figure E5 Appropriate capacity of escape stairs (cont.).

If an escape route from a storey is by a combination of stairs and other escape routes the appropriate capacity for the storey is that proportion of the occupancy capacity of the storey which the total width of all escape routes discharging to the stairs bears to the total width of all escape routes. **E2.76**

Zone transfer requirements from point of origin to place of safety

The main requirements for movement to a place of safety are as illustrated in Figure E6. **E2.11 E2.12 E2.13**

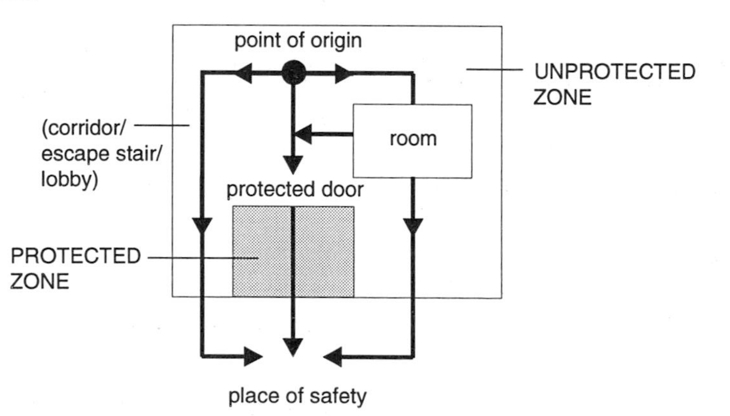

Figure E6 Movement to a place of safety.

E2.14 The following should not form part of an escape route:

(1) lift;
(2) passenger conveyor;
(3) escalator;
(4) revolving door;
(5) turnstile or shutter.

E2.13 ***Additional requirements for escape routes***

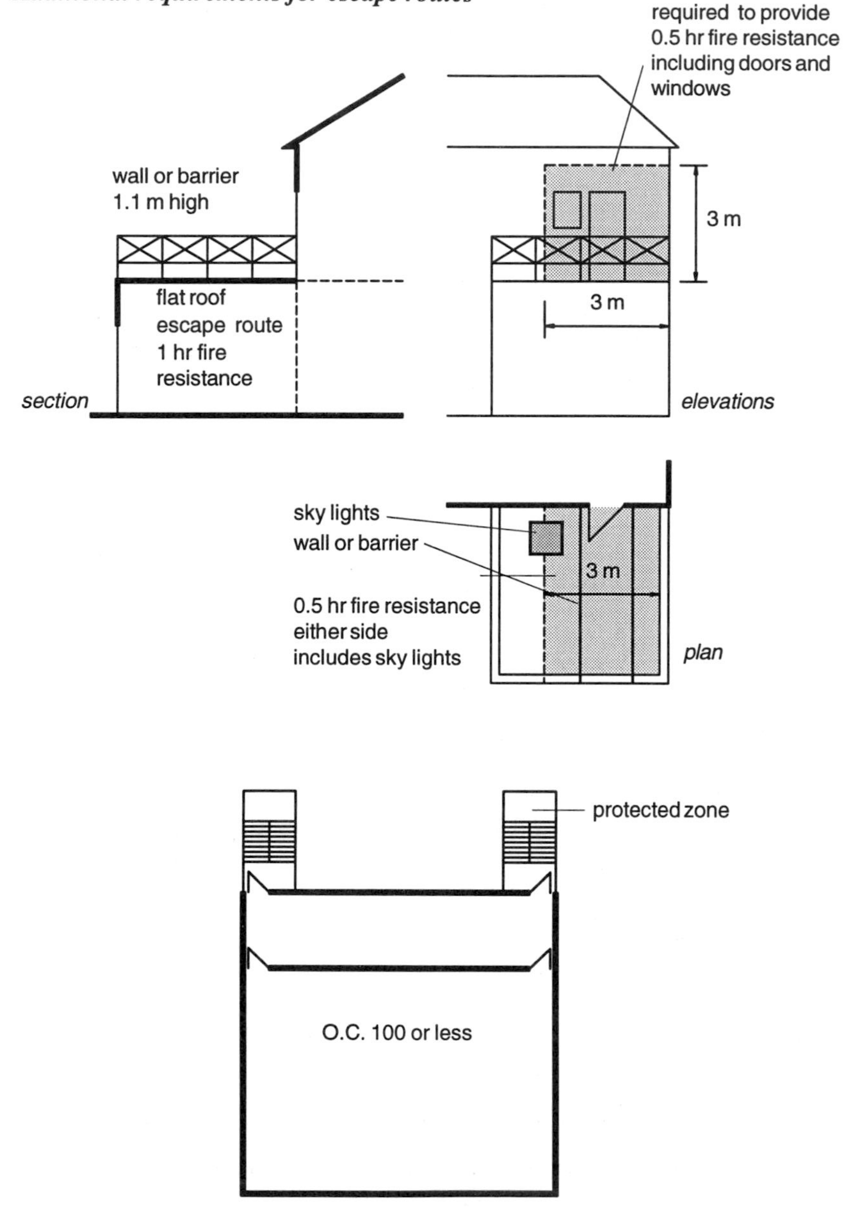

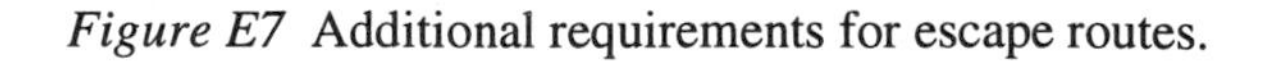
Figure E7 Additional requirements for escape routes.

Requirements for buildings with two or more escape routes
A building with two or more stair enclosures must be designed so that from any point of origin there are at least two escape routes.

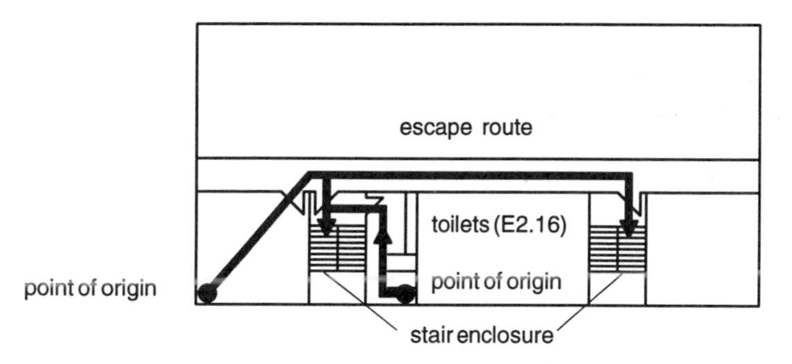

Figure E8 Two or more stair enclosures.

For corridors that provide two directions of travel as illustrated in Figure E8, the length of the corridor between the two protected doors must not exceed the distance given in Table E4. If the corridor exceeds the specified distance then it must be sub-divided by self closing fire doors in order to reduce the travel distance and to prevent smoke spread. **E2.17** **E2.18**

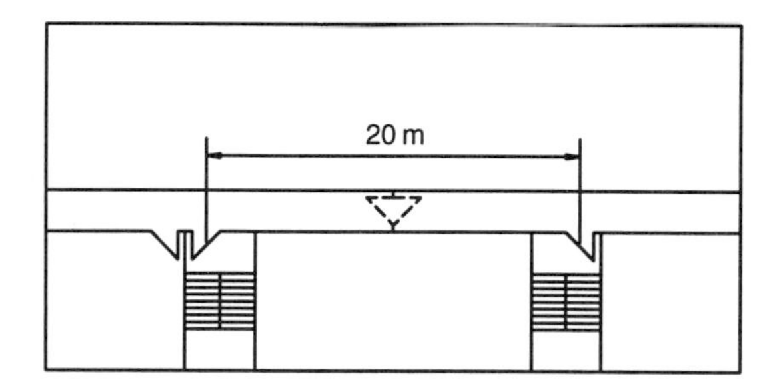

Figure E9 1A purpose group (4.5 m or above – no pressurisation).

Table E4 Maximum distance between fire doors

Purpose group or purpose sub-group	Maximum distance(m)
1A In a storey at a height of not more than 4.5 m; or any storey where a suitable system of pressurisation is installed	no limit
1A In a storey at a height more than 4.5 m where no suitable system of pressurisation is installed	15
2A (hospitals only), 3, 6, 7B or 7C	45
2A (except hospitals), 2B, 4, 5 or 7A	32

If a pressurisation system is installed then the distance between the protective doors can be of unlimited length as there will be no smoke ingress into the corridor. The pressurisation installation must comply with BS 5588: Part 4: 1978. **DTS E2.18**

A section of corridor which provides only one direction of escape must be separated by a fire door from any part which provides two directions of escape if it is more than 4.5 m long.

Openings in walls of escape routes

E2.22 Openings in escape routes adjacent to external walls must be protected by a barrier at least 1.1 m in height. The barrier does not require to be of solid construction; metal railings, glass blocks or toughened glass can be used. Windows must also be 1.1 m above floor level.

DTS E2.22

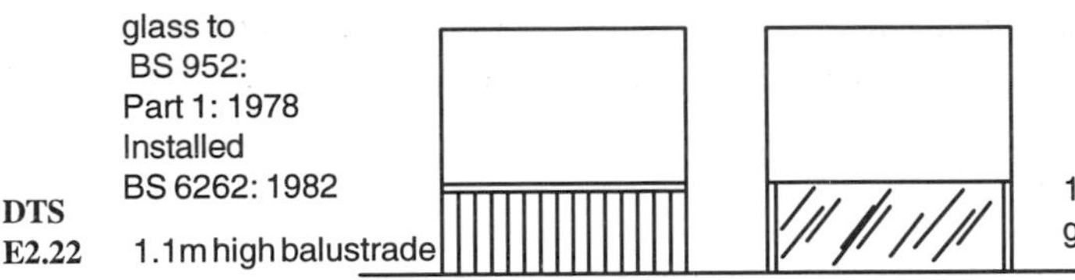

Figure E10 Openings in walls.

Doors across escape routes

E2.23 Doors within escape routes must generally open in the direction of escape. Sliding doors across an escape route may only be used if access to the area is normally restricted. The sliding door must be clearly marked 'slide to open'. Self closing fire doors must also be clearly marked on both sides. The words 'fire door keep shut' should be in letters not less than 5 mm high. The exceptions are fire doors within a dwelling, fire doors to a bedroom or fire doors which can remain open by use of an approved electric device. In a limited number of situations fire doors may be held in the open position by electro-magnetic or electro-mechanical devices designed to respond to smoke. This is achieved by using smoke detectors which can activate the closing device. The smoke detectors should be positioned on each side of the door.

DTS E2.2

E2.27

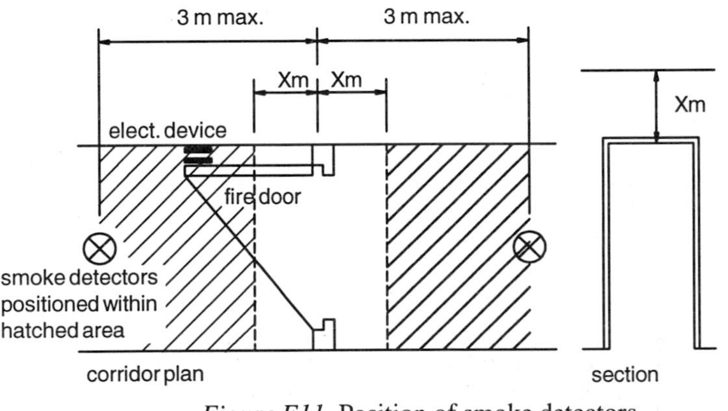

Figure E11 Position of smoke detectors.

E2.26 Self closing fire doors leading onto stair enclosures should not normally be fitted with hold open devices particularly for buildings of purpose group 2 and for fire fighting stair enclosures.

If an exit door has to be secured for unauthorised entry when the building is occupied it must be easily opened from the inside. If the secured exit door is within a building of purpose groups 4 or 5 (shops, assembly and recreational buildings) the door must have a securing device such as a push bar that can be opened when pressure is applied to it. There should also be a notice explaining how the device operates displayed on the inside of the door. **E2.28**

Enclosure of escape routes

Protected zone

The structure surrounding a protected zone must be non-combustible and have a minimum period of fire resistance of 60 minutes. Exceptions include: **E2.31** **E2.32**

(1) fire doors of timber construction;
(2) a glazed enclosure;
(3) an external wall at least 1 m from the boundary;
(4) a roof which is not part of an escape route;
(5) the floor of the lowest storey.

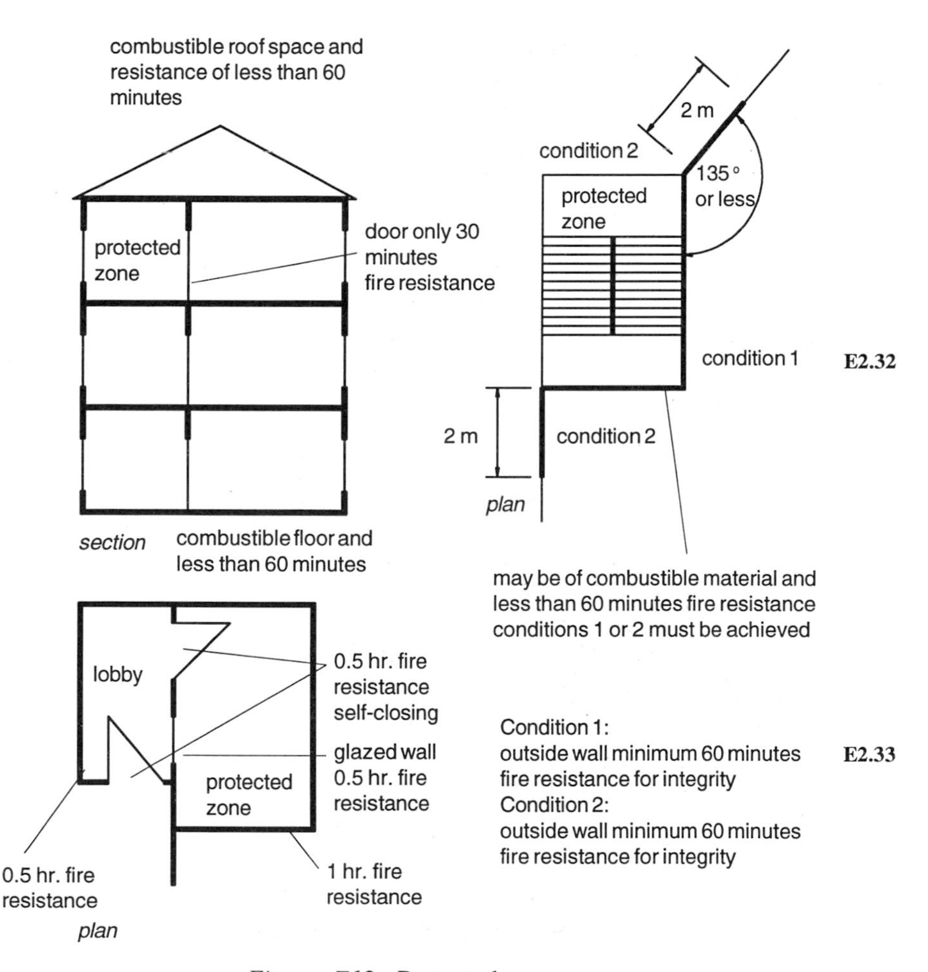

E2.32 **E2.33**

Figure E12 Protected zones.

E2.34 An escape stair enclosure should only contain a stair or ramp and one or more of the following:

(1) a space giving access only to the stair or ramp;
(2) a lift well;
(3) a water closet compartment or washroom (if the enclosure does not serve a storey above 18 m);
(4) a ticket or enquiry office or reception area, kiosk or porters lodge where the area is less than 10 m^2, the wall and ceiling surfaces are class 0, the maximum storey height is 18 m and the building has at least one other escape route.

E2.35 The escape stair enclosure must give access directly to a place of safety at ground level or to an escape route across a flat roof or podium. Not all stairs or ramps have to comply with these requirements. The exceptions are:

(1) a private stair within a dwelling;
E2.29 (2) a stair or ramp to a gallery;

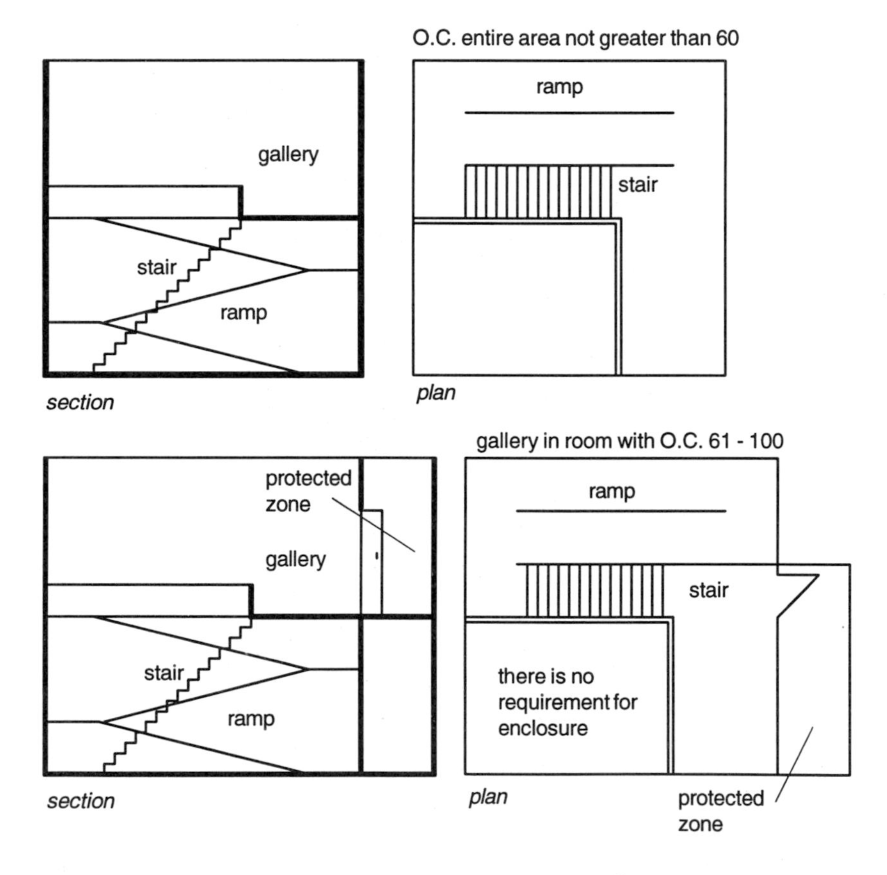

Figure E13 Stair or ramp to a gallery.

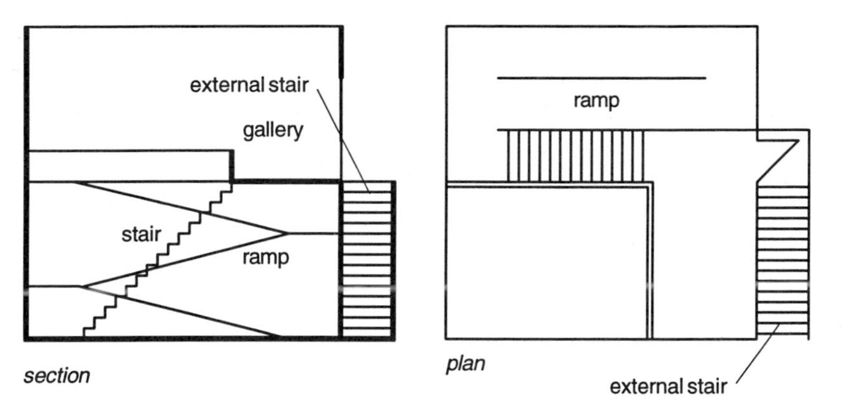

Figure E13 Stair or ramp to a gallery (cont.).

(3) An external stair or ramp. **E2.29** **E2.30** **E2.41**

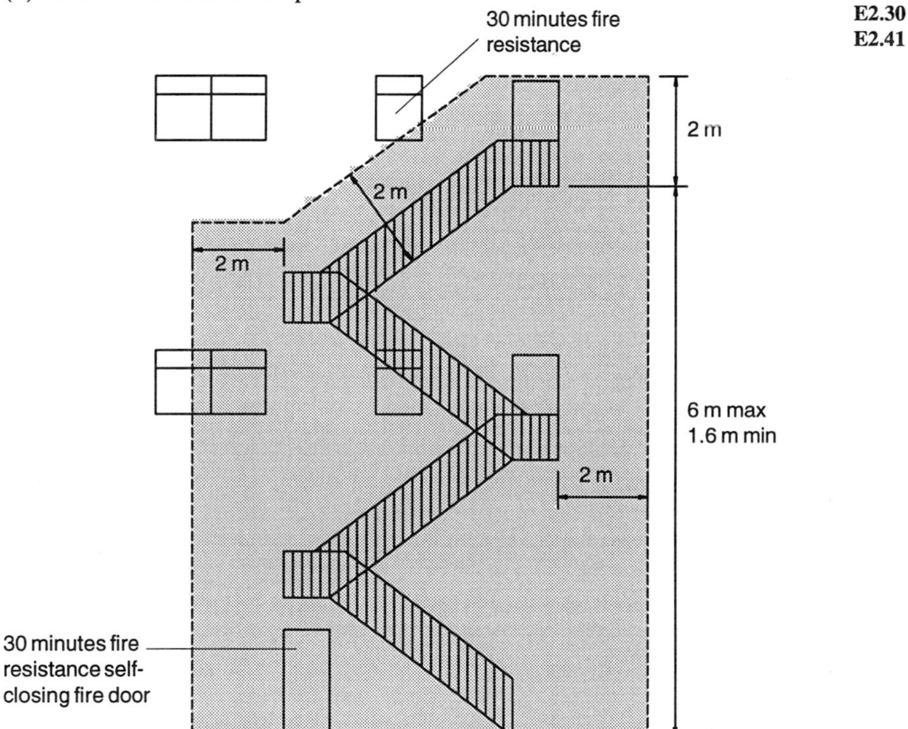

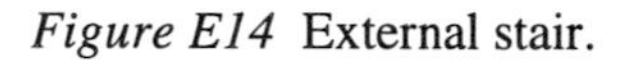
Figure E14 External stair.

Where two escape stairs share a common enclosure wall the access between them must be via a protected lobby. Escape stairs contained within a central core must be designed to ensure that the exits are as far apart as is practicable. If only one escape route is available from a storey, any room on that storey or below that **E2.36** **E2.37** storey which provides access onto the escape stair must normally be separated from it by a protected lobby.

Tall buildings

E2.39 For buildings with storey heights above 18 m the escape stair enclosures must be separated from other accommodation by protected lobbies on each floor except the topmost. In these circumstances the escape stair enclosures will meet the requirements for fire fighting stairs which are described later in this section. In buildings where the only means of escape include an escape stair or a ramp that extends to basement level a means of preventing evacuating persons from

E2.41 bypassing the ground floor and arriving in the basement must be designed. One way to achieve this is to erect a wall at ground floor level with the same fire resistance as the stairway enclosure and containing a self closing fire door.

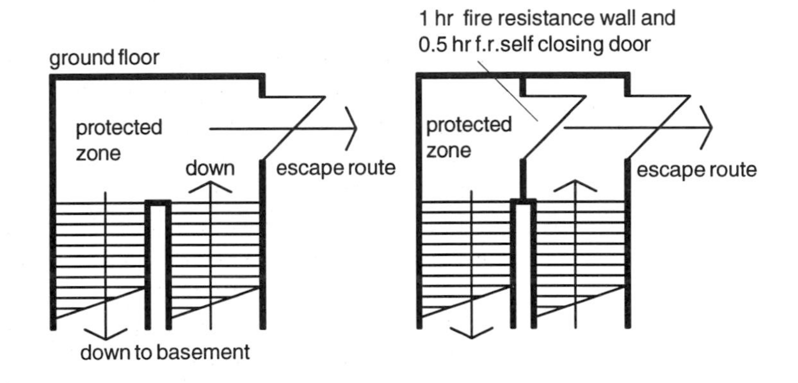

Figure E15 Ground floor of tall building with basement.

Accommodation stairs and escalators

Accommodation stairs and escalators which pass through a compartment floor must comply with the conditions set out in D2.13. The integrity of the compartment during a fire may be maintained by use of a fire shutter to enclose the escalator opening. If the escape route passes close to the escalator then the shutter to the escalator must be automatically activated in response to smoke detectors to prevent people evacuating via the escalator.

Where an accommodation stair or an escalator passes through a floor which is not a compartment floor, an enclosure or suitable shutter with 30 minutes fire resistance must be provided.

Lighting of escape routes

E2.44 Adequate artificial lighting must be supplied to any part of an escape route from the door to a flat or maisonette or from the door to a room in any other building except dwellings and buildings covered by the cinematographic (safety) Scotland Regulations 1955: Part 1. The artificial lighting supply must be electrical with a protected circuit. Within escape stairs and protected lobbies a separate circuit from that in other parts of the escape routes is required. Alternatively emergency lighting complying with the requirements in Table E5 can be provided.

Table E5 Emergency lighting

Purpose group or purpose sub-group	Part of building requiring emergency lighting
any	1. A corridor or protected zone in a building in which any storey is more than 18 m above the level of the adjoining ground. 2. A room with an occupancy capacity of 60 or more, and any corridor or protected zone serving such a room. 3. A public area, corridor or protected zone serving an underground car park where less than 30% of the perimeter of the car park is open to external air. 4. A basement storey required to have two or more escape routes and any corridor or protected zone serving such a storey. 5. A place of special fire risk (other than one only requiring access for the purpose of maintenance) and any corridor or protected zone serving it. 6. Any part of an air supported structure, other than one ancillary to a dwelling.
2.	1. A room with an occupancy capacity of 10 or more and any corridor or protected zone serving such a room. 2. A corridor or protected zone serving a storey required to have two or more escape routes other than, subject to 1 above, a storey in a building not more than two storeys high with a floor area not more than 300 m^2. 3. A protected zone or corridor in a single stair building with two storeys or more with an occupancy capacity of 10 or more.
4.	In shop premises, a corridor or protected zone serving a storey required to have two or more escape routes.
5.	A corridor or protected zone serving, (a) a storey required to have two or more escape routes (b) any storey in a non-residential school of more than one storey.
6.	A corridor or protected zone serving a storey required to have two or more escape routes.
7A.	A corridor or protected zone serving a storey required to have two or more escape routes other than a single storey building with a floor area of not more than 500 m^2.

Fire doors

DTS E2.45 The requirements for fire doors are set out in the deemed-to-satisfy provisions of the Technical Standards. Particular care should be taken when interpreting the requirements relating to buildings with pressurisation installations.

Mechanical ventilation systems

E2.69 When the building has a mechanical system it must comply with the following:

(1) it must be designed to ensure that air movement is directed away from protected and unprotected zones.

(2) if the system uses recirculated air:

- (a) smoke detectors must be provided within all extract ducts, which will if activated cause the recirculation air to stop and direct all the extract air to the outside of the building;
- (b) it must have a suitably designed emergency control system. This can be relaxed for a mechanical system totally within a dwelling.

Additional requirements for escape routes

Places of special fire risk

E2.67 A place of special fire risk must not normally communicate directly to an escape route used by members of the public or to an escape route which is the only escape route from the building.

However, if the escape route is a protected zone serving another part of the building and there is more than one route available, then access from the place of special fire risk can be allowed through a protected lobby.

Fixed storage areas

E2.66 Escape routes between fixed storage areas require the width to be not less than 530 mm. For buildings storing spirituous liquor the gangway may be reduced to 400 mm.

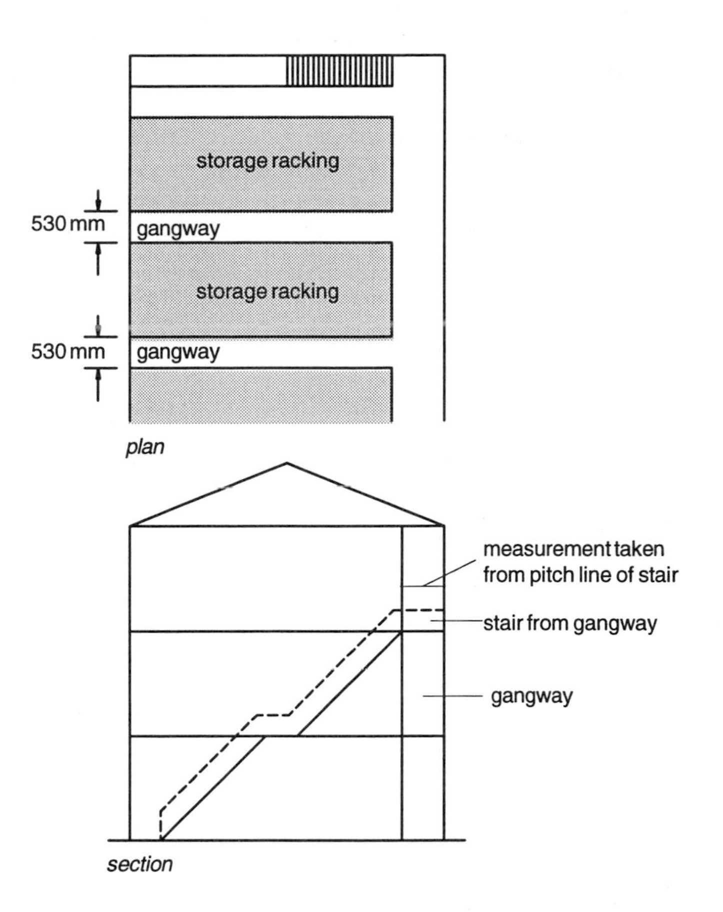

Figure E16 Width of escape routes within fixed storage areas.

Fixed seating

Fixed seating within a room or space can present difficulties, especially when the free movement of people is required in a fire. **E2.63**

The minimum widths of the escape routes are specified in Table E6.

Table E6 Minimum width of escape routes with fixed seating

Situation	Minimum width or effective width (mm)
(a) gangway	1100
(b) seatway serving	
(i) up to 11 seats in one row, gangway one side	400
(ii) up to 22 seats in one row, gangway both sides	400
(iii) more than 22 seats in one row, gangway both sides	500
(c) gangway and seatway combined	1350

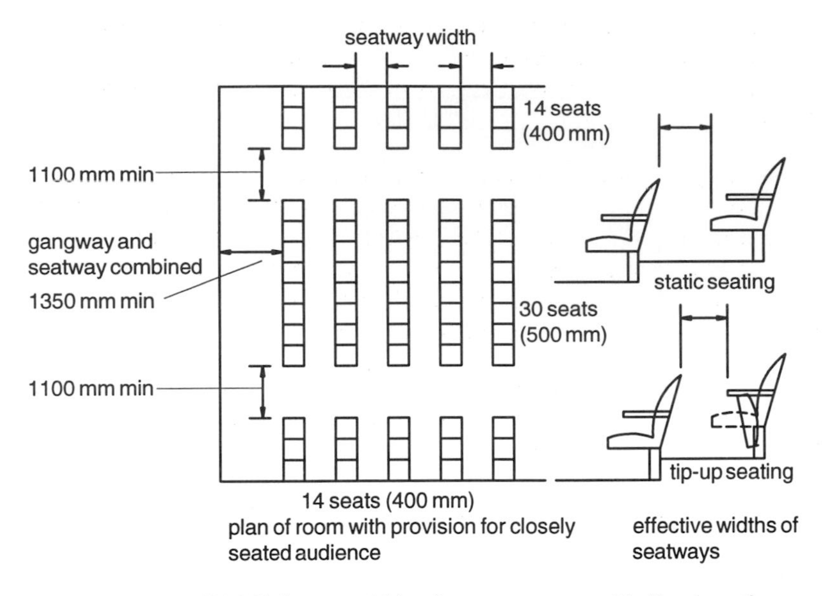

Figure E17 Minimum width of escape routes with fixed seating.

When a room requires more than one exit, at least one of these must be provided not less than two thirds of the distance from the stage, screen or performance area, to the back of the room or space.

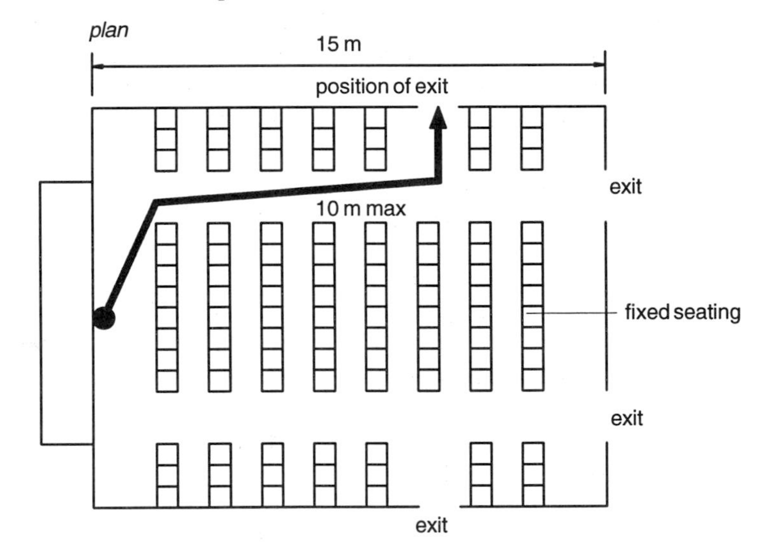

Figure E18 From stage area exits must be within 0.66 of the distance to the back wall (in this case 0.66 x 15 m =10 m).

Stages

E2.64 Within a building that has a stage for the sole purposes of presenting performances, the ventilation system must be designed to ensure that during any performance the air movement is always from the auditorium towards the stage. This is to prevent smoke obscuring exit routes in the event of a fire as all exits are away from the stage area.

Where the auditorium has seating capacity of more than 500, any separated stage, or stage area, and this includes scenery dock, workshop, under-stage area, staff or orchestra room and any ancillary part of the stage, must comply with the following: **DTS E2.65**

(1) the stage or stage area must be separated from the remainder of the building, excluding the proscenium opening, by a wall or screen or other type of construction providing at least 30 minutes fire resistance for integrity;
(2) an outlet must be provided at a high level over the stage, and be of sufficient size to allow the escape of smoke and hot gases;
(3) a suitable safety curtain is required at the proscenium opening.

If a safety curtain is used then the stage area must have its own means of escape and therefore be independent from the auditorium. The separate stage area excludes the proscenium opening.

Special requirements for buildings of purpose group 1A (flats and maisonettes):

General

All flats and maisonettes regardless of position and size are classified as purpose group 1A and are subject to the conditions set out in E2.47 to E2.59. The exceptions which were mentioned previously are certain flats and maisonettes which are to be considered in the same categories as houses. **E2.47 to E2.59**

These are – flats and maisonettes entered from the open air at ground level and with no storey at more than 4.5 m, which should be considered as purpose group 1C; and maisonettes, entered only from the open air at ground level and with a storey at more than 4.5 m, which should be considered as purpose group 1B.

Entrance door: flats and maisonettes

The entrance door to the dwelling must be a self closing fire door. **E2.47**

Inner room: flats and maisonettes

No apartment at a storey height of more than 4.5 m may be an inner room. The terms are: **E2.48**

An apartment is a room in a dwelling; it does not include a kitchen, store or utility room.

An inner room means a room in a dwelling which does not have direct access to a circulation area, (e.g. a hall or passageway).

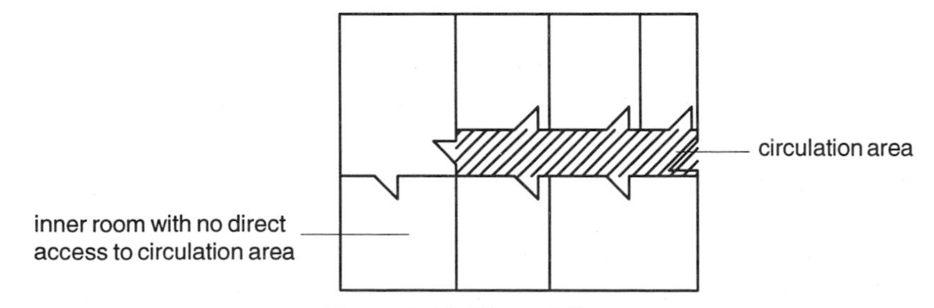

Figure E19 Plan of flat.

Emergency window: flats and maisonettes

E2.49 Emergency windows should be located in every storey of a dwelling and in every apartment which is an inner room at a height between 4.5 m and 11 m, to facilitate the use of ladders for evacuation.

Emergency window requirements

DTS E2.49 The requirements for emergency windows are met by a window:

(1) situated in an external wall or roof;
(2) not situated in a kitchen or stair enclosure;
(3) having a clear opening part at least 850 mm high and 500 mm wide, with the bottom of the opening not more than 1.1 m above the floor and without any built-in obstruction in front of it;
(4) which, if it is in the roof, is:
  (a) either a dormer window or specially designed window to fit in the plane of the roof, with or without an upstand;
  (b) not more than 2 m from the eaves to the nearest part of the window opening measured along the plane of the roof;
  (c) if it is not a dormer, and the roof pitch is less than 40°, the window is inclined at not less than 40° to the horizontal; and
  (d) where necessary, it is counterbalanced to remain in the open position in an emergency; and
(5) provided with an access, either directly or by way of a balcony or flat roof which is in turn directly accessible, from an area of cleared ground which:
  (a) is at least 2 m x 2 m in area with its centre not less than 2.5 m and not more than 3 m from the building;
  (b) has no overhead wires or similar obstructions between it and the window it serves and
  (c) if it serves a building of more than two storeys, has its centre line at least 2.7 m from the face of the building and is either part of, or has an access suitable for, the carriage of a fire service ladder from a public road or footpath.

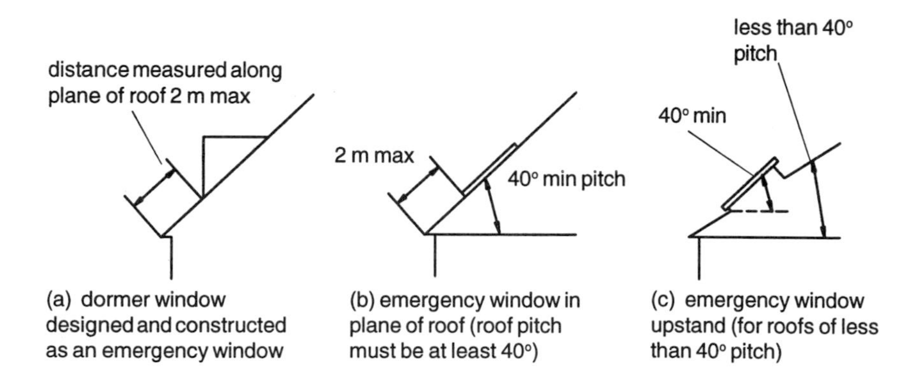

Figure E20 Emergency windows in roof.

Internal layout of flats entered at accommodation level

A flat, and this does not include a maisonette, with the apartments on the same level as the front door, and the storey height of more than 4.5 m must be designed to comply with one of the following three options: **E2.50**

(1) All the apartments must be directly off a private entrance hall of at least 30 minutes fire resistance for integrity and the doors to the apartments must not be more than 9 m from the entrance door.

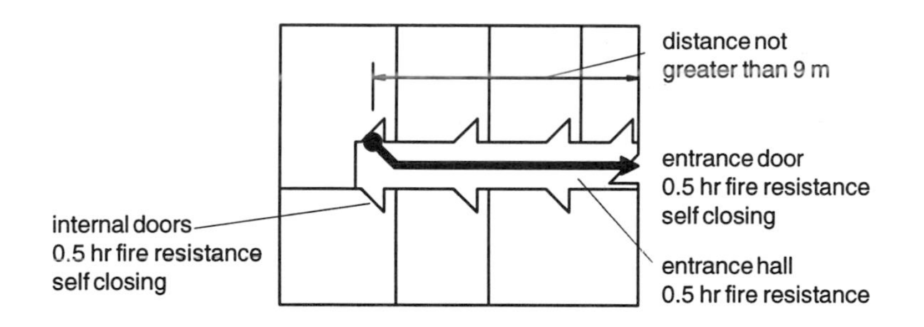

Figure E21 Plan of flat.

(2) No point within thc flat should be more than 9 m from the entrance door, with the cooking area as far away as possible from the front door.

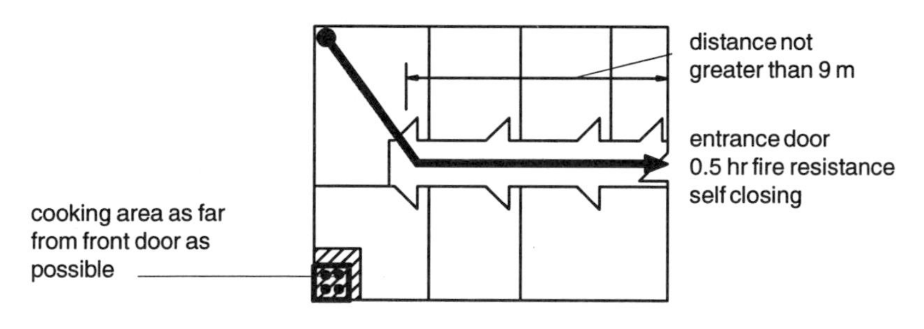

Figure E22 Plan of flat.

(3) The living and sleeping accommodation must be separated by a wall providing at least 30 minutes fire resistance for integrity, and if the flat is more than 11 m above ground level then no point within the flat must be more than 15 m from the entrance door. If it is more than 15 m then an alternative escape route must be provided.

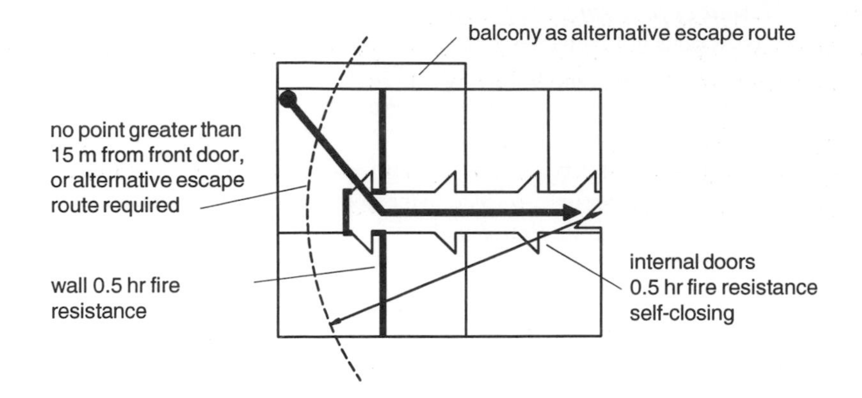

Figure E23 Plan of flat over 11 m above ground level.

Internal layout of flats entered from below accommodation level

E2.51 If the flat or flats are at a storey height above 4.5 m with the entrance door below the apartments, then one of the following requirements must be met:

(1) the apartments lead directly onto a private entrance hall which provides at least 30 minutes fire resistance for integrity and the doors to the apartments are no more than 7.5 m to the head of the stair;

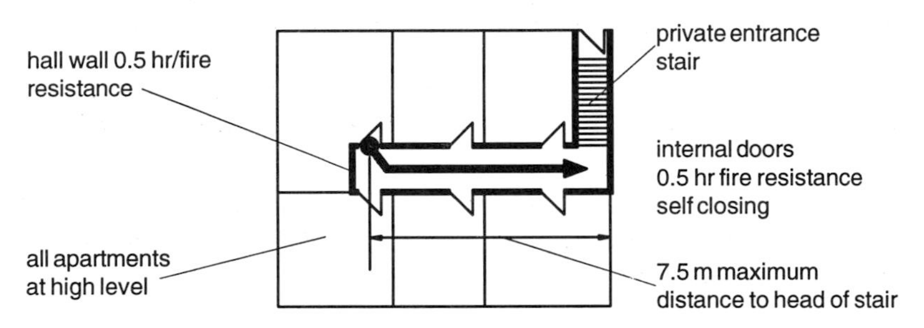

Figure E24 Plan of flat.

(2) no point within the flat must be more than 9 m from the head of the private stair, and with the cooking area as far as possible from the head of the private stair;

(3) as requirement 3 for flats entered at accommodation level.

Internal layout of flats entered from above accommodation level

Two design options are available. The flat or flats must be at a height of more than 4.5 m with the entrance door above the apartments, and either:

E2.52
E2.53

(1) The living and sleeping apartments separated from the rest of the accommodation by a wall providing at least 30 minutes fire resistance for integrity. If any storey is more than 11 m above ground level then an alternative escape route from each floor is required, other than the entrance floor; or

(2) An alternative escape route from each bedroom.

Single stair access to flats and maisonettes

Buildings of purpose group 1A, that have a single access to the dwelling, must be designed to meet one of the following criteria: **E2.54**

(1) The storeys should not contain more than eight dwellings per storey and if they are reached by a common access corridor, the corridor should be divided into each serving not more than four dwellings;

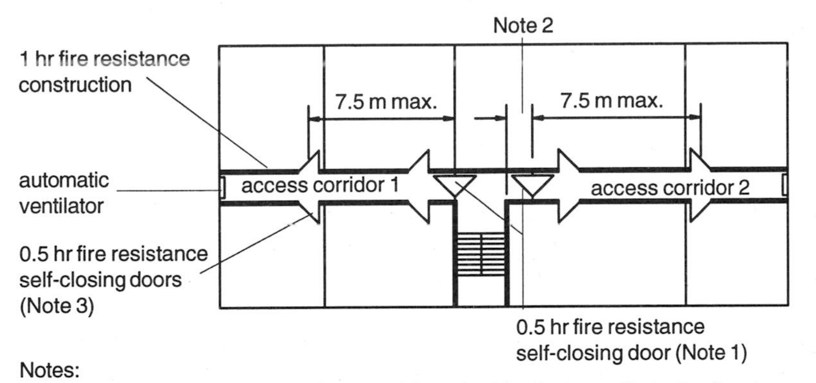

Notes:
1. Access corridor 1 separated from corridor 2 by 30 minutes self-closing fire doors.
2. Stair enclosure extended as shown to accommodate maximum permissible travel distance.
3. Entrance door to each flat/maisonette to be 30 minutes self-closing fire door.

Figure E25 Plan of storey at any height with eight dwellings maximum served by two access corridors, each serving four dwellings corridor to eight dwellings.

(2) If the dwellings are entered via a protected lobby constructed to provide at least 60 minutes fire resistance for integrity, then the maximum number of flats per storey is four;

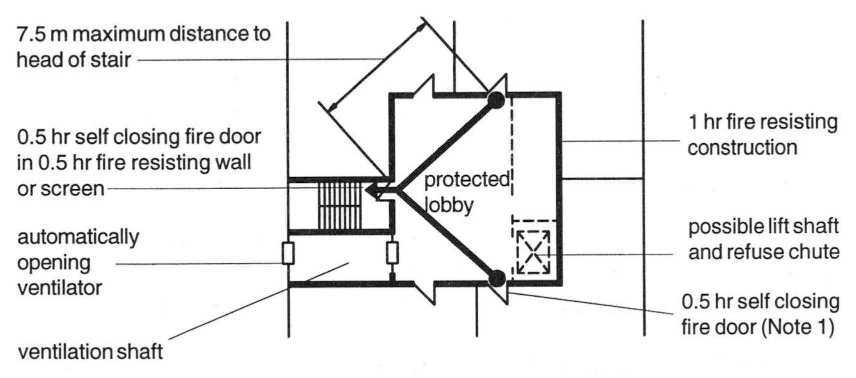

Notes:
1. Entrance door to each flat/maisonette to be 30 minutes self closing fire door.

Figure E26 Plan of storey at any height with four dwellings max., served by a protected lobby with provision for ventilation and a travel distance of 7.5 m maximum.

The common access corridor or protected lobby must be separated from the stair enclosure by utilising either a wall or a screen which provides a minimum of 30 minutes fire resistance.

Between the corridor or protected lobby and the external air, a ventilator is required that is activated by smoke detection. The opening area must be at least 1.5 m^2 and there should be a manual override system.

The conditions stated above can be relaxed if a storey is at a height of no more than 11 m. Either:

(1) the ventilator can be omitted if there are no more than six dwellings on each floor and these are entered directly from a protected lobby and the travel distance from each entrance door to the dwellings and the door to the stair enclosure does not exceed 4.5m; or

(2) A protected lobby can be omitted if there are no more than two dwellings on each floor entered directly from the stair enclosure, and each dwelling has a private entrance hall providing at least 30 minutes fire resistance for integrity. Doors leading from the entrance hall must have 20 minutes fire resistant and be self closing.

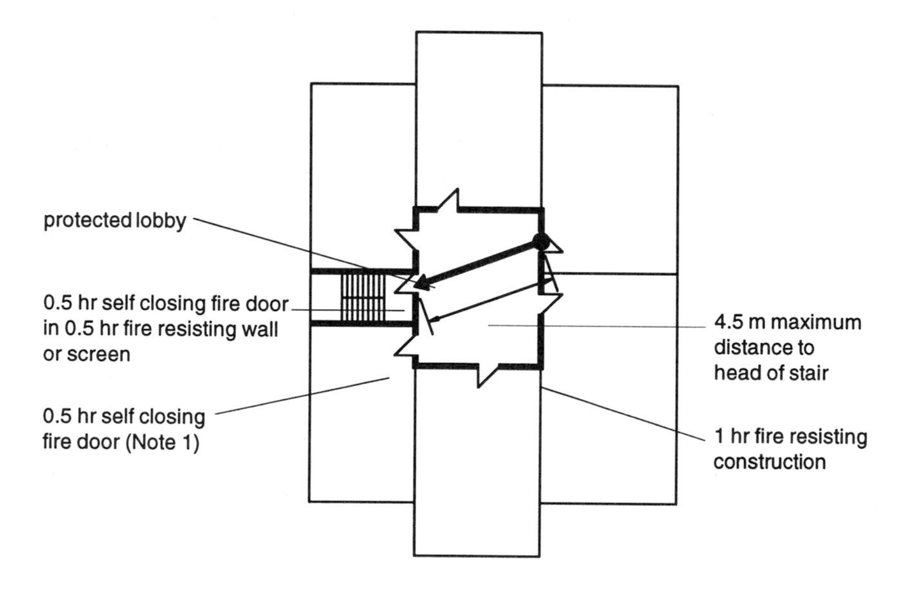

Notes:
1. Entrance door to each flat/maisonette to be 30 minutes self closing fire door.

Figure E27 Plan of storey at 11 m maximum height with 6 dwellings maximum served by a protected lobby without provision for ventilation: travel distance of 4.5 m maximum.

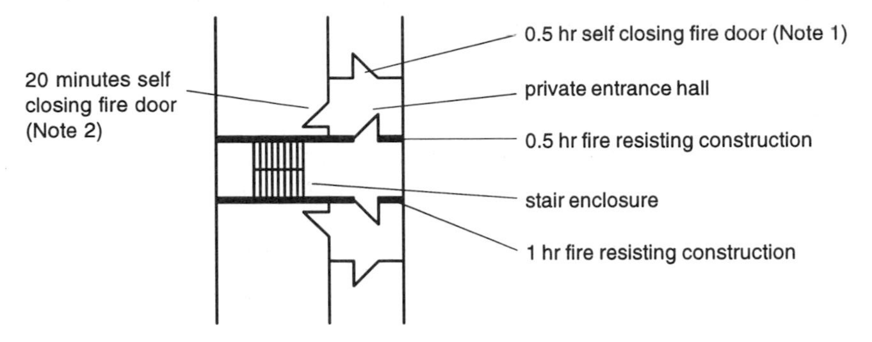

Figure E28 E2.54 exception (b) – Storey at 11 m max. height with two dwellings maximum: no protected lobby.

More than one access to flats and maisonettes

Within a building of purpose sub-group 1A, that has more than one access stair, an openable ventilator must be positioned at the end of each common access corridor. The area of the ventilator must be at least 1.5 m^2 and connect directly to the outside air. **E2.55**

Access balconies and decks to flats and maisonettes

The requirements for protected lobbies and corridors will not apply if access to the flats or maisonettes is by way of an open access balcony. **E2.56**

The definition of open access balcony is, one that has openings to the external air extending over four-fifths of its length and at least one-third of its height.

Wide access balconies and decks

If the access balcony or deck is over 2 m wide, then a downstand must be placed from the underside of the soffit. This must be in line with the separation between the flats at right angles and projecting out from the building line to the edge of the soffit of the balcony of the deck. The dimension of this downstand requires to be 300 mm below either the soffit of the balcony or the soffit of any beam. **E2.57**

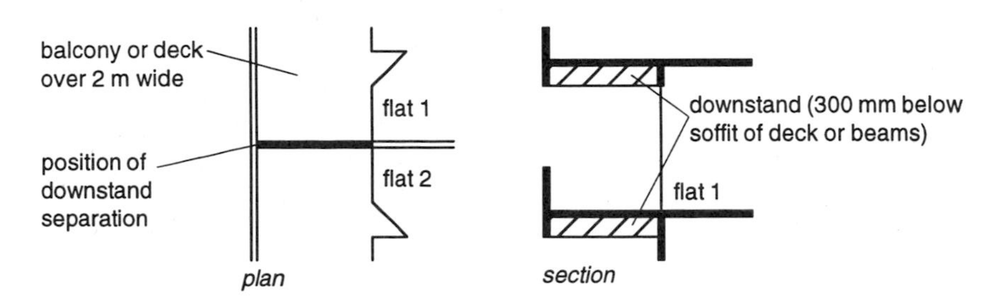

Figure E29 Balcony or deck over 2 m wide.

Flats and maisonettes with access common to other occupancies

E2.58 Within a building of mixed purpose groups a stair providing access to dwellings must not provide access to different purpose groups on storeys below the dwellings, for example flats above shops.

However there are two exceptions to this rule which would allow access at a lower level:

(1) if the storey height of the dwellings is not 4.5 m and a protected lobby providing at least 60 minutes fire resistance is positioned between each occupancy and the stairwell;
(2) if the flats and maisonettes are above 4.5 m, then:
   (a) a protective lobby is required between each occupancy and the stair;
   (b) each dwelling must have an alternative escape route and the building must have a resident caretaker.

Ducted warm air heating in flats and maisonettes

E2.59 Dwellings at storey height of over 4.5 m with a system of ducted warm air heating must comply with the following:

(1) transfer grills must not be fitted between any apartment (room) and the entrance hall or stair;
(2) supply and return grilles must not be more than 450 mm above floor level;
(3) if warm air is ducted to an entrance hall or stair, then the returned air must be ducted back directly to the heater;
(4) if any part of a duct passes through the wall, floor or ceiling of an entrance hall or stair all joints between the duct and the construction must be sealed;
(5) a room thermostat must be positioned within the living room at a height between 1.35 m and 1.85 m and having a maximum setting of 27° centigrade;
(6) if the system recirculates the air, then smoke detectors must be positioned in every extract duct, to terminate the air flow and redirect all extract air to the outside of the building in the event of a fire.

Special requirements for buildings of purpose sub-group 1B

DTS E2.60 Private dwellings with a storey height above 4.5 m and classified as 1B must comply with the following:

(1) no apartment at a storey height more than 4.5 m may be an inner room; and
(2) every storey at a height of more than 4.5 m must have at least one suitably constructed and located emergency window; and
(3) every bedroom at a height of more than 4.5 m must have an alternative escape route. This can be relaxed if the following conditions are met:
   (a) the stair to the upper storeys is an enclosure providing at least 30 minutes fire resistance for integrity;
   (b) every door between the enclosure and a room or cupboard or circulation space is a self-closing fire door, providing 30 minutes fire resistance;

(c) if the stair also serves a basement, then the basement storey must be separated from the remainder of the stair by a wall or screen containing a self closing fire door, both of which must provide at least 30 minutes fire resistance for integrity;
(d) if the dwelling has a system of ducted warm air heating, then the system must comply with the requirements set out in E2.59 of the Technical Standards.

Special requirements for hospitals (purpose sub-group 2A)
Within a hospital the following requirements must be met: **E2.61**

(1) every storey used for bed-patient care requirements must have at least two escape routes;
(2) every upper storey must:
 (a) form a separate compartment; and
 (b) if it is used for bed-patient care then the storey must be divided into at least two compartments to provide for horizontal evacuation of each compartment;
(3) if the storey requires to be divided into compartments, then the horizontal evacuation can be via an adjoining compartment provided that:
 (a) the adjoining compartment has at least one other independent escape route, or
 (b) communicates with another compartment having an independent escape route.

Special requirements for buildings of purpose group 2 other than hospitals
In an upper storey of a building of purpose group 2, other than a hospital, where the building has only one escape route, every bedroom must have a suitable emergency window. **E2.62**

Special requirements for air supported structures
An air supported structure is defined as: **E2.68**
'a structure which has a space-enclosing single-skin membrane anchored to the ground and kept in tension by internal air pressure so that it can support applied loading', and must comply with the following:

(1) exits must be provided to comply with the following:
 (a) the travel distance in one direction must not be more than 9 m;
 (b) every exit is at least 3 m or 0.25 of the smallest plan dimension, whichever is less, from any corner of the structure, refer to Figure E30;
(2) each exit door must have a rigid supporting framework and be clearly marked EXIT and be illuminated by a sign;
(3) inflation equipment must have a standby power system which:
 (a) will start automatically on any failure of the main power supply;
 (b) is independent of the main power supply;
 (c) includes non-return dampers in the ducts, and the equipment outside the structure is suitably weather protected.

(4) if the occupancy capacity exceeds 100, or in the case of a swimming pool, 50, an emergency support system must be provided which will comply with the following:
(a) support the membrane in a deflated state under short term loading;
(b) provide clear emergency escape routes with a minimum headroom height of 2.5 m for at least 10 m adjacent to each exit.

DTS E2.68 (5) if the air supported structure encloses a swimming pool, then steps from the water must be adjacent to the exits.

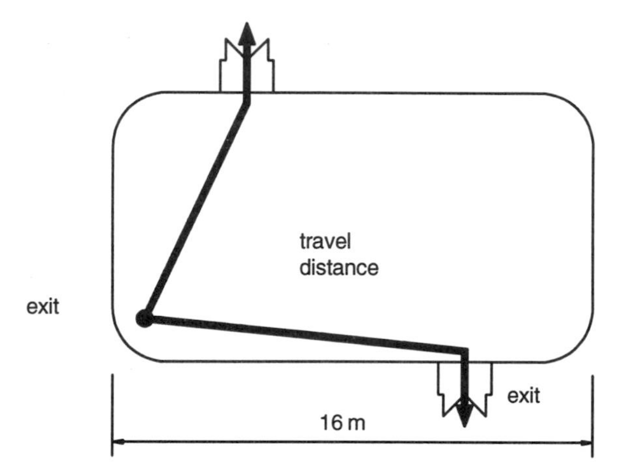

travel distance = 3 m, or 0.25 x smallest plan dimension, whichever is less
therefore 16 m x 0.25 = 4 m

Figure E30 Plan of air supported structure.

Facilities for fire fighting

Access

E3.1 This stipulates that a building must be provided with suitable access for fire-fighting purposes. The deemed-to-satisfy requirements are set out in three tables: the dimensions, the access provisions with wet fire mains, and access provisions (without wet fire mains). Table E7 correlates the type of appliance required and the access position with the cubic capacity and the height of the building for buildings without wet fire mains. Table E8 sets out the minimum dimensions for access routes to be used by fire-fighting vehicles. For buildings fitted with wet fire mains, access is for a pumping appliance to within a suitable entrance on the ground storey to the stair enclosure in which the main is situated (or, if the main is not in a stair enclosure, within 18 m of the foot of the nearest stair enclosure to the main) and within sight of the inlet for the emergency replenishment of the suction tank serving the main.

DTS
E3.1

Table E7 Access provision to buildings not fitted with wet fire mains

Cubic capacity of building (m^3)	Maximum height of building above ground level (m)	Type of appliance	Position of access
up to 7000	9	Pumping	to within 45 m of a suitable entrance.
up to 7000	NL	Turntable or hydraulic platform	along the face of 16% of the total length of all perimeter walls.
7001 to 28 500	9	Pumping	along the face of 16% of the total length of all perimeter walls.
56 501 to 85 000	NL	Turntable or hydraulic platform	along the face of 75% of the total length of all perimeter walls
over 85 000	NL	Turntable or hydraulic platform	along the total length of all perimeter walls.

Note: Consultation with the Fire Authority is advised on all matters concerning fire access.

Table E8 Minimum dimensions of access routes for fire-fighting vehicles

Type of appliance	Width (m)	Clearance height (m)	Turning circle (diameter kerb to kerb) (m)	Turning circle (swept diameter wall to wall) (m)	Width of gateways etc (m)	Laden weight (tonnes)
Pumping appliance	3.7	3.7	16.8	19.2	3.1	12.5
Turntable ladder or hydraulic platform	3.7	4.0	26.0	29.0	3.1	18.6

Ground hydrants

Buildings where the floor area of any storey or compartment exceeds:

E3.2

(1) in purpose group 7A, 230 m^2; or
(2) in any other purpose group, 900 m^2;

must be provided with suitably positioned and constructed ground hydrants.

The deemed-to-satisfy section requires that the hydrants are positioned:

(1) externally, for each building or where common to more than one building
  (a) within 70 m of an entrance to the building;
  (b) not more than 150 m apart;
  (c) so as to be accessible for use at all times; and
  (d) so that there is convenient route for the fire hose between the hydrant and the building; or
(2) internally
  (a) at ground level;
  (b) within 4.5 m of and visible from, an entrance to the building which has a notice clearly indicating the presence of the hydrant; and
  (c) so as to be accessible for use at all times and constructed in accordance with BS 750: 1984.

E3.3 Ground hydrants must be connected to a water service pipe of at least 100 mm diameter supplied from:

(1) a water main vested in a local authority; or
(2) a supply provided under the Fire Services Act 1947;
(3) in certain cases:
  (a) a charged static water tank of adequate size; or
  (b) a spring, river canal, loch or pond, to which access and space are available for a pumping appliance, capable of providing an adequate supply of water for fire-fighting purposes.

Fire mains

E3.4 A building must have at least:

(1) if it has a storey at a height of more than 18 m but not more than 60 m, one wet fire main or dry fire main; or
(2) if it has a storey at a height of more than 60 m, one wet fire main; with all its associated outlets, inlets and fittings, for every 900 m^2, or a part thereof, for every upper storey.

E3.5 A fire main must be enclosed in either:

(1) a protected zone; or
(2) a duct which provides:
  (a) an imperforate surround (except for an opening for access); and
  (b) has at least 60 minutes fire resistance from each side.

E3.6 A wet fire main must be:

(1) of an adequate diameter;
(2) be accessible from a suitable ground hydrant;
(3) be suitably equipped, including pumping and stand-by power sources capable of providing the required water pressures; and
(4) be frost resistant.

An inlet for a dry fire main must be within: **E3.7**

(1) 18 m of and in sight of a parking space for a pumping appliance; and
(2) be 12.5 m horizontally from the vertical main.

Fire-fighting outlets

Fire-fighting outlets must be provided so that no point on any storey is further from the outlet than: **E3.8**

(1) one storey in height; and
(2) 60 m measured along a route suitable for a hose.

If the building has a fire fighting lift, then there must be an outlet from a fire main within 45 m of each entrance hall to such a lift. **E3.9**

A fire-fighting outlet must be in a protected zone, or a protected lobby or an open access deck. **E3.10**

Fire-fighting stairs

A building with a storey height of 18 m or more above ground level or 9 m or more below ground level must have one or more suitably located fire-fighting stairs. These stairs can also be used as escape stairs. **E3.11**

The number of such fire-fighting stairs is determined by the area of the storey. The stair should service an area that does not exceed 900 m^2; if the area does exceed this then extra stairs are required. However this requirement does not apply in a building of purpose sub-group 1A.

The fire-fighting stair must be provided with a protected lobby with an area of at least 5 m^2 at each storey served by the stair. **E3.12**

A protected lobby to a fire-fighting stair on a ground storey or upper storey must have an opening, or openable window or ventilator of at least 1.5 m^2 to the external air. **E3.13**

A protected lobby to a fire-fighting stair on a storey below ground must have a smoke extract which complies with the following: **E3.14**

(1) is independent of any other extract;
(2) has a cross-sectional area of at least 1 m^2; and
(3) discharges directly to the open air at a point at least 3 m measured horizontally from any part of an escape route.

The enclosure to a fire-fighting stair must have: **E3.15**

(1) fire resistance for stability, integrity and insulation of at least:
  (a) between the protected lobby and the rest of the building, 60 minutes;
  (b) between the stair enclosure and the protected lobby, 30 minutes;

(c) between the stair enclosure and a lift shaft, 30 minutes; and
(d) between the protected lobby and the lift shaft, 30 minutes; and

(2) suitable provision for the control of smoke which conforms to BS 5588: Part 5:1986, Clause 8.

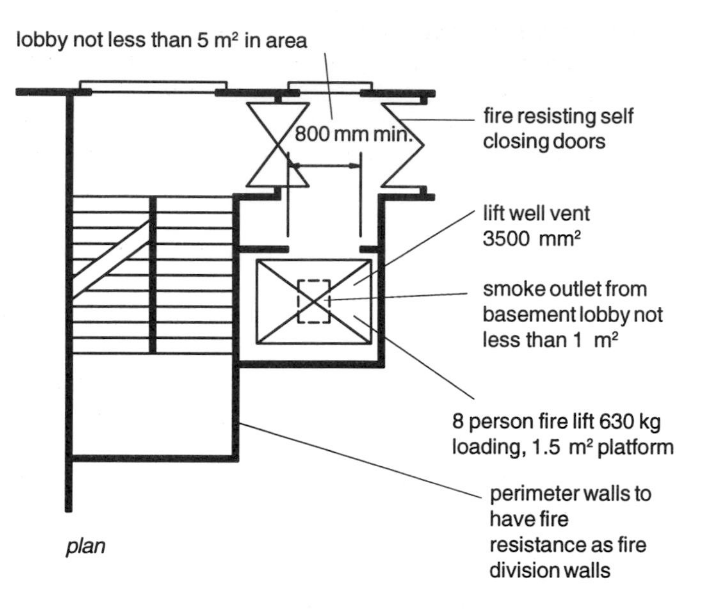

Figure E31 Protected lobby to fire-fighting stair.

Fire-fighting lifts

E3.16 If a building has a storey height of more than 18 m above the ground level, or 9 m below the ground level, then it must have at least one or more suitable fire-fighting lifts serving every storey, except for the following situations:

(1) a storey in a building of purpose sub-group 1A which has no entrance to a dwelling; or

(2) the top storey of a building which is equipped with fire mains, where:
(a) there is access by stair from the storey below; and
(b) the foot of the stair is within 4.5 m of a fire-fighting lift.

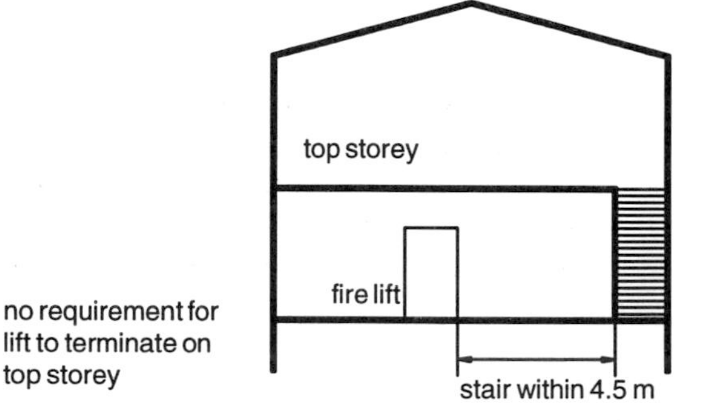

Figure E32 Fire-fighting lift.

Each lift is required to serve an area that does not exceed 900 m^2; if the area does exceed this, then again extra lifts are required.

A fire-fighting lift must comply with the following: **E3.17**

(1) a suitable electrical supply;
(2) a load-carrying capacity of at least 630 kg at a speed enabling it to complete its full travel range in one minute;
(3) internal plan dimensions must be at least 1.4 m x 1.1 m;
(4) power operated doors giving a clear opening width of at least 800 mm; and
(5) a suitable fire control system which:
   (a) enables firemen to take control of the lift without interference from landing control points;
   (b) enables the lift doors to remain open when the lift is under fire control; and
   (c) has a fire switch at the landing call station at ground level, housed in a lockfast glass fronted box marked 'FIRE SERVICE'.

The deemed-to-satisfy requirements will be met where the fire-fighting lift conforms to Clauses 10-16 of BS 5588: Part 5:1986 and the electrical supply conforms to BS 5655: Part1:1986.

A fire-fighting lift must be entered only from: **E3.18**

(1) an open access balcony; or
(2) a stair enclosure meeting the requirements of E2.31 to E2.32; or
(3) a protected lobby.

Automatic detection in dwellings (the July 1993 Amendment)
As a result of this amendment there is a change to Regulation 13 to introduce a requirement to provide a means of warning of fire in dwellings. This has been done by inserting, between Regulations 13(1) and 13(2), the new requirement 13(1A) which reads 'Every dwelling shall be provided with means of warning the occupants of an outbreak of fire'.

This requires consequential additions to both the standards and the deemed-to-satisfy provisions.

Within the standards, the newly inserted E4.1 requires that a dwelling must be provided with a suitable fire detection and alarm system. The exception is for cases where the shortest route from any point in a room to any point in a room on the same storey is not more than 30 m, and in these cases, one or more suitable mains-operated smoke alarms may be located on each storey. For the majority of dwellings the exception would apply. **E4.1**

With regard to the deemed-to-satisfy provisions, the requirements of E4.1 will be met by an automatic fire detection and alarm system which complies with **DTS E4.1**

BS 5839: Part 1: 1988: Type L3.

In cases where compliance can be met by the provisions of a suitable smoke alarm system, the relevant standard is BS 5446: Part 1: 1990. The deemed-to-satisfy provisions also set out installation requirements which inform that smoke alarms must be located:

(1) in circulation spaces;
(2) within 7 m of the door to a living room or kitchen and within 3 m of the door to the bedroom;
(3) where circulation spaces are greater than 15 m in length, within 15 m of another smoke alarm in the same storey;
(4) for ceiling mounting, at least 300 mm distance from any wall or light fitting;
(5) for wall mounting, not less than 150 mm and not more than 300 mm below the ceiling;
(6) not less than 300 mm distance from any heater or air conditioning outlet and not placed directly above it;
(7) on a surface which is normally at the ambient temperature of the surrounding space.

Where it is necessary to have more than one smoke alarm installed in a dwelling (e.g., two storey terrace house, semi-detached or maisonette), these must be interconnected so that detection of a fire by any alarm operates the alarm signal in all of them.

There is also a requirement that smoke alarms must be permanently wired to an electrical circuit which is protected at the consumer unit and has no other equipment connected to it other than a regularly used local lighting circuit.

Smoke alarms installed in dwellings in residential accommodation where there is supervision (e.g. warden) are required to be connected to a central monitoring unit so that in the event of activation of the alarm, the warden or supervisor can identify the dwelling concerned.

Means of escape for the disabled (the June 1994 Amendment)
Means of escape from fire and facilities for fire fighting

E37 DTS E5.1 In every building which requires the application of Part T, suitable means of escape for disabled people must be provided.

The deemed-to-satisfy section stipulates two sources of information:

(1) BS 5588: 1988 Part 8 Clause 6.2, Section 2.
(2) BS 5810: 1979 Clause 7.6, Section 2.

These refer to the design and location of refuge areas for wheelchair bound disabled people.

HEAT PRODUCING INSTALLATIONS AND STORAGE OF LIQUID AND GASEOUS FUELS

PART F — Regulations 14 & 15

Building Standards

Regulation

14 Heat producing installations

Every fixed heat producing installation incorporating an appliance designed to burn solid fuel (including wood and peat) or gaseous or liquid fuel shall be so constructed and installed that:

(a) it operates safely;

(b) its operation does not cause damage by heat or fire to the building in which it is installed;

(c) the products of combustion are not a hazard to health; and

(d) it receives sufficient air for its safe operation.

15 Storage of liquid and gaseous fuel

(1) Subject to paragraph (2), every fixed storage tank for:

(a) the storage of fuel oil used principally to serve an appliance providing space or water heating, or cooking facilities, in any building; or

(b) the storage of liquefied petroleum gas serving any appliance in a building of purpose group 1 or 2, shall be so constructed and installed as to minimise the risk of fire spreading to the tank or of the contents of the tank contaminating any water supply, watercourse, drain or sewer.

(2) This regulation shall not apply to:

(a) a fuel oil storage tank of a capacity not exceeding 90 litres;

(b) a liquefied petroleum gas storage tank of a capacity not exceeding 150 litres water equivalent.

Aim

To ensure the safe operation of heat producing appliances and effective discharge of products of combustion to the atmosphere. Also to avoid damage by heat or fire to the building.

TECHNICAL STANDARDS

Scope

The standards apply to all buildings in respect of heat producing installations and to:

(1) oil storage tanks with capacities exceeding 90 litres; and
(2) liquified petroleum gas storage tanks with capacities exceeding 150 litres water equivalent, for purpose groups 1 and 2 buildings.

Requirements

The requirements of this part will be satisfied if the installations or storage facilities are specified and designed to meet the requirements set out in the deemed-to-satisfy section.The basic criteria for heat producing appliances are:

(1) efficient combustion must take place;
(2) the products of combustion must be discharged safely to prevent a health hazard;
(3) the constructional details and materials must prevent damage to the structure, fabric of the building;
(4) appliances must be safely installed.

Background

The health and safety of people in and around buildings is the prime intention of this part of the Regulations. This is achieved by setting minimum requirements. Appliances must be safe and of good quality (for example, have kite marked labels) and be correctly installed by utilising good constructional detailing and materials. Installation must be by a competent person.

For large installations, even though reference to the British Standards is provided, the relevant specialists may well design and specify to other standards. This should be accepted where covered by a design certificate from an appropriate person.

Again for large installations other legislation will have to be consulted. For example, for LPG tanks in excess of 135,000 litres water equivalent capacity the Health and Safety Executive should be consulted.

'Good practice' will move ahead of changes in regulations and British Standards. It is essential, therefore, to monitor related changes via technical updates, etc.

COMPLIANCE

Application of Part F

F1.2 Standards apply to all buildings in respect of heat producing appliances and oil storage tanks with capacities exceeding 90 litres; and to LPG storage tanks to buildings of purpose groups 1 or 2 with capacities exceeding 150 litres water equivalent.

Specialist advice should be taken when an open flued device is used in conjunction with a fitted extract fan. **F1.3**

When an appliance is installed that can function on more than one type of fuel, then the most onerous requirements of each must be taken into account. **F1.4**

Heat-producing installations

Large installations

Large installation appliances having an output rating exceeding 45 kW must be constructed to prevent ignition of the building, chimney, flue-pipe or adjacent building, and the emissions through smoke, grit, dust etc. must be kept to a minimum. Manufacturers' instructions on installation and fitting must be strictly adhered to. **F2.1**

Appliances

Appliances must be constructed to operate safely. The deemed-to-satisfy section stipulates the relevant standards that should be met, for example:

solid fuel – BS 8303: 1986
oil fired – BS 5410: Part 2:1978
gas-fired – BS 5258: 1975-1989,
BS 5314: 1976-1982 or
BS 5386: 1976-1988. **DTS F2.2**

An appliance fitted in a garage with a floor area less than 60 m^2, must be of a room sealed type. This infers that if the garage is over 60 m^2, then the requirement for a sealed appliance does not apply. **F2.3**

An incinerator together with its chimney, flue-pipe or hearth, regardless of type of fuel used, should comply with the following:

If the chamber capacity exceeds 0.08 m^3, then the requirements of F2.1 should be met; if the chamber capacity exceeds 0.03 m^3 but does not exceed 0.08 m^3 then the requirements set out in F3.1 to F3.19 should be met; and if the chamber does not exceed 0.03 m^3 then the requirements of F5.1 to F5.11 should be met. **F2.4**

Removal of combustion products

Flue outlets must be positioned away from obstructions and away from flammable or vulnerable materials. The outlets must also be of a height to prevent flowback of combustion products into buildings. Reference is made to the deemed-to-satisfy section which highlights shaded areas on roofs where the flue outlet should not terminate, as illustrated in Figure F1.

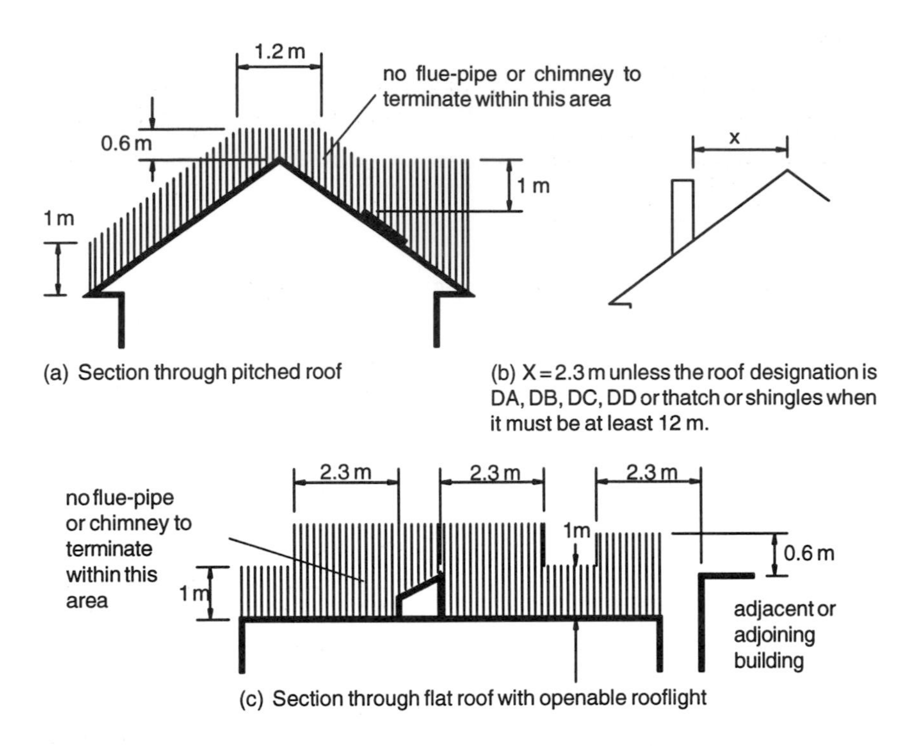

DTS F2.5

Figure F1 Extract from deemed-to-satisfy section.

Solid fuel burning installations

Small installations

Output rating not exceeding 45 kW is classified as a small installation. The associated chimney, flue-pipe or hearth should be constructed in accordance with the standards set out in F3.2-F3.19.

F3.1

Supply of combustion air

A fireplace or a free standing appliance requires to have a fixed opening area. The deemed-to-satisfy section produces a table (Table to F3.2) which provides the standard opening area required for various types of fires to satisfy the relevant British Standards, as shown in Table F1.

Table F1
Supply of combustion air

Type of appliance	Minimum ventilation (Note 1)
Open fire of nominal size 450 mm or less (Note 2) installed in accordance with BS 8303: 1986	an opening or openings with a total free area of 15000 mm^2.
Open fire either of nominal size greater than 450 mm (Note 2) or not installed in accordance with BS 8303:1986	as above, together with any additional openings which may be required by the manufacturer's instructions in respect of the particular appliance.
Any other solid fuel burning appliance	an opening or openings with a total free area of 550 mm^2 for each kW of appliance rated output over 5 kW (an appliance with an output rating of 5 kW or less has no minimum requirement).

Notes:
(1) If a draught stabiliser is fitted to a heating appliance, or to a chimney or flue-pipe in the same room as an appliance, additional ventilation must be provided with a free area of at least 300 mm^2/kW of the appliance rated output.
(2) Fire size is related to the free opening at the front of the fireplace recess when firebricks are in place.

Table F1 also highlights a factor that is frequently ignored, especially when modernisation or refurbishment takes place. This is the provision of further ventilation to allow the fire within the fireplace to operate adequately through convection and combustion.

DTS
F3.2

Removal of combustion products
A solid fuel burning appliance should be connected individually to a flue. An exception is incinerators which can have a shared flue.

F3.3

The cross-sectional flue area must be adequate for the appliance. The table to F3.4 in the deemed-to-satisfy section provides minimum areas for various appliances as illustrated by Figure F2.

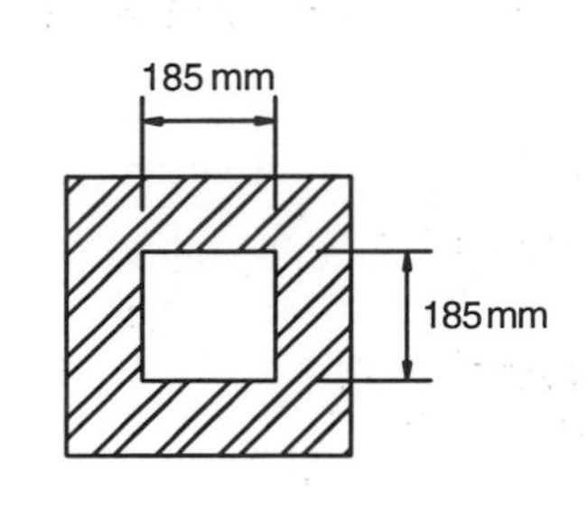

Open fire, room heater or stove within a fireplace recess capable of accepting an open fire of nominal size 450 mm or less

185 mm x 185 mm square, or 225 mm diameter (the diameter may be reduced to 200 mm for a factory made insulated chimney).

DTS F3.4

Figure F2 Extract (removal of combustion products).

Bends in the flue-pipe must not exceed 45° from the vertical. An exception is allowed for back-entry appliances, as illustrated in Figure F3.

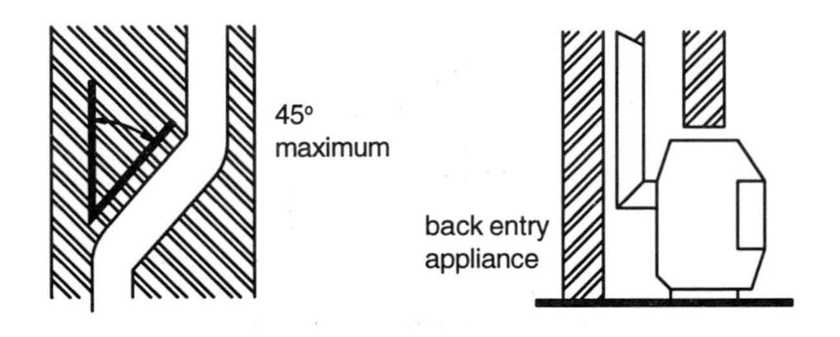

F3.5

Figure F3 Back-entry appliances.

Flues should have no intermediate openings from the appliance to the termination point in the open air. The standard allows for three exceptions to this rule:

(1) if a draught stabiliser has been fitted but only where easy access can be gained for maintenance (in other words within the room), for example, as indicated in Figure F4;

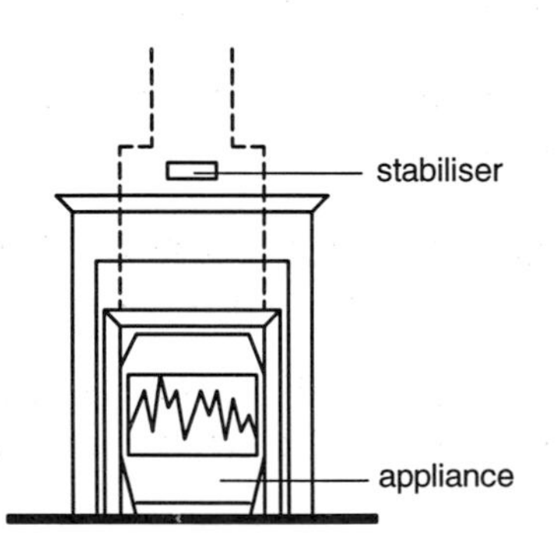

Figure F4 Access to stabiliser.

(2) an explosion door leading into a sootbox has been fitted, as indicated in Figure F5;

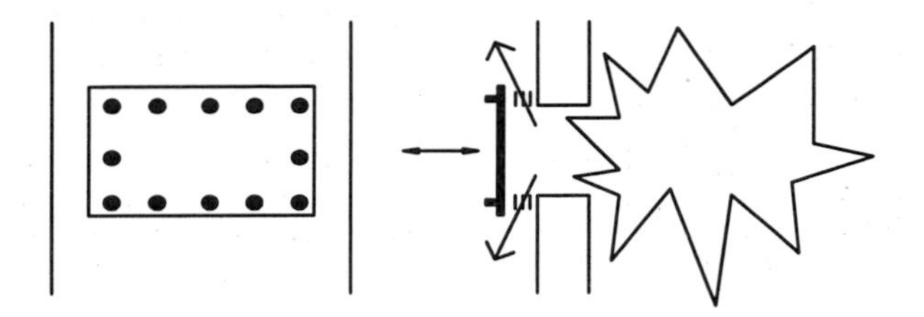

Figure F5 Explosion door to sootbox.

(3) an inspection cover for cleaning can be fitted if the door is of gas-tight construction as illustrated in Figure F6.

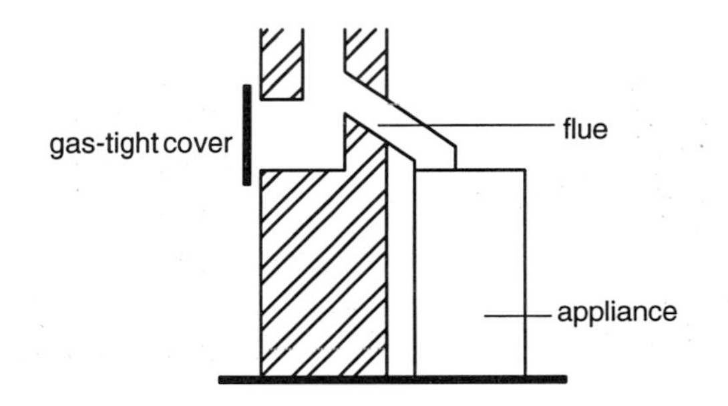

Figure F6 Inspection cover.

F3.6
F3.7

Protection from combustion products

Protection can be split into two functions: firstly to carry off the gases through the building (the flue) and secondly to expel the gases away from the building safely (the chimney). The materials used for the construction of the chimney should be:

(1) masonry, with a lining or heat resistant material; or
(2) factory made insulated chimney.

Dimensions and manufacturing requirements for the materials can be found in the deemed-to-satisfy section.

DTS
F3.8

Although the requirement in F3.8 stipulates that a factory made insulated chimney can be used, the following restrictions must be followed:

(1) it must not pass through or be attached to any part of a building in different occupation, for example Figure F7;

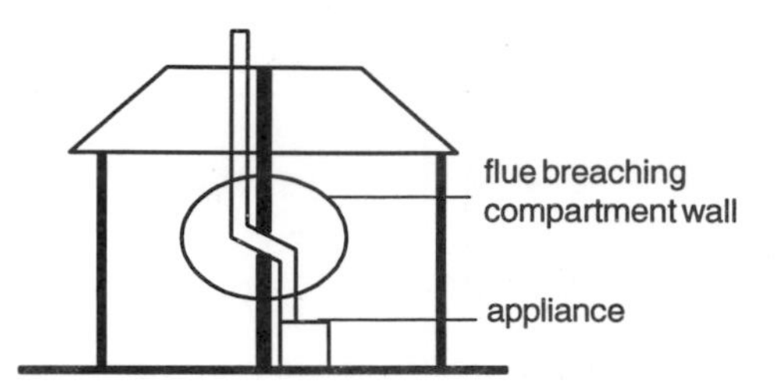

Figure F7 Section through semi-detached dwelling.

(2) it must not pass through any storage space with one exception (that is, if the chimney is separated from the storage area by a removable imperforate casing, refer to the deemed-to-satisfy requirements for the casing positioning).

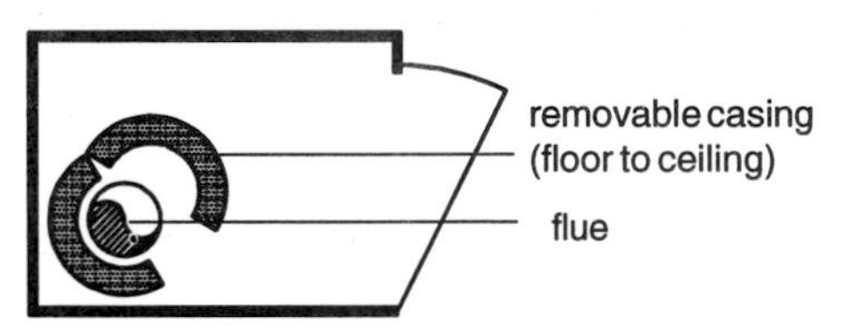

Figure F8 Flue with removable casing.

DTS F3.13

Flues must be individually separated and surrounded by non-combustible material (Figure F8). The material should extend from the top of the fireplace opening to the top of the chimney stack and be capable of withstanding 1100°C without significant change in the properties of the materials. One combustible material is allowed in the construction of the stack however and that is the damp proof course. The inner surface of the chimney or flue-pipe should be constructed with a uniform smooth, impermeable and chemical resistant material. The deemed-to-satisfy requirements provide methods of achieving this. If a flue-pipe is used it must also be capable of withstanding temperatures of 1100°C without significant change to the materials; again reference should be made to the deemed-to-satisfy requirements.

F3.9 DTS F3.10 F3.11

Flue-pipes can only be used to connect a heating appliance to a chimney, and should not pass through a roof space or through an internal wall (allowed connection to a chimney with non-combustible material) or through a ceiling or floor (except if the construction is non-combustible and the flue-pipe discharged into a chimney).

F3.12

Relationship to combustible materials

Flue-pipes should be separated from combustible materials by at least three times the diameter of the flue-pipe. The distance can be reduced to 1.5 times the diameter if there is a non-combustible shield between the flue-pipe and the combustible material, the length of which should extend 1.5 times the diameter on both sides of the flue-pipe. The air space between the shield and the combustible material must be at least 12 mm, as illustrated in Figure F9.

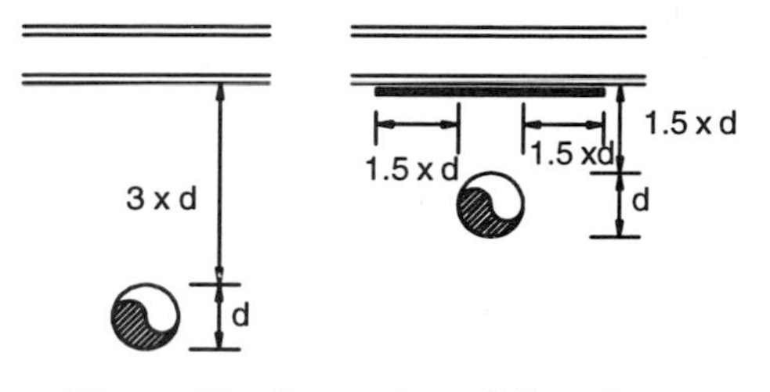

Figure F9 Separation of flue-pipes.

F3.14

All combustible material must be at least 200 mm from the inner surface of a flue or fireplace opening, the exceptions being a damp proof course firmly bedded in mortar, dooks or similar fixings at least 150 mm from the inner surface of the flue or fireplace opening or material under a hearth as described in F3.18. Metal fastenings in contact with combustible material must be at least 50 mm from the inner surface of the flue or the fireplace opening.

F3.15

F3.16

The construction of a hearth should be carried out in such a way that it can easily accommodate an appliance and also prevent ignition through direct radiation, conduction or falling embers. Simple sketches can be found in the deemed-to-satisfy section to F3.18, which provides dimensions for the construction. Under the hearth there should be no combustible materials, except as indicated in Figure F10.

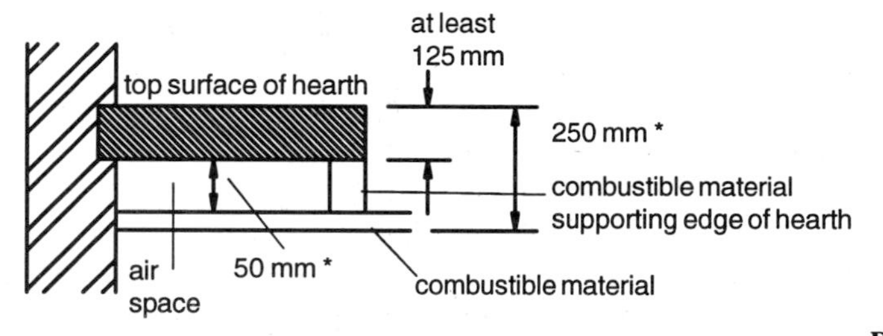

* Note:
Combustible material (not supporting edge of hearth) should not be nearer than either (but not both) of these distances.

DTS
F3.17
F3.18

Figure F10 Section through hearth.

The recess for a fireplace must be constructed from solid non-combustible material. The walls and jambs should be at least 38 mm in thickness and built of fireclay. Refer to the deemed-to-satisfy table for the dimensions of the fireplace material. The exceptions are, if the appliance has a fireclay lining of at least 38 mm or a purpose made chamber, when the building of a non-combustible surround is not required.

DTS
F3.19

Oil-fired installations

Small installations

As in F3.1 the definition for a small appliance is that the output rating should not exceed 45 kW, and if the temperatures of the base and sides of the appliance exceed 100°C, that is under normal working conditions, the requirements of F3.14 to F3.19 should apply. If the appliance does not exceed these temperatures then the requirements of F4.3 to F4.14 should be met.

DTS F4.1

If the appliance is designed to operate with a flue, then the associated chimney or flue-pipe must meet the requirements of F4.3 to F4.15. If the gas temperatures exceed 260°C under normal working conditions then the requirements of F3.3 to F3.18 should be met. The method of measuring the temperatures stated above can be found in BS 4876:1984.

DTS F4.2

Supply of combustion air

A room or space containing an oil fired heating appliance must be provided with permanent ventilation direct to the open air. The area of ventilation must be at least 550 mm² for every kW above 5 kW. This ventilation can be achieved by direct ventilation on an outside wall or via a duct. For an appliance less than 5 kW, if designed, ventilation is also required, but there is no requirement for a minimum area.

F4.3

Removal of combustion air

Every appliance should be connected individually to a flue, with the exception illustrated in Figure F11.

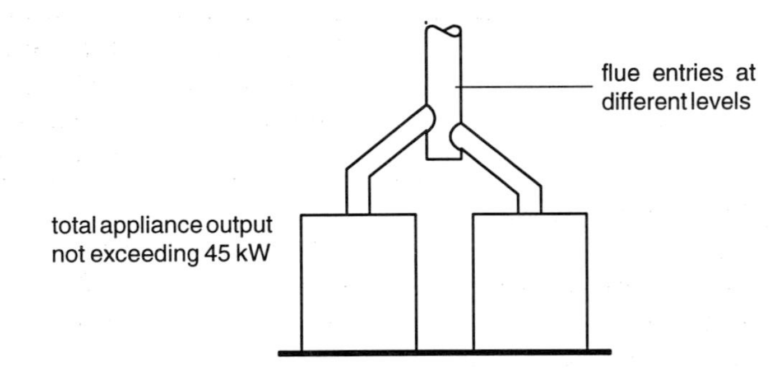

Figure F11 Connection to flue.

A balanced flue for an oil fired appliance is illustrated in Figure F12.

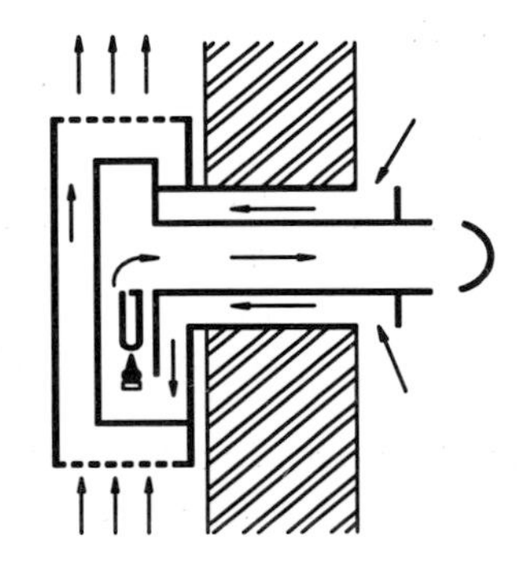

F4.4

Figure F12 Oil fired balanced flue.

Flues should be positioned safely on the exterior of a building and protected by a cover to facilitate air movement, restrict alien matter entering the flue and accidental damage due to collision. The deemed-to-satisfy section refers to BS 2869: Part 2: 1988 where class D fuel is used. Figure F13 shows the appropriate positioning for the flue outlets with other fuels; their terminals should be so positioned as to avoid products of combustion entering openings into buildings.

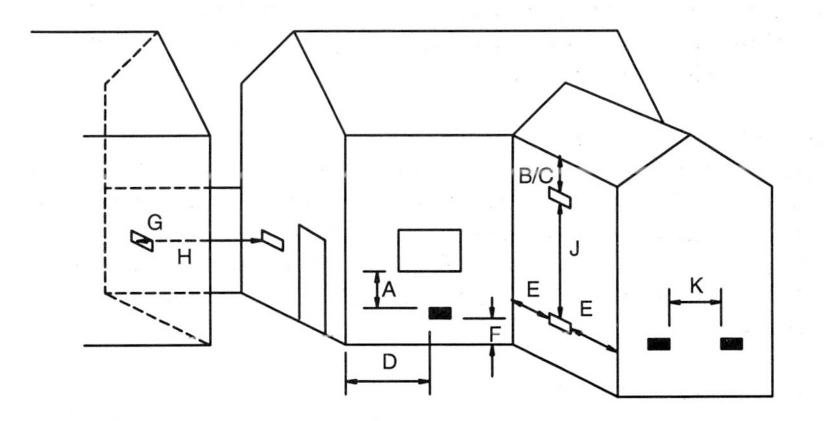

Location	Distance to a balanced flue terminal	Distance to a low level discharge flue terminal
A Directly below an opening, air brick, window etc.	600	600
B Below a gutter, or sanitary pipework	600	1000
C Below eaves or balcony	600	1000
D From vertical sanitary pipework	600	1000
E From an internal or external corner	600	600
F Above ground, or balcony level	600	600
G From a surface facing the terminal	600	600
H From a terminal facing the terminal	600	600
J Vertically from a terminal on the same wall	1500	1500
K Horizontally from a terminal on the same wall	600	600

Figure F13 Flue terminal positions for oil fired appliances.

DTS F4.5 F4.6

Flues should not exceed 45° from the vertical as illustrated earlier in Figure F3 (solid fuel). No intermediate openings are allowed in the flue.

F4.7

Protection from combustion products

The flue within a chimney must be surrounded and separated from any other flue by non-combustible material and must extend from the top of the fireplace to the top of the chimney stack. The flue must also be able to withstand a temperature of 260°C without alteration of the properties, with the exception that a damp proof course material can be utilised if built into the chimney stack.

DTS F4.8

The inner lining of the chimney or flue-pipe must be uniformly smooth, impermeable and resistant to chemical attack. The requirements for the above can be satisfied with reference to the deemed-to-satisfy section which highlights the relevant British Standards and materials that can be used in the construction.

DTS F4.9

Flue-pipes must be non-combustible and be able to withstand temperatures of 260°C without a change to their properties. The deemed-to-satisfy requirement stipulates materials and the Standards that would comply with this requirement.

DTS F4.10

An insulated metal chimney or a double-walled flue-pipe must not pass through a cupboard, storage or roof space.The exception is where the flue pipe can be designed as in Figure F14.

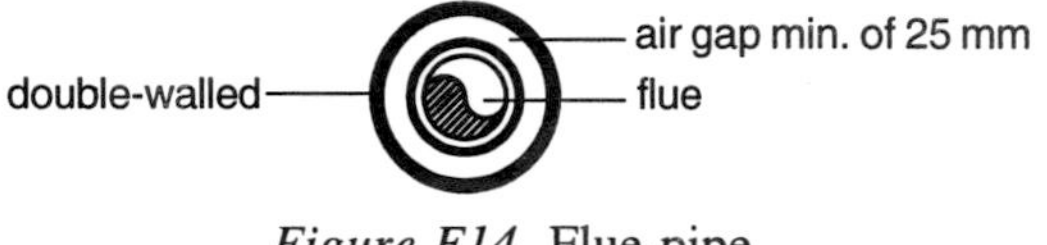

Figure F14 Flue-pipe.

F4.11 Single-walled flue-pipes must be situated on a building as illustrated in Figure F15.

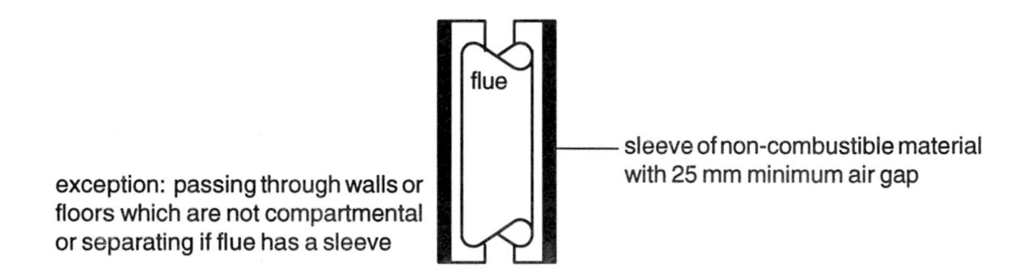

F4.12 *Figure F15* Restrictions for single-walled flue-pipe.

The flue-pipe must be protected to prevent accidental damage. The only part that is excluded from protection is in an area containing the appliance. Reference to the deemed-to-satisfy section relates to clause 42.4 of BS 5410: Part 1:1977 for details of protection.

DTS F4.13

Relationship to combustible materials

A flue-pipe in close proximity to combustible materials should be positioned as illustrated in Figure F16.

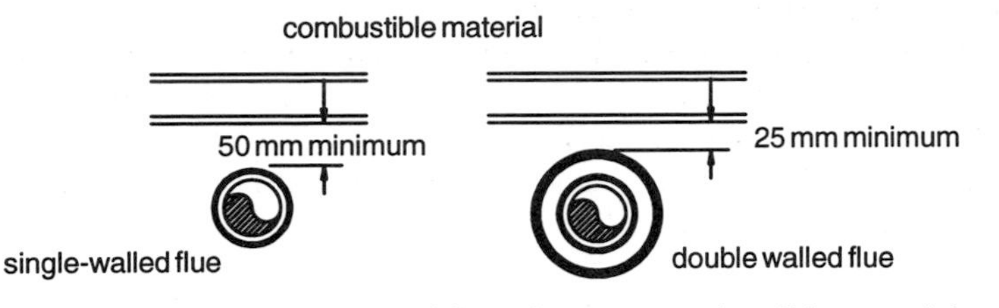

Figure F16 Positioning of flue-pipe near combustible materials. **F4.14**

The appliance must also be separated from combustible materials by either providing a non-absorbent imperforate base or the appliance itself can incorporate a rigid non-combustible base. **F4.15**

Gas-fired installations

Small installations

An appliance with an output rating not exceeding 60 kW is classified as a small appliance and in this respect the requirement of F5.2 to F5.11 will apply, with the following two exceptions:

(1) a water heater without a flue can be installed without having to meet the standards but must be fitted in accordance with the manufacturer's instructions. BS 5546:1979 provides details for fitting such an appliance;

(2) a decorative fuel-effect gas appliance (e.g. gas fire) can be installed without reference to the standard or can be fitted to comply with F3.1 to F3.19 and F5.3, if required, depending on the manufacturer's instructions. BS 6714:1986 provides details for the installation of such an appliance. **DTS F5.1**

Appliances

A gas appliance can be fitted into a room or a space containing a shower or a fixed bath but must be of the balanced flue design as illustrated in Figure F17.

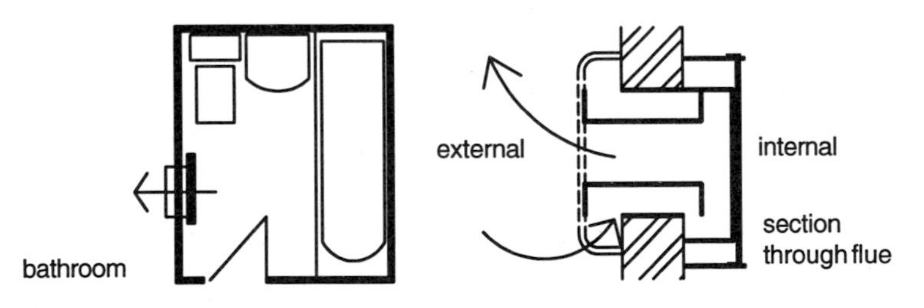

Figure F17 Balanced flue fitted to bathroom. **F5.2**

Supply of combustion air

A gas installation fitted into a room or space must be provided with sufficient permanent ventilation to operate the appliance safely. Reference to BS 5440: Part 2: 1989 provides methods of achieving these ventilation requirements.

DTS F5.3 Ventilation must be direct to the outside air or to an adjoining space which is also permanently ventilated to the outside air.

Removal of combustion products

Each appliance must have its own separate flue with the following exceptions:

(1) incinerators may share the same flue;

DTS F5.4 (2) where a shared flue or duct has been specially designed in accordance with BS 5440: Part 1:1990.

DTS F5.5 A chimney or flue-pipe which is connected to an appliance must be designed in accordance with BS 5440: Part 1:1990 or DM2:*Guide to Gas installation in Timber Frame Housing*, published by British Gas in 1984.

If balanced flues are used then the outlets must:

(1) be situated externally and away from obstructions or combustible material, as illustrated in Figure F18. For flat roof positioning of flues, refer to Table F2 and Figure F19;

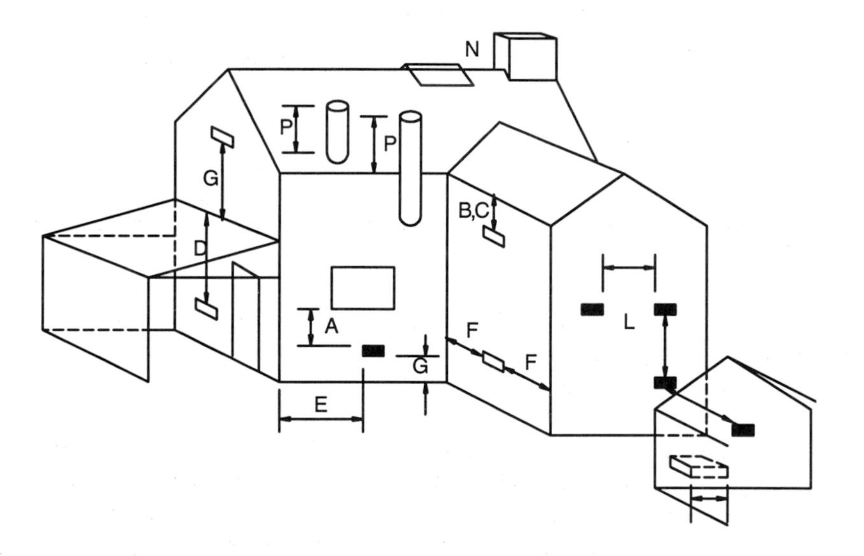

Figure F18 Flue terminal positions.

Table F2 : Minimum height P to the base of the terminal for roof mounted individual natural draught open flue systems

Type of roof	Where N ≥ 1.5 metres	Where N < 1.5 metres
	where the flue system is internal	all flue systems internal and external
Flat roof	where the roof has a parapet 600 mm	
	where the roof has no parapet 1 m	
Pitched roof	where the roof pitch exceeds 45° 600 mm	

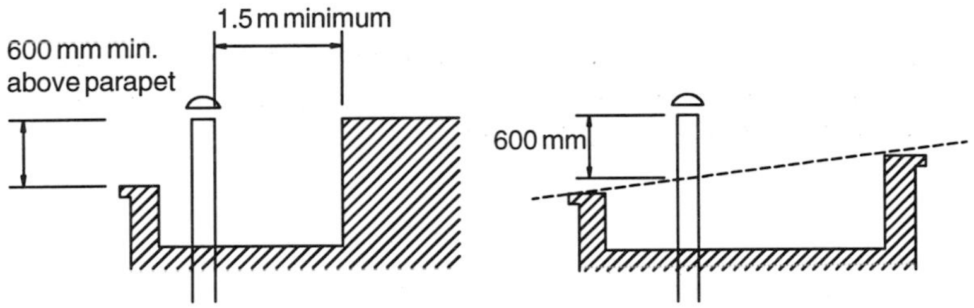

Figure F19 (Table 2) Extract flat roof.

(2) the flue must be constructed to prevent matter entering and causing an obstruction and must be protected by a terminal guard. Reference to BS 5258: Part1:1986 will provide details. **DTS F5.6**

As with flues for solid and oil appliances, the gas flues must have no intermediate openings. Reference to the requirements stated in F3.6 applies also to flue-pipes for gas appliances. **F5.7**

Flue-pipes must be protected to prevent damage or danger to people where they pass through a room or an accessible space. This can be achieved by providing sleeves for the flue. The only part of the flue that does not require protection is where it connects to the appliance within the room. The methods of protection are stated in BS 5440: Part 1:1990. **F5.8**

Relationship to combustible materials

Flue-pipes should be positioned in relation to combustible materials as shown in Figure F20.

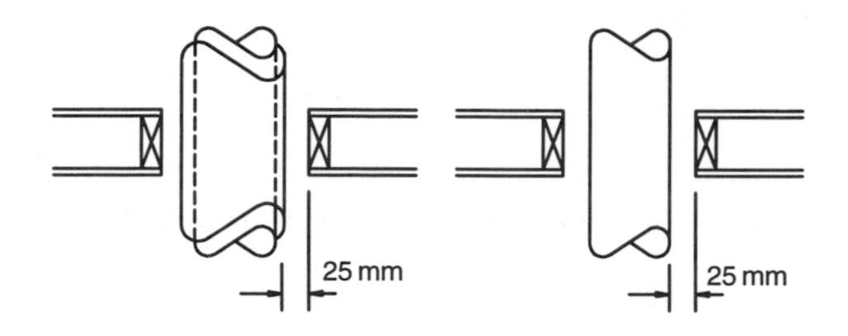

F5.9

Figure F20 Flue-pipes in relation to combustible materials.

The gas appliance must be provided with a suitable hearth as illustrated in Figure F21.

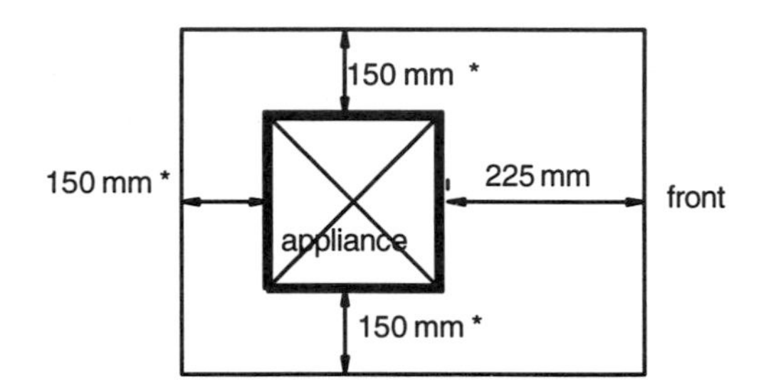

* Note:
The 150 mm minimum does not apply where the back or side of the hearth either abuts or is carried into a solid non-combustible wall which extends above the height of the appliance, nor does it apply where a gas appliance back boiler is built into a fireplace opening suitable for solid fuel appliances.

DTS
F5.10

Figure F21 Hearth details.

The gas appliance and associated draught diverters must be separated from combustible material as illustrated in Figure F22 or the appliance can be installed in accordance with BS 5258:1975-1989 or BS 5386:1976-1988.

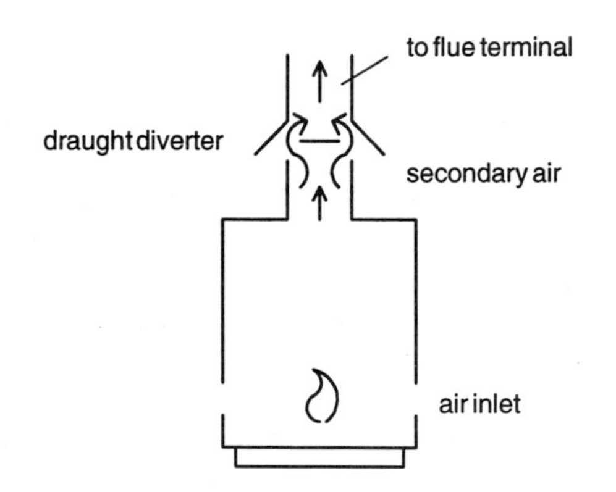

DTS
F5.11

Figure F22 Separation of appliance/diverters.

Storage of liquid and gaseous fuels

Oil storage

Tanks with capacities exceeding 90 litres must be:

(1) constructed adequately and safely, (reference to BS 5410: Part 1:1977 or BS 5410: Part 2:1978, will provide constructional details);

(2) separated from any building (same occupation) or separated from the boundary. Catchpits if required should be designed to retain the contents in the event of leakage, for example refer to diagram and table to F6.1.

The requirements of F6.1a will be met where the tank is constructed in accordance with BS 5410: Part 1:1977 or BS 5410: Part 2:1978.

The requirements of F6.1b and c for a tank with a capacity not exceeding 3500 litres will be met when the tank is located in accordance with the table to the specification in Table F3.

Table F3 : Oil storage tanks

Location of tank	Catchpit required	Protection from fire in building	Protection from fire in relation to boundary
Within a building	yes	within chamber	n/a
External, above ground	yes (2)	1.8 m from building, or building wall is fire resisting, or barrier is provided	750 mm from boundary, or barrier is provided
External, wholly below ground	no	no requirement	no requirement

Notes:

(1) **CATCHPIT** means a pit, without a drain, which is capable of containing the contents of the tank, plus 10%;
CHAMBER means a fully enclosed ventilated space, bounded by construction of not less than 60 minutes' fire resistance, including a self closing door wholly above catchpit level;
FIRE RESISTING in relation to a building wall means that any part of the external wall within 1.8 m of the tank must have not less than 30 minutes' fire resistance from the inside; and
BARRIER means a wall or screen having not less than 30 minutes' fire resistance extending 300 mm above and beyond the ends of the tank.

(2) A tank not exceeding 1250 litres does not require a catchpit.

DTS F6.1 The requirements of F6.1b and c for the location of a tank with a capacity exceeding 3500 litres will be met where the tank is located in accordance with BS 5410: Part 2: 1978 (clauses 45.2 and 45.3).

Liquefied petroleum storage

Tanks with a capacity exceeding 150 litres water equivalent must:

(1) be constructed of steel with the requirements for pressures being met by the following extract from the deemed-to-satisfy section:

The requirements of F6.2a will be met by a tank constructed of steel suitable for use at the safe operating temperatures and pressures given in the table to this specification:

	Propane	**Butane**
Maximum safe operating pressure	14.5 bar gauge	4.83 bar gauge
Minimum safe operating pressure	0 bar gauge	480 millibar absolute
Minimum safe operating temperature	–40°C	–18°C

Where a tank is for use with both gases the more onerous criteria apply.

(2) be separated from a building or boundary in the same occupation and from any other storage tank or group of tanks. The maximum number of tanks that can be grouped together is six. The capacity of an individual tank should not exceed 135 000 litres water equivalent and if grouped together should not exceed 450 000 litres water equivalent. Refer to Table F4 extracted from the deemed-to-satisfy section.

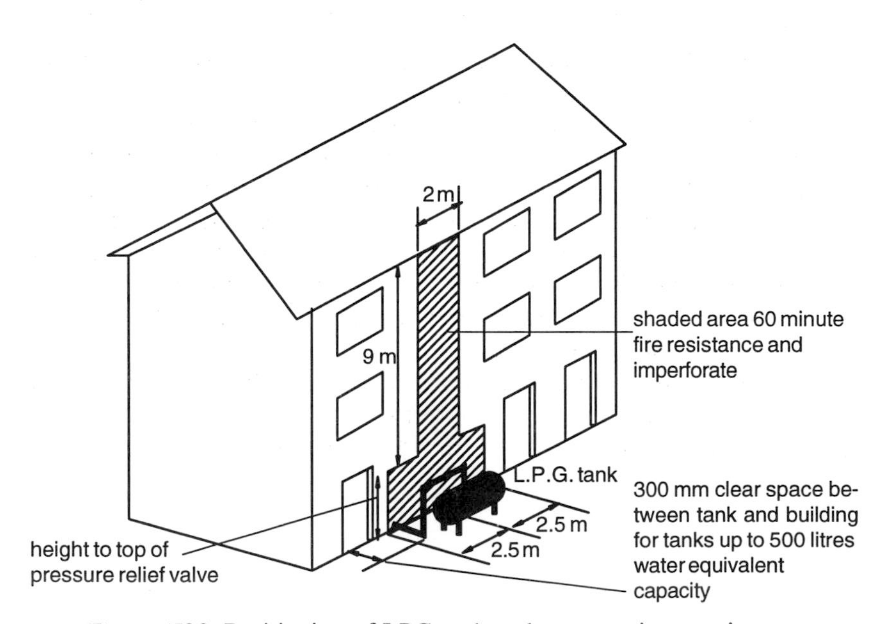

Figure F23 Positioning of LPG tank and construction requirements.

Table F4 : Liquefied petroleum gas storage

		Minimum separation distance (metres)		
Capacity of any single tank in litres (water equivalent) not exceeding (Note 2)	Maximum capacity of any group of tanks in litres (water equivalent)	For above ground tanks, from building, boundary or fixed source of ignition		
		Without barrier	With barrier	Between tanks
500		In accordance with Figure F23		
2500	7500	3.0	1.5	1.0
9000	27 000	7.5	4.0	1.0
135 000	450 000	15.0	7.5	1.5
		For buried or mounded tanks		
		From building, boundary or fixed source of ignition	Between valve assembly and building	Between tanks
500		In accordance with Figure F23		
2500	7500	1.0	3.0	1.5
9000	27 000	3.0	7.5	1.5
135 000	450 000	3.0	7.5	1.5

Notes:

(1) **MOUNDED** means partly or wholly above ground and covered by a mound of earth or similar inert material;

BARRIER means a wall or screen having not less than 60 minutes' fire resistance, located between 1 m and 3 m from the tank and extending:

(a) longitudinally: so that the distance specified above without a barrier is maintained when measured around the ends of the barrier, and

(b) vertically: 2 m or the height of the top of the tank, whichever is greater.

VALVE ASSEMBLY means the manhole and pressure relief valves.

(2) For tanks exceeding 135,000 litres water equivalent the Health and Safety Executive must be consulted.

(3) Where the capacity of a tank is between the capacities shown in the table, separation distances can be interpolated.

(4) The number of tanks in a group should not exceed six.

DTS F6.2

PREPARATION OF SITES AND RESISTANCE TO MOISTURE

PART G **Regulations 16, 17 & 18**

BUILDING STANDARDS

Regulations

16 (1) Subject to paragraph (3) a site and ground immediately adjoining a site shall be so prepared and treated as to protect the building and its users from harmful effects caused by:
(a) harmful or dangerous substances;
(b) matter in the surface soil;
(c) vegetable matter.

(2) Subject to paragraph (34) a site and ground immediately adjoining a site shall be so drained or otherwise treated as to protect the building and its users so far as may be reasonably practicable from harmful effects caused by:
(a) ground water;
(b) flood water;
(c) existing drains.

(3) Paragraphs (1)(b), (1)(c) and (2)(c) shall not apply to a limited life building of purpose groups 2 to 7 inclusive.

(4) In paragraph (1)(a) 'harmful or dangerous substances' includes deposits of faecal or animal matter and any substance or mixture of substances which is or could become corrosive, explosive, flammable, radioactive or toxic or which produces or could produce any gas likely to have any such characteristic.

17 (1) Subject to paragraph (2), a building shall be so constructed as to protect the building and its users from harmful effects caused by:
(a) moisture rising from the ground;
(b) precipitation.

(2) This regulation shall not apply to a building where penetration of moisture from outside will result in effects no more harmful than those likely to arise from use of the building.

18 A building of purpose group 1 shall be so constructed as to protect the building and its users, so far as may be reasonably practicable, from harmful effects caused by condensation.

Aim

The purpose of this part is to ensure that adequate measures are taken to protect both people and the fabric of the building from any harm which could arise from site conditions (e.g. harmful substances in the soil) or from moisture in various forms such as rising damp, rain penetration into the building, surface and interstitial condensation.

TECHNICAL STANDARDS

Scope

Generally the standards apply to all buildings with the exceptions indicated in regulation 16 (3) in relation to buildings having a limited life.

They deal with matters of site preparation and resistance to moisture entering the building from the ground, from precipitation, and via the mechanisms of surface and interstitial condensation.

Requirements

In terms of site preparation there is a requirement to remove or make safe any substance considered to be harmful or dangerous. Furthermore with respect to the potential for accumulation of ground or flood water there is a requirement to drain the site area and if necessary any land ajoining the site.

In terms of the design of the buildings there is a requirement to use appropriate detailing and specifications which will preclude rising damp, evaporation from the ground within basement areas, water penetration arising from precipitation, and surface and interstitial condensation.

Background

For moisture to condense on a room surface the vapour present in the room, air must be cooled below its dew point temperature. Any cold bridge present in the building fabric would facilitate such cooling but Part J regulations require that no part of the external fabric (other than glazing) will have a U-value greater than 1.2 W/m^2C. This reduces the risk of such cooling occurring anywhere other than on a window surface.

The same physical process of condensation of vapour can occur within the building fabric. Generally, the levels of moisture in the air and the corresponding vapour pressure are higher within a building than outside. This is especially so in winter, when low outside temperatures inhibit moisture levels in the outside air. The vapour pressure difference across the building fabric is the driving agent which causes vapour flow. Any vapour within the interstices of the fabric, if cooled below its dew point temperature, will condense, thus wetting the fabric. Materials such as timber or insulants if wetted in this manner could deteriorate.

The application of a vapour check on the warm side of any applied insulation reduces the risk of interstitial condensation. In cases such as flat or pitched roofs, ventilation of voids reduces the vapour pressure and lowers the dew point temperature, thereby reducing the risk.

COMPLIANCE

Introduction

The three themes covered in Part G, preparation of sites and resistance to moisture from the ground, resistance to precipitation and means of precluding condensation, will be considered in turn.

This will include a review of deemed-to-satisfy provisions and tables of harmful and dangerous substances.

Preparation of sites

The site and any ground immediately adjoining the site must have any harmful or dangerous substances either removed or made safe. Such substances include faecal or animal matter, or material which is or could become corrosive, explosive, flammable, radioactive or toxic or which produces or could produce any gas likely to have such characteristics. **G2.1**

When planning application is made to develop a site, records or local knowledge may identify the presence or possible presence of harmful or dangerous substances and the resultant planning permission will then normally be subject to conditions. Such circumstances could well arise from the range of possible sites identified in Table G1 as follows:

Table G1 **Part G Appendix**

Land likely to contain contaminants
Asbestos/chemical/gas works
Industries making or using wood preservatives
Landfill and other waste disposal sites
Metal mines/smelters/foundries/steel works/metal finishing works
Munitions productions and testing
Nuclear installations
Oil storage and distribution
Paper and printing works
Railway land/large sidings/depots
Scrap yards
Sewage works/sewage farms/sludge disposal sites
Tanneries

On the the other hand the presence of possible contaminants may not be detected until work has started on site. Once there is the certainty or even the suspicion of the presence of a contaminant the local authority should be told immediately. Environmental health officers would offer advice in dealing with such matters and Table G2 provides assistance on how to identify and deal with certain contaminants. The courses of action provided in Table G2 are based on the assumption that the building will have at least 100 mm concrete cover *in situ*. Expert advice may be required especially where contaminants are present in large quantities or where there is considerable health risk.

Part G Appendix

Table G2
Possible contaminants and actions

Signs of possible contamination	Possible contaminant		Probable remedial action required
vegetation (absence, poor or unnatural growth)	metals, metal compounds		none
	organic compounds, gases		removal
surface materials (unusual colours and contours may indicate wastes or residues)	metals, metal compounds		none
	oil and tarry wastes		none
	asbestos (loose)	*	removal, filling or sealing
	other fibres		none
	organic compounds including phenols		removal or filling
	potential combustibles, coal, coke dust		removal or inert filling
	refuse and waste		removal
fumes and odours (may indicate organic chemicals at very low concentrations)	flammable, explosive, toxic and asphyxiating gases including methane and carbon dioxide		removal: construction must be free from unventilated voids
	corrosive liquids	*	removal, filling or sealing
	faecal, animal and vegetable matter (biologically active)		removal or filling
Drums and containers (full or empty)	various	*	removal with all contaminated ground

* local authority may require removal by specialists

In Table G2:

removal means that the contaminant and any contaminated ground to be covered by the building should be taken out to a depth of 1 m (or less if agreed by the LA) below the level of the lowest floor and taken away to a place to be named by the LA.

filling means that the ground to be covered by the building is to be covered to a depth of 1 m (or less if agreed by the LA) with a material which will not react adversely with any contaminant remaining and will be suitable for making up levels. The type of filling and the design of the ground floor should be considered together.

inert filling means that the filling is wholly non-combustible and not easily changed by chemical reactions.

sealing means that a suitable imperforate barrier is laid between the contaminant and the building and sealed at the joints, around the edges and at the service entries. Note that polyethylene may not be suitable if the contaminant is a liquid such as tarry waste or organic solvent.

Other useful information on site preparation can be obtained from:

BS 5930:1981	Code of practice for site investigations
DoE	Guidance Notes – Inter-departmental Committee on the Redevelopment of Contaminated Land
BS DD175:1988	Code of practice for the identification of potentially contaminated land and its investigation

Drainage provisions where necessary

G2.2 to G2.4

In situations where the proposed site has a high water table (within 0.25 m of the lowest floor in the building) or is liable to accumulate ground or flood water, the ground must be drained to overcome these effects or the building constructed to resist moisture penetration.

Where the building is erected over an existing drain (or field drain), which will remain active, such drains will require attention. The possibilities are to re-route them, if practicable, or as an alternative they may be re-constructed as appropriate (see Part M) below the building. Possible means of dealing with single or multiple subsoil drains are indicated in Figures G1 and G2.

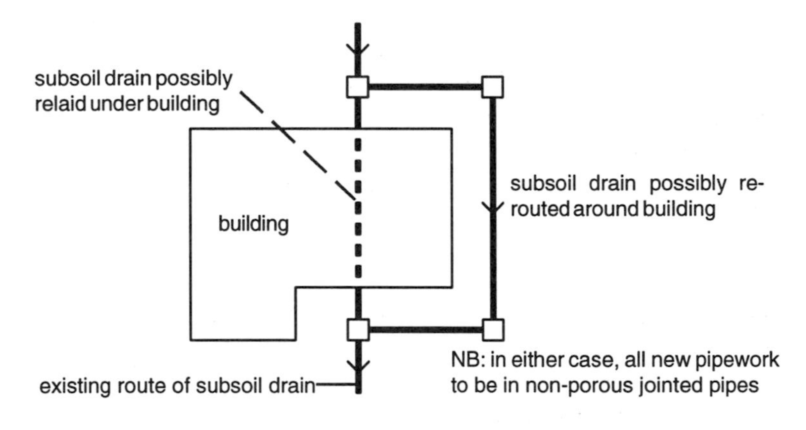

Figure G1 Single subsoil drain.

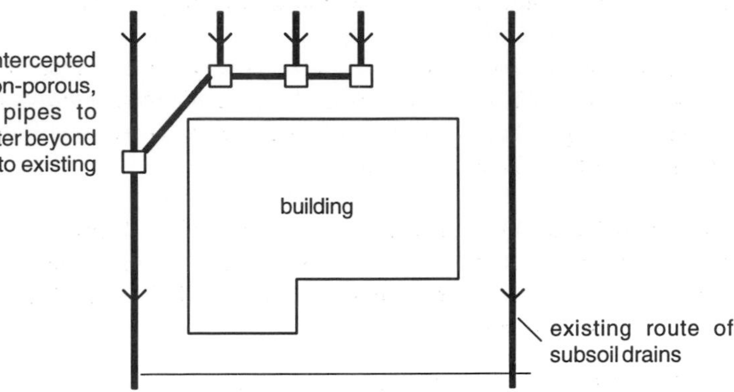

Figure G2 Multiple subsoil drains.

Resistance to moisture from the ground

G2.5

In the construction of the solum of the building, which is the prepared area within the containing walls, measures must be taken to minimise the evaporation of water from the ground (by means of a membrane) which may otherwise provide conditions in which damage to materials such as timber within the building could occur.

G2.6

Floors, walls or any other element adjoining the ground must be treated to prevent moisture from the ground reaching any materials which could deteriorate as a result of the presence of moisture.

Specific means of achieving these functional requirements are indicated within the deemed-to-satisfy provisions which relate to ground supported concrete floors, suspended timber floors and suspended concrete floors.

Ground supported concrete floors

DTS G2.6

The functional requirements can be met by covering the ground with a dense concrete slab which incorporates a damp proof membrane and which is laid upon a hardcore bed set above a level solum.

The hardcore bed should be at least 100 mm thick and of broken brick or similar inert material. Care must be taken in the selection of such material to avoid the possibility of the inclusion of water soluble sulphates in quantities which could damage the concrete slab. The hardcore should be blinded with suitable fine material and consolidated to form a level crack-free surface.

The concrete slab should be at least 100 mm thick and if insulation is required this may be laid either directly above or below the slab. A finishing screed on the top of the floor is optional.

There are three possible locations for the damp proof membrane: above, below or within (sandwich) the concrete slab. In all cases care must be taken to ensure effectiveness of all joints and seals. As an instance of this it is common practice to use at least a 100 mm overlap when joints are being lapped.

When laid below the slab the membrane (especially when polythene is used) should be provided with a uniform support by a material which will not damage the membrane. Upper surface membranes will normally require protection by either a screed or a suitable floor finish.

No matter which location is chosen for the membrane it will require to be sealed to the damp proofing material in the walls, columns or any other elements in contact with the floor in accordance with clause 11 of CP 102:1973. A typical arrangement is shown in Figure G3.

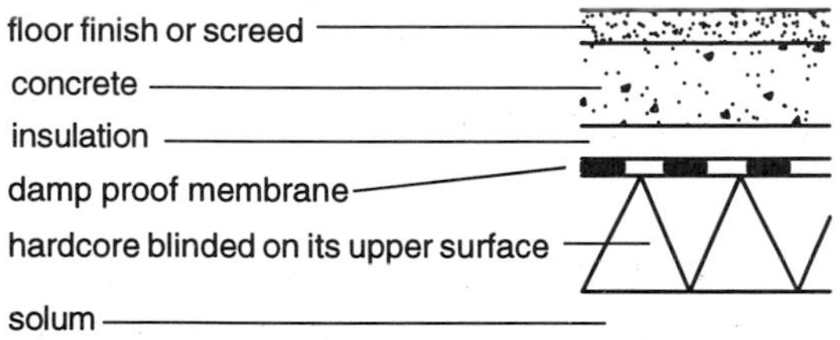

floor finish or screed
concrete
insulation
damp proof membrane
hardcore blinded on its upper surface
solum

Figure G3 Ground supported concrete floor.

Suspended timber floor

DTS G2.6

The functional requirements can be met by supporting the timber floor on wall plates resting on sleeper walls with isolation via damp proofing, laid beneath the wall plates and with the ground being covered with a suitable moisture resistant material, there being provision for adequate ventilation in the space between the top surface of the solum and the underside of the floor to protect the timber.

The solum should be level (ensure that any infilling is by means of a hard dry material). The hardcore bed should be at least 100 mm thick and of broken brick or similar inert material, taking care to avoid the possible inclusion of water soluble sulphates.

Three possible methods may be used to finish the ground cover surface:

(1) a damp proof membrane as per CP 102: 1973; Section 3; or
(2) 50 mm concrete on 1000 gauge polythene sheet; or
(3) 100 mm concrete.

Whichever method is chosen, the finished ground surface must not lie below that of the adjacent ground outside the building.

The suspended timber floor may include thermal insulation in the space between the joists but in any case a clear space must be available between the underside of the joists and the finished ground surface of minimum depth 150 mm for ventilation.

Provision must be made for permanent ventilation of this space by the inclusion of ventilators in two external walls on opposite sides of the building with an opening area in each wall of either:

(1) 1500 mm^2 per metre run; or
(2) 500 mm^2 per m^2 of floor area.

Where internal sleeper walls or any other obstruction interferes with the ventilation arrangements the same opening area is necessary in these elements.

The minimum allowable height from the finished ground surface to the underside of wall plates is 75 mm. A typical arrangement is shown in Figure G4.

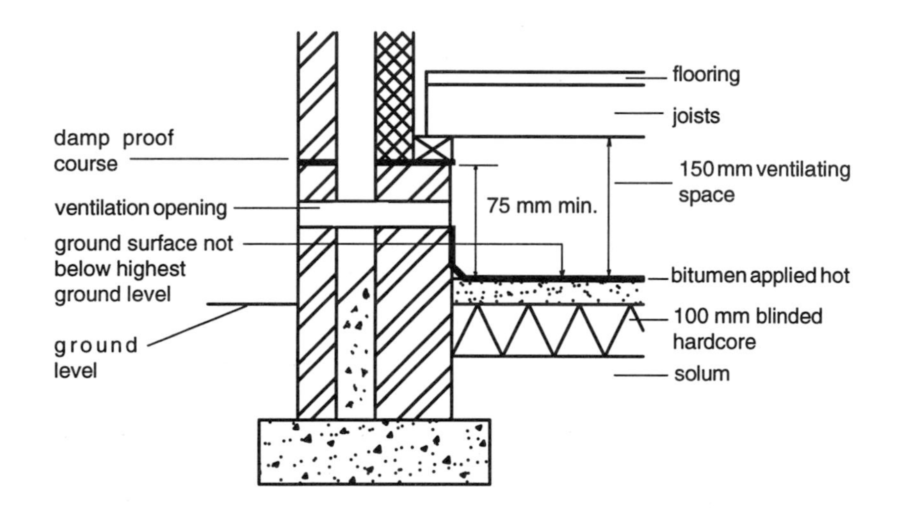

Figure G4 Suspended timber floor.

DTS G2.6

Suspended concrete floor

The functional requirements can be met by the use of *in situ* concrete or precast concrete slabs or beams having concrete or clay infill units. Insulation, if required, is placed above the concrete and provided with a screed or other acceptable floor finish or with boards.

The solum should be level, with any infilling being of hard, dry material.

The permanent ventilation arrangements are as for the suspended timber floor in terms of the 1500 mm^2 / 500 mm^2 requirements and their application to two opposing external walls and any internal sleeper walls or similar obstructions. A ventilating space of vertical height 150 mm from finished ground surface to underside of the floor slab or beams is also required. A typical arrangement is shown in Figure G5.

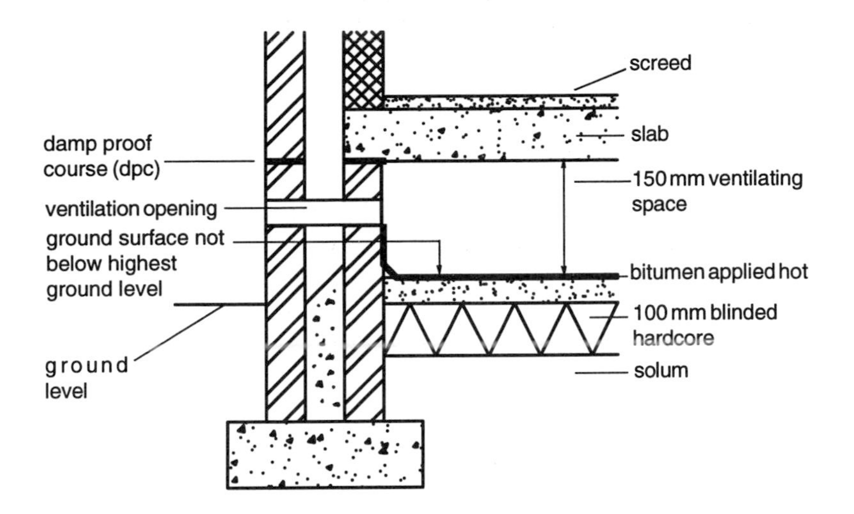

Figure G5 Suspended concrete ground floor.

Protection of walls against ground moisture

DTS G2.6

Wall can also mean pier, column and parapet. It may also include a chimney if it is attached to the building. Windows, doors and other openings are not included.

In constructing a wall (in accordance with clause 10 of CP 102: 1973):

(1) ground moisture must be prevented from reaching the inside of the building;
(2) the wall must not be adversely affected by ground moisture;
(3) the wall must not transmit ground moisture to any other part of the building which could be damaged as a result.

These requirements are normally met by the provision of a suitable damp proof course (dpc) located in the appropriate position, (see examples in Figures G4 and G5). The normal materials used are bituminous sheets, engineering brick or slate laid in cement mortar, polythene or pitch polymers. It is important that damp proofing should be continuous. The damp proof course should be set at least 150 mm above the finished outside ground level.

With regard to the damp proof course in cavity walls, the cavity should be extended by at least 150 mm below the lowest level of the damp proof course as indicated in Figure G6.

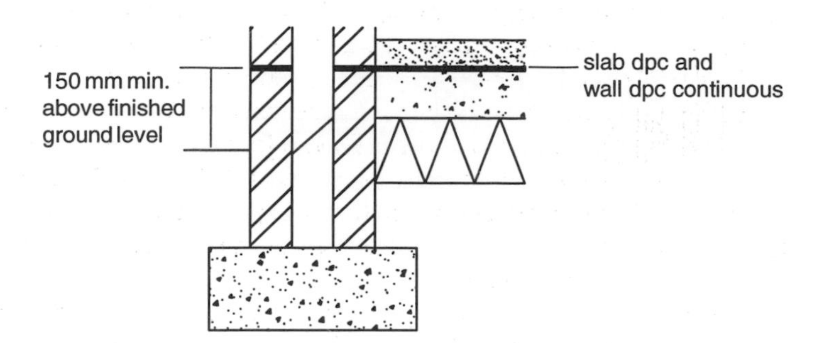

Figure G6 Location of damp proof courses.

It is impractical to apply the 150 mm rule to cavity walls which are supported by either a ground beam or a raft foundation. In such cases the support must be considered to bridge the cavity, and protection should be provided by a flashing or dpc as shown in Figures G7 and G8.

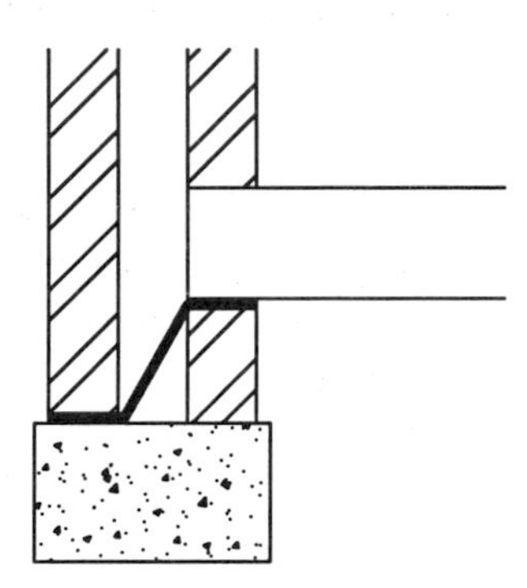

Figure G7 Dpc – ground beam.

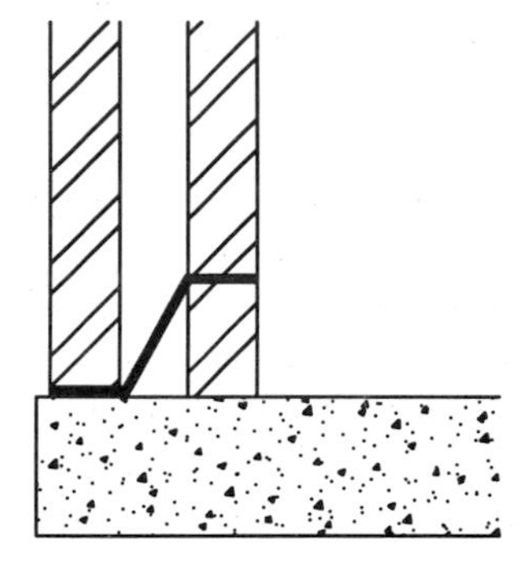

Figure G8 Dpc – raft foundation.

Structures below ground, such as basements, are covered by the relevant clauses in Section 2 of CP 102: 1973.

G3.1 **Weather resistance of walls and roofs**

Item G3.1 of the Standards requires that walls, floors, roofs and other building elements must be constructed in such a manner as to prevent moisture penetration to inner surfaces which could be damaged as a result.

Specified Constructions are given within the deemed-to-satisfy provisions, the thicknesses and other dimensions specified being the minimum acceptable unless otherwise stated.

DTS G3.1 **External walls**

Before looking at some of the Specified Constructions, there are some general points.

(1) masonry walls incorporating dpcs, flashings and other materials and components, to be constructed as per BS 5628, CP for use of masonry, Part 3; 1985, Materials and components, design and workmanship. The construction must be appropriate for the degree of exposure to wind and rain as per clause 21 and as per BSI Draft for Development DD93:1984;
(2) rendered masonry walls must be as per BS 5262:1976 CP External rendered finishes, selected in relation to exposure and types of masonry;
(3) walls of natural stone or cast stone must be as per Section 3 of BS 5390: 1976, CP for stone masonry;
(4) walls incorporating insulating materials must be constructed as per the Agrement Certificate for the material together with the recommendations within the relevant British Standard. For example, where urea formaldehyde foam insulant is used the relevant source would be BS 5617: 1985 and BS 5618:1985. For other materials see the deemed-to-satisfy provisions;
(5) where cladding materials are included in the construction of the wall then construction must be in accordance with the relevant BS. For example, where aluminium is used the relevant source would be CP143: Part 1:1958 and Part 15:1973. For other materials see the deemed-to-satisfy provisions.

Within the deemed-to-satisfy provisions three wall types are given: solid masonry, cavity masonry and framed walls. Solid walls which have sufficient thickness can retain moisture during wet weather and release it to the air again during the next dry spell. The outer leaf of a cavity wall retains moisture in the same manner as the solid wall and the cavity prevents any penetration to the inner leaf. Framed walls may be protected in a variety of ways including brick outer leaf, or various forms of cladding.

Solid masonry walls

DTS G3.1

These require to be at least 200 mm thick and constructed of brick, block, slabs of clay, calcium silicate or cast stone.

Where the wall is insulated on the inside face this is accomplished by creating a cavity at least 25 mm wide between the inner face of the masonry and the insulant; lining is by plasterboard sheet or equivalent. The wall may be rendered externally if required.

Where the wall is insulated on the outer face, the insulant must be protected by a rendering or cladding (such as sheets, tiles, or boarding) with permanent ventilation. The inner surface of the wall may be finished with plaster or plasterboard, if required.

Cavity masonry walls

DTS G3.1

These require two leaves of at least 100 mm thickness each, separated by a cavity of at least 50 mm and constructed of brick, clay block, calcium silicate or concrete. The wall may be rendered externally if required.

Three possible arrangements are available with respect to the use of insulant linings:

(1) when insulating on the internal face of the inner leaf the insulant is applied to the face and lined with plasterboard;
(2) complete fill of the cavity with insulant. The internal face of the inner leaf may be faced with plaster or plasterboard, if required;
(3) partial fill of the cavity with insulant. The internal face of the inner leaf may be faced with plaster or plasterboard, if required. This specified construction is shown in Figure G9;

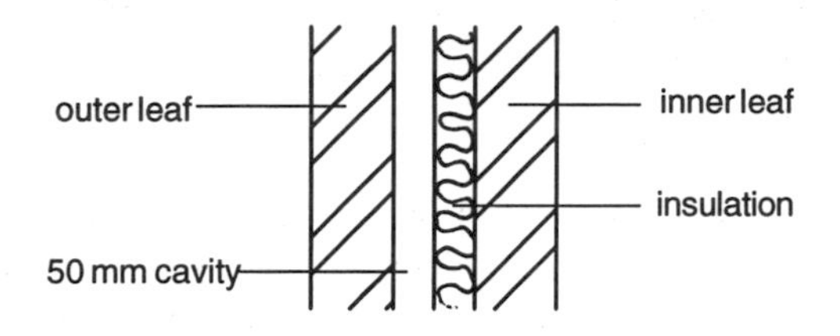

Figure G9 Cavity wall with partial insulant fill.

DTS G3.1 ***Framed walls***

Three specified constructions are offered, each consisting of a framed wall and cladding and these are considered next:

(1) In situations where an external masonry cladding is used this should be of 100 mm thick brick, block, calcium silicate, dense *in situ* concrete, light weight concrete or autoclaved aerated concrete with an externally ventilated cavity in the range 50 mm to 100 mm which separates the masonry leaf from the timber frame. Ventilation is by one open perpend joint per 1.2 m run, at the top and bottom of the wall. The masonry may be rendered if required.

The framed wall is of timber standards and rails with a vapour permeable sheathing to the frame which is covered by a breather membrane. The insulant is applied as a fill within the timber frame and the frame is lined internally with a vapour control layer and faced with plasterboard. The arrangement is shown in Figure G10.

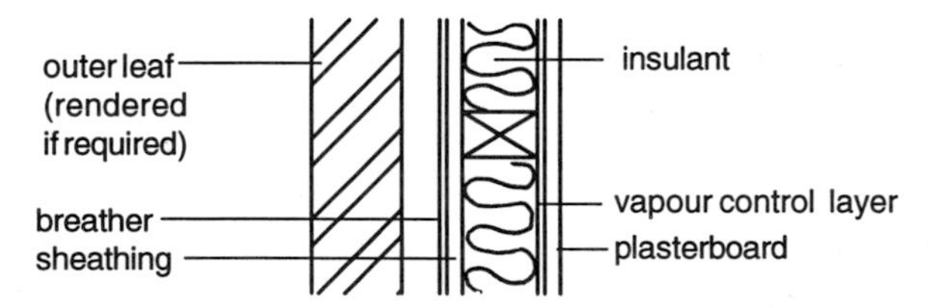

Figure G10 Framed wall with masonry cladding.

(2) Using the timber frame arrangement as indicated in Figure G10, an alternative to the masonry cladding is to use weatherboarding, tiles or slates as a cladding applied to battens/counterbattens as indicated (when

tiles are used) in Figure G11.

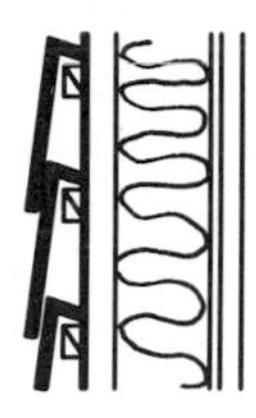

Figure G11 Framed wall with tile cladding.

Where claddings are used which are of sheet or panel form of fibre cement, plastic, metal, GRP or GRC then the framed wall may be of either timber or metal standards and rails.

The insulant can be applied at the inner surface of the framed wall or as an infill within the framing. Permanent ventilation is required behind any impervious cladding. The inner face of the framed wall is lined with a vapour control layer and plasterboard lining. The arrangement is shown in Figure G12.

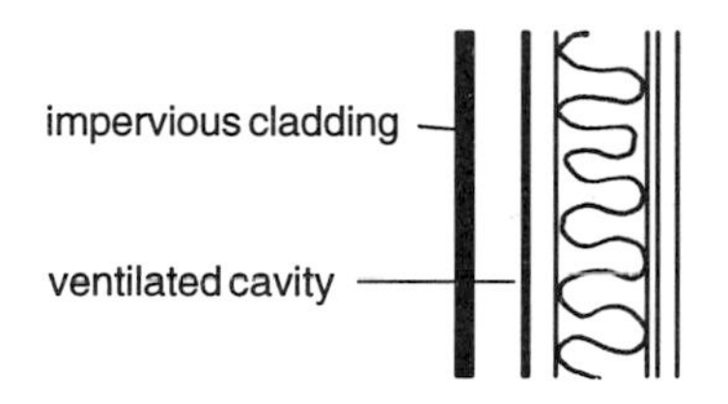

Figure G12 Framed wall with impervious cladding.

Roofs

Before looking at some of the specified constructions, two general points must be considered: **DTS G3.1**

(1) where cladding materials are included in the construction of the roof then construction must be in accordance with the relevant BS. For example, where mastic asphalt is used the relevant source would be CP 144: Part 4: 1970. For other materials see the deemed-to-satisfy provisions;
(2) provision must be made for expansion and contraction of materials such as copper, lead, zinc and other forms of sheeting. Cold roof design has, in the past, resulted in failure caused by condensation within the roof and cold deck specifications are not offered as specified constructions. With warm deck roofs it may be necessary to provide a ventilated air space on the cold side of the insulant in conjunction with a high performance vapour control layer between the insulation and the roof structure in order to minimise the risk of condensation and corrosion. It should be noted that for flat roofs of the warm deck type, sheet metal coverings which require joints to allow for thermal movement, are not suitable.

Flat roofs

Specified Constructions are available for two basic categories of flat roof: the warm roof and the inverted roof type. This is in conjunction with three forms **DTS G3.1** of decking: concrete, timber or metal frame and troughed metal.

The warm roof category has the insulation placed above the roof deck, with the insulant sandwiched between the impermeable (weatherproof) covering to the outside and the vapour control layer to the warm side.

The inverted roof category (sometimes referred to as the upside down roof) is still a class of warm roof but in this case the insulant has a protective outercovering and is sandwiched between this covering and the impermeable (waterproof) membrane on top of the decking. Figure G13 shows examples of these two categories in association with a concrete decking.

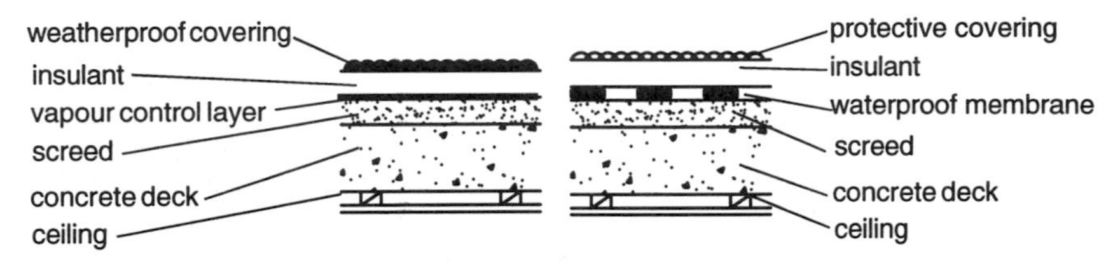

Figure G13 Comparison of 'warm' and inverted roofs.

DTS G3.1 ***Pitched roofs***

Four Specified Constructions are given:

(1) structure of timber or metal frame with an external weatherproof covering of slates or tiles on an underslating felt on sarking boards applied via battens and counterbattens. The insulation is laid between joists on a level ceiling with provision for adequate ventilation of the space between the insulant and the roof structure, as indicated in Figure G14;

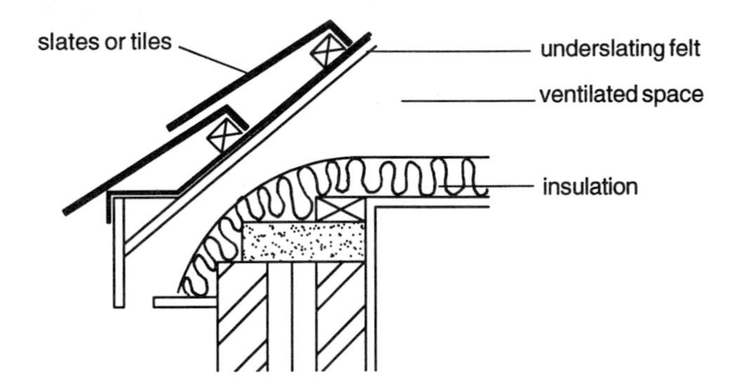

Figure G14 Slate or tiled roof with insulant on level ceiling.

(2) same structural arrangement as (1) except that the insulant and its associated vapour control layer are laid on a sloping ceiling with a ventilated space at least 40 mm deep between the insulant and the sarking, as indicated in Figure G15;

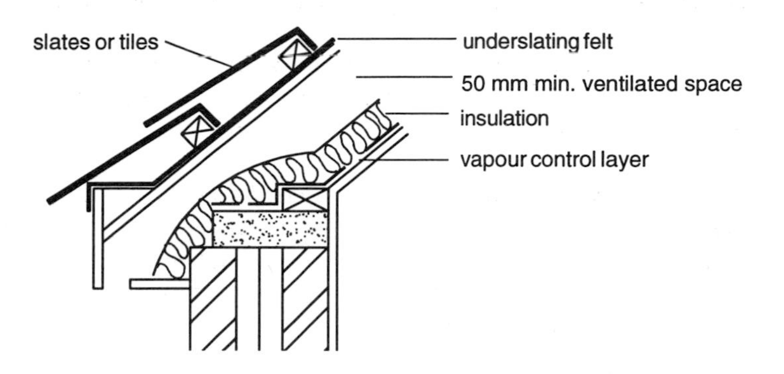

Figure G15 Slate or tiled roof with insulant on sloping ceiling.

(3) same structural arrangement as (1) except that the decking of sarking is replaced by one of low permeability insulation fitted to and between the roof framing. The external weatherproofing is, as before, slates or tiles on battens and counterbattens. A breather membrane is placed on top of the insulant;

(4) same structural arrangement as (1) but with an external weatherproof covering of metal (note that sheet metal coverings which require joints to allow for thermal movement are not suitable) or fibre cement sheet sandwich construction on purlins with the insulant sandwiched between the external and soffit sheeting, with or without a ceiling, as indicated in Figure G16.

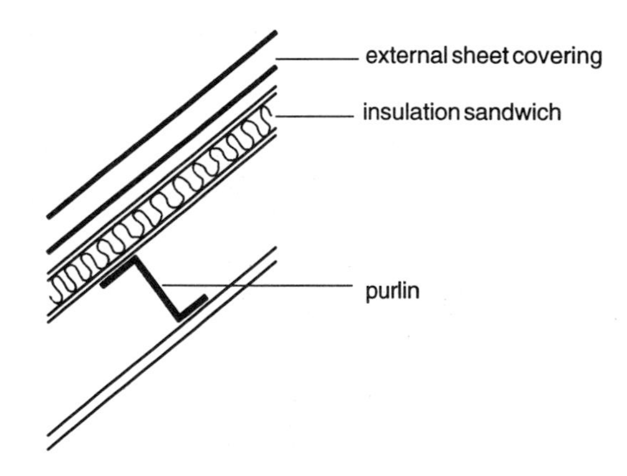

Figure G16 Metal or fibre cement sheet roof with sandwiched insulant.

Condensation

Section G4 of the Standards raises the requirement to minimise the risk of interstitial and surface condensation in relation to floors, walls, roofs or any other building elements. It is interesting to note that a distinction is made in the Standards in that concern about interstitial condensation is with regard to damage to fabric whereas concern about surface condensation is predominately related to the health and wellbeing of the occupants.

G4.1 and G4.2

These requirements can be met by designing and constructing in accordance with BS 5250:1989 Control of condensation in buildings. In the case of interstitial condensation issues the relevant passages are Appendix D and clauses 9.1 to 9.5.5.2. In the case of surface condensation issues the relevant passages are clauses 9.6.1 to 9.6.3.

The general strategy with regard to minimising interstitial condensation is to use vapour control layers on the appropriate (warm side) side of the insulant, ensuring effective sealing of these layers, together with, where necessary, adequate ventilation.

The general strategy for minimising surface condensation is to preclude any elements of construction (other than windows) having a thermal transmittance (U-value) of more than 1.2 W/m^{2}K (see Part J).

Ventilation of pitched roofs

Where the roof has a pitch angle of greater than 15°, cross ventilation should be provided by permanent vents at eaves level on opposite sides of the roof. The required opening area for each vent is equivalent to a continuous vent along each side of the roof of width 10 mm.

Where lean-to or monopitch roofs are to be constructed (for pitch angles greater than 15°) the 10 mm rule still applies at eaves level but in addition high level ventilators are required with an opening area equivalent to a continuous gap of 5 mm width (see example G1).

BS 5250

With low pitched roofs the volume of air in the void is less than with higher pitched roofs and therefore there could be a greater risk of condensation within the void. The same cross ventilation strategy is employed as before except that the equivalent continuous vent should be 25 mm wide rather than 10 mm wide. Again with lean-to and monopitch roofs the 25 mm vent is used in conjunction with the opening area at high level equivalent to a continuous gap of 5 mm width.

Example G1

A dwelling with a monopitch roof measures 12 m by 6 m in plan and has a roof of pitch 30°. It is proposed to supply cross ventilation by means of a continuous strip ventilator of width 12.5 mm at eaves level and two ventilators on the ridge each of opening area 20 000 mm^2 as indicated in Figure G17. Does this arrangement comply?

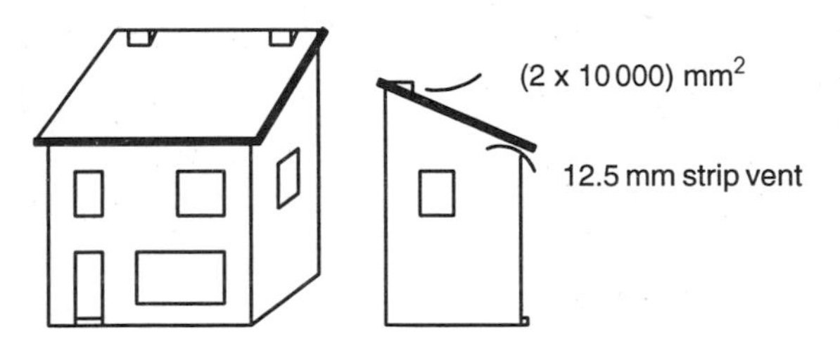

Figure G17 Ventilation of pitched roofs.

Compliance check
Requirement is continuous eaves vent, 10 mm minimum width
Proposal is 12.5 mm continuous eaves vent Complies

Also required is a high level vent of area (12 x 1000 x 5) which is 60 000 mm^2
Proposal is (2 x 20 000) which is 40 000 mm^2 Fails to comply

The addition of a third high level vent would allow compliance

Cold bridges

BS 5250

Projections such as parapets or balconies, junctions of walls and roofs and details around windows such as lintels, jambs or sills are all susceptible to providing cold bridge conditions which would facilitate condensation on inner surfaces. Figure G18 gives a comparison of a poor detail and a reasonable detail for the head of a window where masonry is supported by a steel lintel.

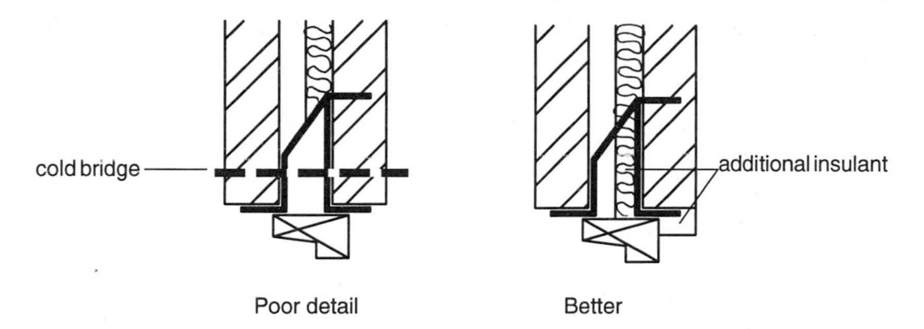

Figure G18 Avoidance of cold bridges at window lintels.

Looking in more detail at the pitched roof, in this case for a dwelling with flat (insulated) ceiling and with walls with insulant in the cavity, the possibility exists for cold bridging at the junction of the wall and the ceiling. Figure G19 gives a comparison of a poor detail and a reasonable detail for this situation.

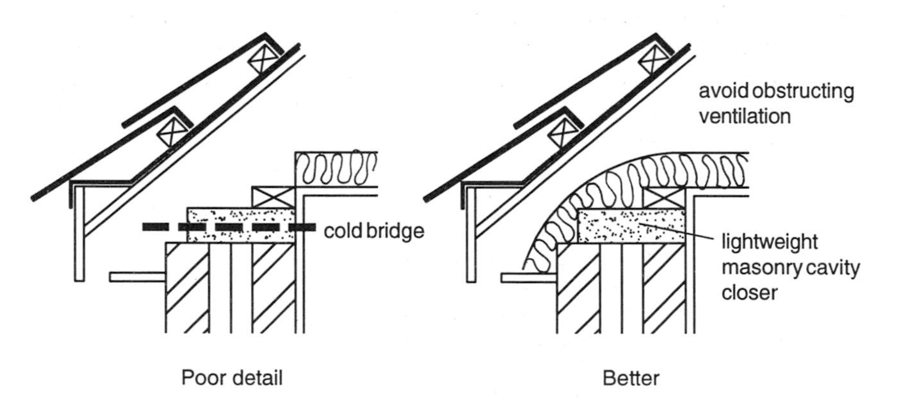

Figure G19 Avoidance of cold bridges at wall and ceiling junctions.

Parapets are commonly used for blocks of flats, maisonettes and other flat roofed buildings. Figure G20 gives a reasonable detail for this situation.

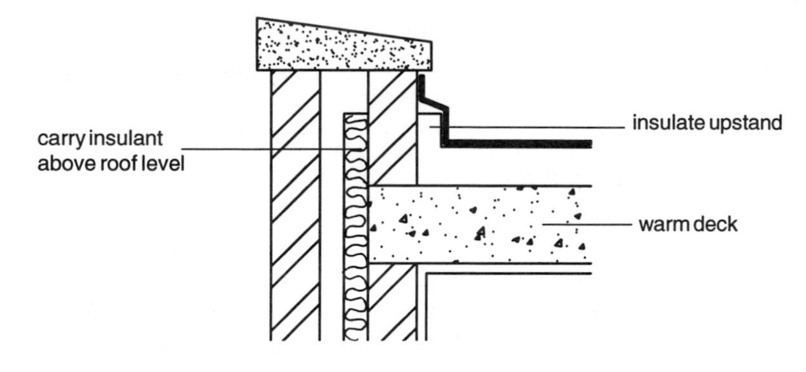

Figure G20 Parapet detail.

Reference

Technical Standards, 1991, Part G, HMSO

RESISTANCE TO TRANSMISSION OF SOUND

PART H — Regulations 19, 20 & 21

BUILDING STANDARDS

Regulations

19 (1) Subject to paragraph (2), every wall which separates a dwelling from another building and, in the case of a dwelling forming part of a building, every wall and floor which separates the dwelling from another part of the building shall provide adequate resistance to transmission of airborne sound.

(2) Paragraph (1) shall not apply to a wall between a dwelling and any area which is open to the external air.

20 (1) Every floor separating a dwelling from any other part of a building above the dwelling shall provide adequate resistance to transmission of sound caused by impact.

(2) Roofs or walkways which are situated directly above a dwelling and to which there is access other than for maintenance purposes shall provide adequate resistance to transmission of sound caused by impact.

21 Regulations 19 and 20 shall not be subject to specification in a notice served under section 11 of the Act.

Aim

To protect the occupants of dwellings from excessive airborne and impact noise which has been generated elsewhere in the building by the provision of reasonable standards of sound insulation by common and economically viable forms of construction.

TECHNICAL STANDARDS

Scope

The standards apply only to dwellings which have separating walls or floors to other dwellings or common access spaces or to other parts or types of buildings.

Requirements

The two ways in which compliance can be achieved are either the use of specified constructions in association with reasonable standards of workmanship or by satisfying the performance standards after completion of construction.

In ensuring reasonable standards of insulation it is important to stress that good and well informed workmanship is just as necessary as good detailing and good element specification in avoiding excessive flanking due to items such as weak air paths, bridging and the bypassing of resilient layers (such as arises when nails penetrate floating floors).

Sound resisting elements

The elements covered by the regulations are the separating walls and floors of all new dwellings, dwellings created by conversions (for example, the development of an old city centre warehouse into flatted properties) and dwellings where a change of use of an element makes its acoustic performance more onerous, (for example, where an existing internal wall of a rehabilitated flat becomes a separating wall between flats). In these current regulations, the concept of sound resisting elements has been extended to include roofs or walkways above dwellings which may be subject to use other than for maintenance.

Background

H1.4 H1.4 defines dwelling to include any part of a building intended to be occupied as a separate dwelling irrespective of the occupancy group of the rest of the building.

Separating elements include walls between a dwelling and another dwelling, but also between a dwelling and a common access corridor or stair, or waste chute or another building.

They also include floors between a dwelling and another dwelling, but also between another building and a dwelling or between a roof or a walkway with access other than for maintenance (for example, for leisure) and a dwelling. These various possibilities for separating elements are illustrated in Figure H1. The sectional plan shows one arrangement and the sectional elevation another arrangement.

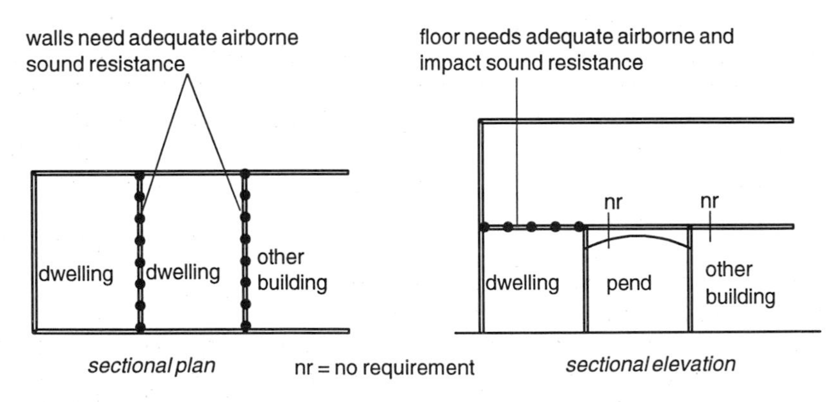

Figure H1 Possible arrangements for separating elements.

Further illustrations of the requirements with respect to sound resisting walls and floors (and in particular, when dealing with refuse chutes and rooftop accessible areas) are shown in Figure H2.

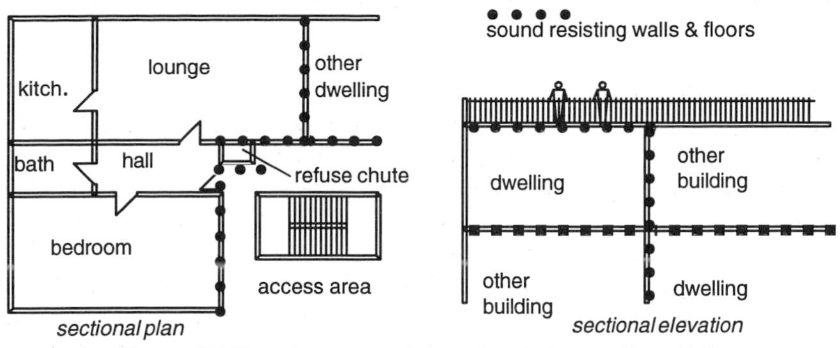

Figure H2 Requirements of sound resisting walls and floors.

The nature of airborne sound

Airborne sound arises from voices, radio, TV, hi-fi and household items such as vacuum cleaners, etc. The sound energy enters the structure of the building via the air and causes vibration of walls and floors which results in acoustic energy being transferred from the air in a room in one dwelling or building to the air in a room of an adjacent dwelling or building. This can occur directly through walls or floors or indirectly as flanking noises as indicated in Figure H3.

H2.1 of the Standards requires that all separating walls and floors must provide adequate resistance to airborne sound. The meaning of *adequate* will be examined later in the text.

H2.1

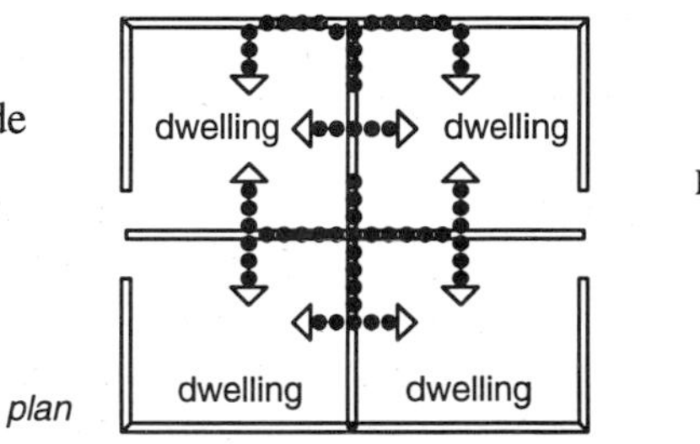

Figure H3 Transfer of airborne noise.

The nature of impact sound

Impact sound arises in buildings from a variety of sources involving energy directly entering the structure through activities such as walking, closing doors, pulling plugs from sockets, vacuuming, etc. The focus for the Standards is associated with the effect of footsteps. Again the energy can pass directly from the floor to the room below or indirectly by flanking noise passing through other elements, as indicated in Figure H4.

H2.2 requires that a floor separating a building above from a dwelling below must provide adequate resistance to impact sound. H2.3 requires that the same conditions must apply to a roof or walkway, provided for purposes other than maintenance, which lies above a dwelling.

H2.2

H2.3

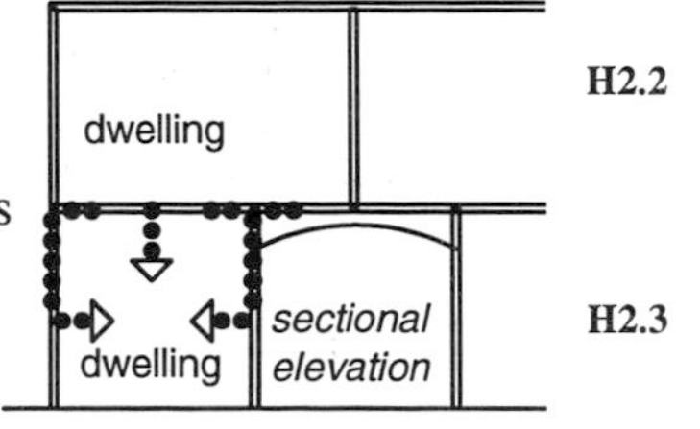

Figure H4 Transfer of impact noise.

COMPLIANCE

Provisions deemed-to-satisfy the Standards

There are two ways in which the Standards can be met. The first way is the use of a form of construction selected from a range offered from points 10 to 17 within the provisions section of the regulations and in association with good practice. The second way is to carry out airborne tests on walls and/or airborne and impact tests on floors all in accordance with points 18 to 29 within the same provisions.

DTS H10

DTS H18

Each of these means of satisfying the Standards is considered next.

Specified constructions

These are provided for four wall types and for four floor types.

Acoustic principles for effective insulation utilise sufficient mass, isolation, and resilience where appropriate and ensure adequate bracing while minimising the existence of weak air paths to reduce flanking effects. These principles are central to the specified constructions and associated precautions to achieve good performance within this section.

Ensuring sufficient mass

Basic principles

Surface density of an element, kg/m^2, (the product of volumetric density kg/m^3 and thickness, m) is the means of checking for compliance of the element with the relevant specified construction. An example will help illustrate the method.

Example H1

The regulations include within the classification Wall Type 1: Solid Masonry the concrete, *in situ* construction shown in Figure H5.

An architect proposes to construct to this specification using 200 mm thick concrete of density 2300 kg/m^3 and does not intend to include a plaster finish. Does the proposal comply in terms of adequate mass?

Specification E/Wall Type 1: Solid Masonry:
Concrete (minimum volumetric density 1500 kg/m^3)
in situ or large panel, plaster optional.
Mass (including plaster if used) 415 kg/m^2

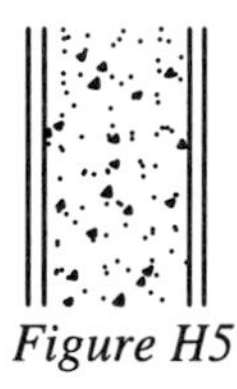

Figure H5

Compliance check

The proposal satisfies the minimum volumetric density criterion of 1500 kg/m^3.

Surface density = volumetric density x thickness
= 2300 x (200/1000) = 460 kg/m^2

The proposal satisfies the minimum surface density criterion of 415 kg/m^2. Thus in terms of adequate mass, the proposal complies.

Composite elements
Where the separating element is constructed of brick or block, account needs to be taken of the surface density of both masonry and mortar. Formulae (each formula is for a given course height) for performing this check, based on the principles indicated in example H1, are set out in the Standards. These formulae make an allowance for the proportional areas of masonry and mortar assuming 10 mm mortar joints and with a mortar density of 1800 kg/m^3 being assumed.

For 200 mm coursing the relevant formula is:

$$M = T(0.93D + 125) + NP$$

where M is the surface density of the element, kg/m^2
T is the thickness of brick or block, m
D is the volumetric density of brick or block, kg/m^3
N is the number of finished faces
P is the surface density of the finished faces, kg/m^2

DTS H5

NB. The finished faces are assumed to be 13 mm thick and the surface densities are assumed to be 29, 17, 10 and 10 kg/m^2 for cement render, gypsum plaster, lightweight plaster and plasterboard respectively.

Example H2
The regulations include within the classification Wall Type 1: Solid Masonry, the concrete block, plastered both sides, construction indicated in Figure H6.

An architect proposes to construct to this specification using 100 mm thick concrete block, with 200 mm coursing, the block having a volumetric density of 1800 kg/m^3 and the wall being 200 mm thick. Facing on each side is to be 13 mm gypsum plaster. Check the proposal for compliance in terms of adequacy of mass.

Compliance check
Step 1
Check the surface density (mass) requirement of the appropriate Specified Construction.
From the regulations:
Specification B/Wall Type 1: solid masonry:
Concrete block, plastered both sides
Mass including plaster, 415 kg/m^2
13 mm plaster each side.
Use blocks which extend to the full thickness of the wall.

Figure H6.

(cont.)

Step 2
Check surface density by formula:

$$M = 0.2\ ([0.93 \times 1800] + 125) + (2 \times 17)$$
$$= 393.8\ kg/m^2$$

This is less than 415 kg/m^2, and thus fails to comply with the mass criterion.

The solution may well be to use a denser block but if a cement render was used rather than plaster, with a surface density of 29 kg/m^2 then the increase in M would be $[(29 \times 17) \times 2] = 24\ kg/m^3$ giving a value of 417.8 kg/m^3. In terms of adequacy of mass, this would secure compliance.

Separating walls

In addition to the need to follow the specified constructions and the associated points of detail there are some overriding conditions as follows:

openings the only opening permitted is a doorway between a dwelling and a stairway (or a passageway) in which case the doorway must be protected by a door having adequate fire resistance.

services (pipes and ducts) the only permitted penetration by services pipes or ducts is between a dwelling and a stairway (or passage) or between a dwelling and a duct. As with openings there remains the need to ensure adequate fire resistance.

DTS
H8

services (chases) chases are allowed in the leaves of masonry walls with a proviso that the depth of the chase must not exceed 1/6th leaf thickness (horizontal) or 1/3rd leaf thickness (vertical). Chases are not permitted back to back on a single leaf. Thus for example on a 112.5 mm brick or block leaf, the maximum allowable depth of a chase would be 18 mm (horizontal) and 37 mm (vertical).

Where walls are of masonry core with freestanding (isolated) panels on each side of the core the service may pierce the panel (providing any gaps are sealed with tape or caulking) but not impinge on the core. Since the minimum distance between panel and core is 25 mm this effectively allows a maximum chase depth of (25 + 12.5) giving 37 mm. No chases are allowed in separating walls of timber-frame construction.

flues only permitted when the flues are of masonry construction; not permitted on separating walls of masonry core with freestanding panels.

Specified constructions for walls

1. Solid masonry

The mass of the wall provides the resistance to the direct transmission of airborne sound. It is important to ensure that bricks and/or blocks of the correct density are used and that masonry joints are properly filled. Where bricks are used they should be laid frog up.

Junctions

Care must be taken at junctions to minimise indirect (flanking) transmission of airborne sound as follows:

DTS H10

Junctions of solid masonry walls

Junction	Comments
with roof	seal the joint between the top of the wall and the underside of the roof with an adequate fire-resisting material. 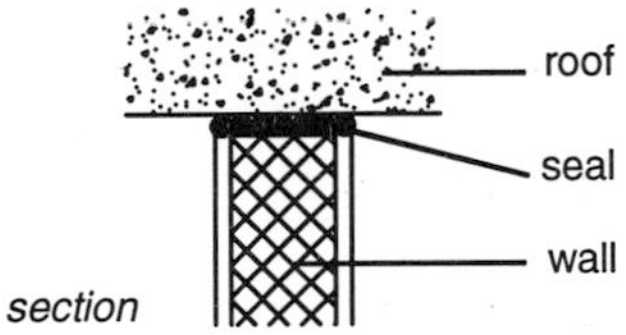
with ceilings directly below roof spaces	lightweight ceilings – no change to wall mass; heavy ceilings with sealed joints (12.5 mm plasterboard or equivalent) – the mass of the wall above ceiling level may be reduced to 150 kg/m^2. 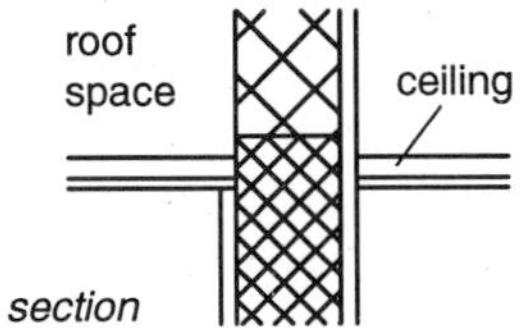 NB Where lightweight aggregate blocks are used to reduce mass above ceiling level, seal one side with cement paint or plaster skim.
intermediate and ground floors	timber floors – support with joist hangers concrete floors – permitted to bear on wall
internal walls	there are no restrictions

external walls outer leaf of cavity walls – no restrictions

solid masonry/masonry inner leaf:
masonry must be bonded at junction or butted and tied at 300 mm centres vertically.

timber inner leaf:

timber leaf must abut separating wall and be tied at 300 mm centres vertically, joints taped or caulked

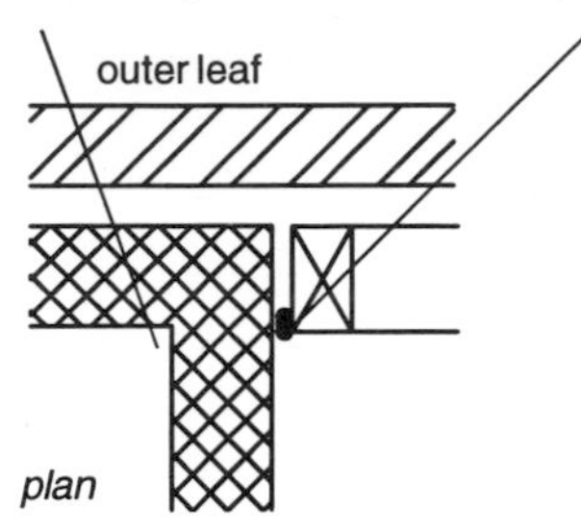

*mass of inner leaf at least 120 kg/m² unless there are doors or windows on either side at least 1 m high and not more than 700 mm from the faces of the separating wall

*flanking in external walls is reduced significantly by the reduction of their length by openings such as doors or windows, if not too close to the separating wall.

The basic specifications for the five Specified Constructions are as given in Table H1

Table H1 Summary for solid masonry walls

subtype	core	facings	comments
A	brickwork 375 kg/m² (with plaster)	13 mm plaster each side	bricks laid in a bond to include headers
B	blockwork 415 kg/m² (with plaster)	13 mm plaster each side	blocks to extend full thickness of wall
C	brickwork 375 kg/m² (with lining)	12.5 mm plaster-board each side, any normal fixing	bricks laid in a bond to include headers
D	blockwork 415 kg/m² masonry alone (with lining)	12.5 mm plaster-board each side, any normal fixing	blocks to extend full thickness of wall
E	concrete, *in situ* or large panel 415 kg/m²	plaster optional	joints between panels filled with mortar

Example H3

Front and rear elevation sketches shown in Figure H7 are part of a proposal for the construction of semi-detached dwellings where:

(1) the separating wall is solid masonry blockwork, of surface density 420 kg/m² including plaster, bonded to the lightweight inner leaf of the external cavity wall, of density 80 kg/m²;
(2) the separating wall within the attic space is of reduced thickness, with surface density 165 kg/m², the ceilings of the upper rooms being 9.5 mm plasterboard;
(3) door positions and window positions are as indicated in the sketches.

Check items (1), (2) and (3) for compliance.

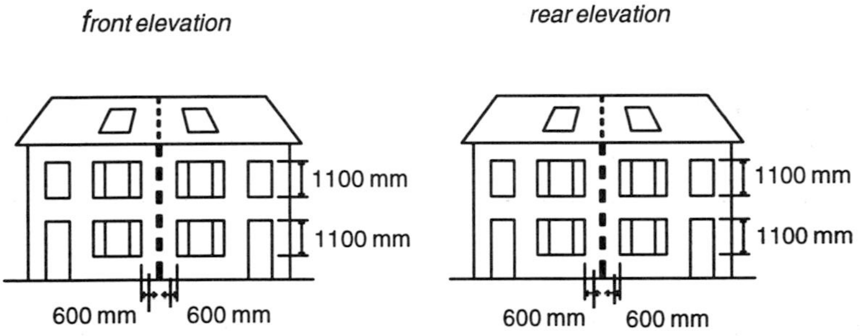

Figure H7 Proposal for semi-detached buildings.

Compliance check

(1)	Proposed surface density 420 kg/m²	
	Minimum acceptable surface density 415 kg/m²	Complies
	Bonding arrangements	Complies(see (3))
(2)	Proposed ceiling is lightweight. No change in wall mass is thus allowed within attic space	
	Proposal to change mass to 165 kg/m²	Does not comply
(3)	Proposed doors/windows greater than 1 m high	
	Proposed doors/windows within 700 mm of wall	Complies

2. Cavity masonry

DTS H11

As with solid masonry walls, resistance to the direct transmission of airborne sound depends on the mass of the wall. Additionally the degree of isolation between the leaves also contributes to this resistance. The cavity should be carried up to the underside of the roof. The earlier comments about filling of mortar joints also hold here.

Wall ties

Only butterfly type to be used, no more than 900 mm apart horizontally and 450 mm apart vertically (see comments in BS 5628: Part 3: 1985 limiting their use to cavities in the range 50 mm to 75 mm using a minimum leaf thickness of 90 mm).

Risks of bridging

Care must be taken during construction to avoid bridging effects which can arise when mortar droppings catch and hold on the ties.

Also in situations where the cavities of external walls are filled with insulants ensure none of these are allowed to enter the cavity of the separating wall.

Junctions

Care must be taken at junctions to minimise indirect (flanking) transmission of airborne sound as follows:

Junctions of cavity masonry walls

Junction	Comments
with roof	seal the joint between the top of the cavity wall and the underside of the roof with an adequate fire resisting material 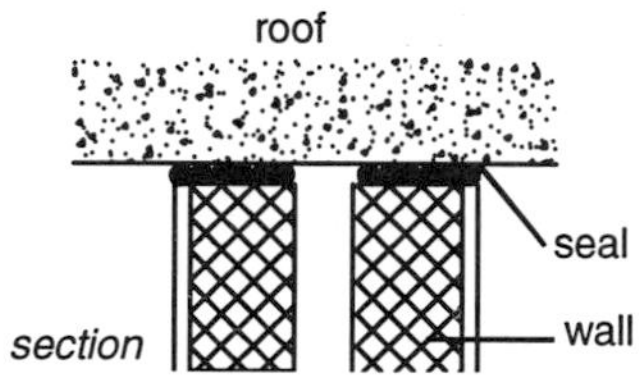
with ceilings directly below roof spaces	lightweight ceilings – no change in wall mass heavy ceilings with sealed joints (12.5 mm plasterboard or equivalent) – the mass of the wall above ceiling level may be reduced to 150 kg/m^2 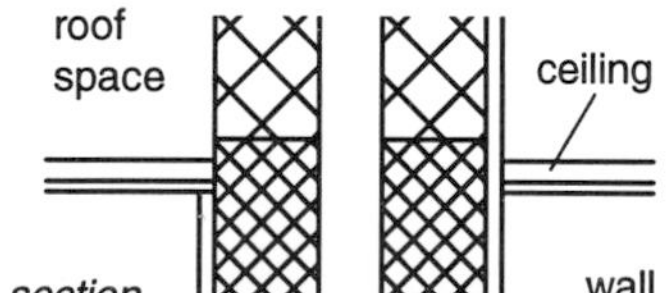 NB Where lightweight aggregate blocks are used to reduce mass above ceiling level, seal one side with cement paint or plaster skim.
intermediate and ground floors	timber floors – support with joist hanger concrete floors – may bear on one leaf only maintaining isolation a concrete slab on the ground may be continuous.
internal walls	no restrictions

external walls	outer leaf of cavity walls	no restrictions
	masonry inner leaf:	timber inner leaf:
	masonry must be bonded at junction or butted and tied at 300 mm centres vertically	timber leaf must abut separating wall and be tied at 300 mm centres vertically, joints taped or caulked
		mass of inner leaf at least 120 kg/m² unless the specified construction is a type B wall as per Table H2

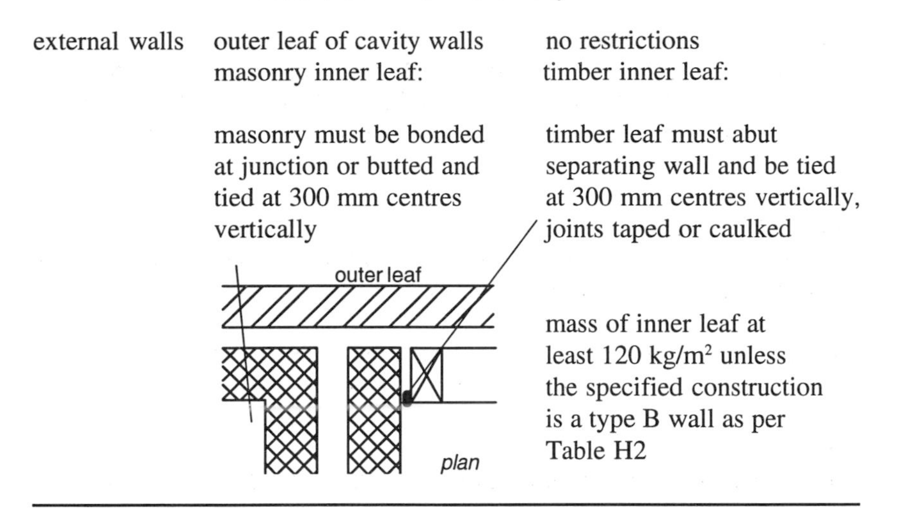

The basic specifications for the four Specified Constructions are as given in Table H2.

Table H2 Wall type 2 : Cavity masonry

subtype	core	facings	comments
A	brickwork 415 kg/m² (with plaster)	13 mm plaster each side	50 mm cavity
B	blockwork 415 kg/m² (with plaster)	13 mm plaster each side	50 mm cavity
C	*blockwork 415 kg/m² masonry alone (with lining)	12.5 mm plasterboard each side, any normal fixing	50 mm cavity
D	*blockwork 250 kg/m²	13 mm plaster each side	75 mm cavity seal the faces (with plaster) of blockwork through the full width and depth of intermediate floors

* These two constructions can only be used where there is a step or a stagger of at least 300 mm as follows:

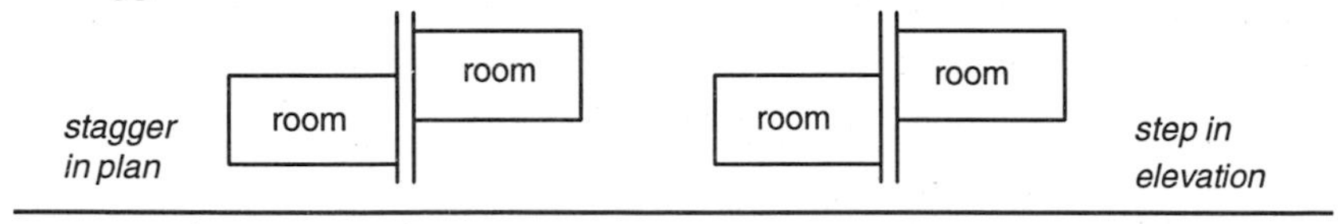

3. Solid masonry between isolated panels

DTS H12 Resistance to direct transmission of airborne sound depends on the mass of the core and panels and also the isolation produced by the panels. As before, care must be taken in filling mortar joints.

Fixing the panels

It is important to ensure that the panels do not touch the core; fixing of these panels should be to the floor and ceiling.

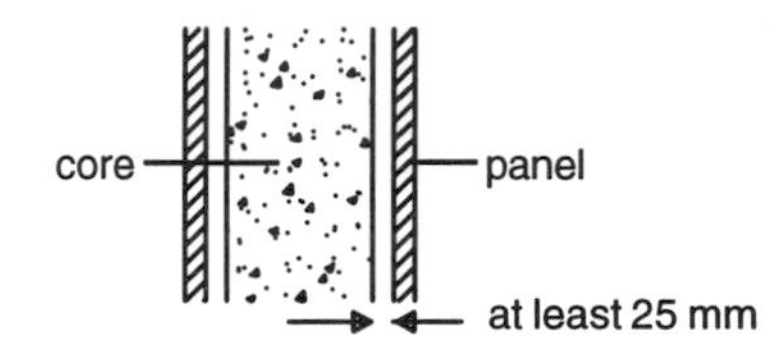

Figure H8 Panels must not touch core.

Junctions

Junction of solid masonry/isolated panel walls

Junction	**Comments**
with roof	seal the joint between the top of the masonry core and the underside of the roof with an adequate fire resisting material
with ceilings directly below roof spaces	lightweight ceilings – no change in wall mass heavyweight ceilings, sealed joints (12.5 mm plasterboard or equivalent) – the mass of the core may be reduced to 150 kg/m² and the panels may be omitted within the roofspace roof space / ceiling / section NB Where lightweight aggregate blocks are used to reduce mass above ceiling level, seal one side with cement paint or plaster skim also, seal the junction between the ceiling and freestanding panels with tape or caulking
intermediate and ground floor	timber floors – support with joist hangers and use dwangs to close floor at joist ends at wall concrete floors – may be carried through the core if of sufficient mass (365 kg/m² or more) seal the junction between ceiling and panel with tape or caulk
internal walls	masonry partitions are not permitted. loadbearing type – fix to masonry core through a continuous pad of mineral fibre quilt non-loadbearing type – should be fixed to the freestanding panels all joists between partitions and panels sealed

external walls outer leaf of cavity walls – no restrictions
inner leaf of cavity walls – must have isolated panels (25 mm gap) as per separating walls except in cases where the core of the separating wall is dense brick or block, in which case plaster or dry lining (10 mm gap), joints sealed and taped, may be used. Thermal insulants may be added to the internal finishes

where dry linings are used the leaf must be masonry of at least 120 kg/m² and butt jointed to the separating wall core, tied at 300 mm centres vertically

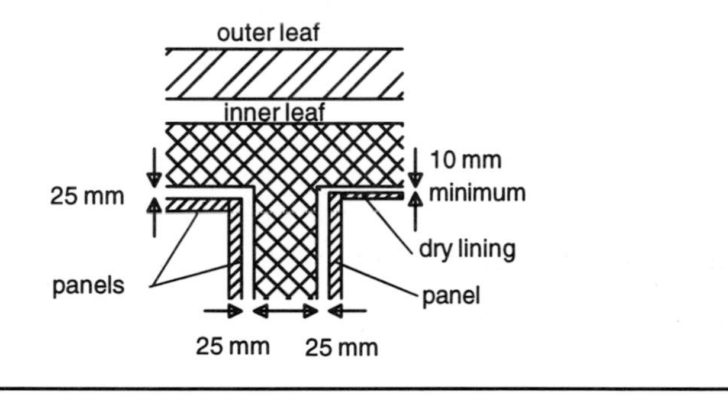

The basic specifications for the four Specified Constructions and for panels are as given in Tables H3 and H4 respectively.

Table H3 Wall type 3 : Solid masonry between isolated panels

subtype	core	facings
A	brickwork (300 kg/m²)	subtype panels E or F
B	blockwork (300 kg/m²)	subtype panels E or F
C	lightweight aggregate blockwork (200 kg/m²)	subtype panels E or F
D	autoclaved aerated blockwork (160 kg/m²)	subtype panels E or F

Table H4
Freestanding panels

subtype	specification	comments
E	two sheets of plasterboard joined by cellular core, mass 18 kg/m^2	panel joints taped, panels fixed at ceiling and floor only, minimum separation from core, 25 mm
F	two sheets plasterboard staggered joints, framed or unframed	each sheet 12.5 mm for framed panels, total thickness 30 mm for unframed panels

4. Timber frames with absorbent curtain

DTS H13 The resistance to direct transmission of airborne sound depends primarily on the isolation of the pair of frames plus absorption in the intervening airspace. The limitation on its use is to dwellings up to three storeys in height (see Part D for other constraints).

The basic construction is a pair of timber frames with a 200 mm gap between the facings, with an absorbent curtain suspended within the cavity. The facings will be two or more plasterboard sheets with an overall thickness of 30 mm with joints staggered to minimise airpaths. The curtain is unfaced mineral fibre quilt, density in the range 12-36 kg/m^3 or equivalent. If fixed to one frame then a 50 mm quilt is used, whereas when suspended clear within the cavity a 25 mm quilt is used.

A masonry core is an option but its role is then primarily one of structural support. Where masonry cores are used then a 200 mm gap should still be maintained between the facings, and the frames should be at least 5 mm clear of the core. It may be useful to consider the use of masonry cores for structural reasons for cases where there are steps or staggers.

Fixing the panels
Minimise connections between frames. If connections are necessary, use 40 mm by 3 mm metal ties just below ceiling level and 1.2 m apart. Masonry cores can only be connected to one frame.

Junctions
Junctions of timber framed walls

Junction	Comments
with roof	seal the joint between each frame (and the core if present) and the underside of the roof with an adequate fire resisting material.

with ceilings directly below roof spaces	unlike the previous specifications, for this case the complete construction must be carried up to the underside of the roof. There are no restrictions on the type of ceiling.
intermediate and ground floors	with timber floors care must be taken to avoid weak sound paths through the floor and between the joists into the wall cavity. Use dwangs to provide a solid edge to timber floors.
internal walls	there are no restrictions
external walls	if the wall is of cavity construction, the cavity should be sealed between the ends of the separating wall and the outer leaf of the external wall to minimise air gaps. The internal finish to the external walls must be 12.5 mm plasterboard, which may also include thermal insulation.

Separating floors

There is a need to provide adequate airborne and impact resistance to the transmission of sound. A range of four floor types is offered which employ a variety of measures, among them mass, soft top covering, isolation. As with walls, it is important during construction to minimise the likelihood of weak air paths, particularly around the perimeter of the floor. It is also important to avoid nailing through floating floors, which defeats the purpose of isolation.

Specified constructions for floors

1. Concrete base with soft covering

The mass of the concrete base provides the resistance to direct transmission of airborne sound together with minimising weak air paths. The soft top covering to the base offers resistance to direct transmission of impact sound. In situations where concern only relates to airborne sound transmission (e.g. a dwelling above a shop) the soft covering may be omitted.

DTS H14

It is important to fill and seal all joints to minimise weak air paths. This requires particular attention when the floor consists of beams or planks. Care must also be taken regarding sealing and isolation of services which penetrate floors (see later).

Soft covering

A soft covering is any resilient material with an overall uncompressed thickness of 4.5 mm bonded to the floor.

Junctions

Care must be taken at junctions to minimise indirect (flanking) transmission of airborne and impact sound as follows:

Junctions of concrete base floors (soft covering)

Junction	Comments
with external walls or with separating walls of cavity type	Openings (windows/doors) reduce flanking effects in walls. Thus for opening areas greater than 20% of external wall area there is no restriction on the mass of the wall. Where such opening areas are less than 20%, the mass of the wall leaf adjoining the floor must be at least 120 kg/m^2 (including any plaster). The floor base must pass through the leaf but care must be taken not to bridge the cavity.
with internal walls or with separating walls of solid type	In most cases the concrete floor base (but not the screed) will pass through the walls. Where the wall mass is less than 355 kg/m^2 this must be the case. Where the wall mass is greater than 355 kg/m^2 then either the wall or the floor may pass through. When the wall passes through the floor the base requires to be tied to the wall and the joint grouted. In both cases the figure of 355 kg/m^2 includes plaster (if any).
penetrations by services	Any duct or pipe passing through the floor must be within an enclosure either side of the floor. The walls of the enclosure require to be at least 15 kg/m^2 (e.g. double sheeted plasterboard). To aid isolation, the service should be wrapped with 25 mm mineral fibre. See also Part D for fire protection.

The four forms of floor bases (which can also be used with type 2 floors) are as given in Table H5.

Table H5 Floor types 1 and 2 : bases

subtype	specification	comments
A	*in situ* solid concrete slab 365 kg/m^2 (with screed/ceiling)	floor screed and/or ceiling finish optional
B	solid concrete slab with permanent shuttering 365 kg/m^2 (with shuttering, screed/ceiling)	floor screed and/or ceiling finish optional
C	concrete beams with infill blocks 365 kg/m^2 (with screed/ceiling)	floor screed must be used, but ceiling finish is optional

D	concrete planks, solid or hollow 365 kg/m^2 (with screed/ceiling)	floor screed must be used ceiling finish is optional

Example H4

In a particular project (see Figures H9 and H10) for a block of flats it is proposed to use: 100 mm blockwork of density 400 kg/m^3 for the inner leaf of external walls, minimum ratio of window area/external wall area for any rooms is 25%; floor of concrete planks supported on the inside leaf of the external walls and an internal wall, and butted against an internal wall, (both internal walls of density 200 kg/m^2) and against a solid separating wall of surface density 415 kg/m^2. Once in place, the floor is to be screeded.

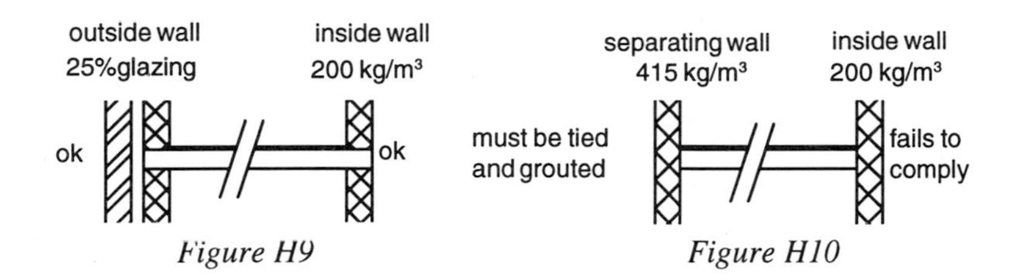

Figure H9 *Figure H10*

Compliance check

Mass of inner leaf of outside wall = 0.1 x 400 = 40 kg/m^2. However since openings > 20% there is no restriction on mass of wall. A solid separating wall of 415 kg/m^2 may pass a floor but it must be tied and grouted to the floor.

Floor base must rest on internal walls which are less than density of 355 kg/m^2.

2. Concrete base with floating layer

The mass of the concrete base together with the mass of the floating layer provides the resistance to direct transmission of airborne sound together with minimising weak air paths.

DTS H15

The resilience of the material which isolates the floating layer from the floor base is the key to the provision of resistance to direct transmission of impact sound. In situations where concern only relates to airborne sound transmission (e.g. a dwelling above a shop) the full construction still requires to be used.

Good detailing and good workmanship are critical to minimise effects of flanking. Thus careful sealing of the base at the perimeter and between beams and planks of subtypes C and D is crucial (a screed is advised to remove irregularities for these two types of base). Equally important is care in laying the resilient material and ensuring that the floating layer really does float (e.g. avoid any direct contact between services such as radiator pipes and the floating floor or between skirtings and the floating floor.

Junctions

Care must be taken at junctions to minimise indirect (flanking) transmission of airborne and impact sound as follows:

Junction of concrete base floors (floating layers)

Junction	Comments
with external walls or with separating walls of cavity type	The same restrictions apply as for concrete floors with soft coverings with respect to the mass of the inner leaf of outside walls and the need to support the floor on both these walls. Note the isolation detail for the resilient layer at the skirting (or leave a 3 mm gap) 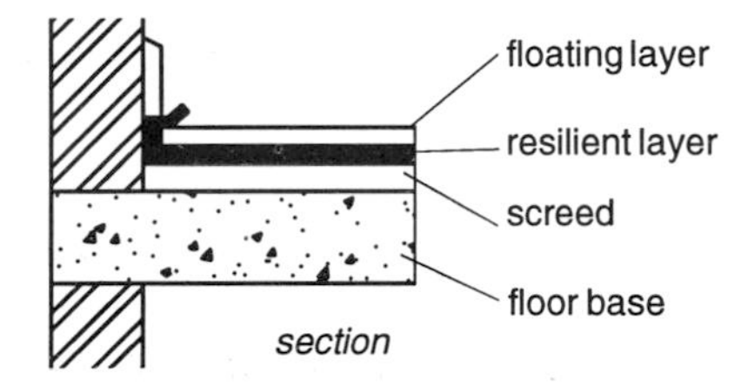
with internal walls or with separating walls of solid type	The same restrictions apply as for concrete floors with soft coverings with respect to structural connections to walls.
penetration by services	The same restrictions apply as for concrete floors with soft coverings with respect to minimising weak air paths via openings in the floor for services. Ducts to house services provide additional opportunities for bridging from the floating layer and consequently (as with skirtings) a 3 mm gap should separate the duct from the floating layer. (See also Part F regarding flue-pipes which penetrate floors.)

The four forms of floor bases (which can also be used with type 1 floors) are as given earlier in Table H5.

The floating layer

This may be a timber raft or a screed. The timber raft consists of 18 mm thick tongued and grooved boards secured by 45 mm by 45 mm battens, the raft being supported entirely by the resilient layer (note earlier comments regarding bridging). The screed consists of 65 mm cement sand with a mesh underlay, the purpose of the mesh being to protect the resilient layer during laying of the screed.

The resilient layer
The norm is 25 mm mineral fibre of density 36 kg/m^3 or equivalent and it is important to ensure that isolation of the floating layer occurs. Under a screed the upper face must be paper faced to prevent screed material entering the interstices of the mineral fibre. Under a timber raft, the mineral fibre may be paper faced on its underside.

Where the battens of the timber raft are of the acoustic type, being lined on the underside with a closed cell resilient foam, the mineral fibre layer can be reduced to a 13 mm thickness.

Alternative resilient layers which can be used under screeds are:

(1) 13 mm pre-compressed polystyrene boards of impact sound duty grade;
(2) 5 mm extruded, closed cell, polyethylene foam, of density in the range 30 to 45 kg/m^3.

Example H5
Figure H11 shows a proposed detail for dealing with a steel heating pipe which requires to penetrate a separating floor of concrete base with a floating layer. Check for compliance.

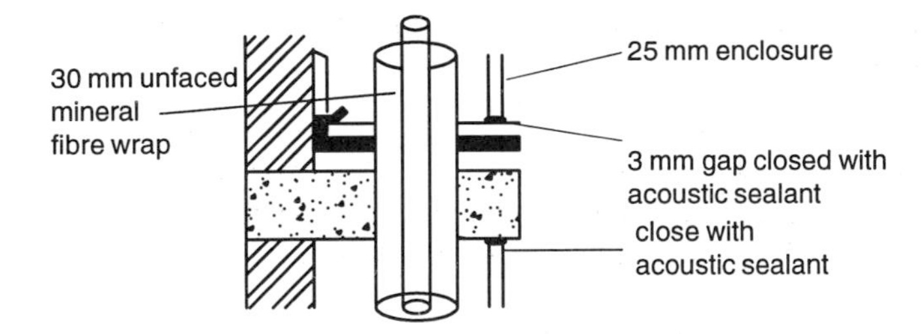

Figure H11 Steel heating pipe.

Compliance check
The enclosure is on both sides of the floor as required; 30 mm unfaced mineral fibre is to be wrapped around the pipe. At least 25 mm thickness is required (either lining the enclosure or the pipe) and thus this requirement is satisfied.

The enclosure has a gap of 3 mm between it and the floating layer and the gap is caulked with an acoustic sealant as required. Hence the detail complies.

3. Timber base with floating layer

DTS H16

The mass of the timber base and associated deafening/absorbent blanket, provides the resistance to direct transmission of airborne sound. The resilience of the material which isolates the floating layer from the timber base and the surrounding construction provides the resistance to direct transmission of impact sound. In situations where concern only relates to airborne sound transmission the full construction still requires to be used.

Its use is limited to buildings up to four storeys in height and the fire resistance must meet criteria set out in Part D. It is a useful construction for converted properties.

As with concrete floating floors, it is important to ensure that bridging does not occur at the perimeter (chipboard flooring will expand after laying) between flooring and skirtings and where services penetrate the floor.

Junctions

Care must be taken at junctions to minimise indirect (flanking) transmission of airborne and impact sound as follows:

Junction of timber base with floating layer

Junction	Comments
with timber frame wall	Seal gap between wall and flooring with a resilient strip glued to the wall, leaving a 3 mm gap between skirtings and flooring which may be closed with a flexible sealant. To minimise weak air paths block off space between the wall and the nearest parallel joist within the floor base; also when joists are at right angles to the wall, close the space between the ends of the joists and the wall. The junction of wall linings with ceilings should be taped or caulked.

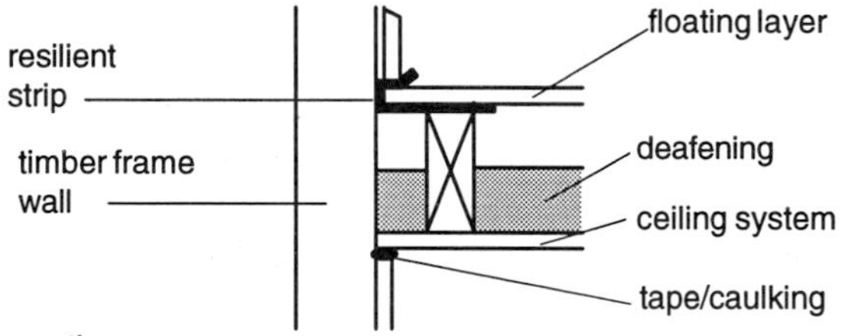

Junction	Comments
with heavy masonry leaves	These are leaves of density 355 kg/m^2 or more. Same isolation arrangements as for timber framed walls. To minimise weak air paths block off the space between the nearest parallel joist and take the deafening hard against the wall where the joists are at right angles to the wall. The junction of wall linings with ceiling should be taped or caulked.

with light masonry leaves	These are leaves of density less than 355 kg/m^2. There is a risk of flanking with lighter walls and these require free standing panels set at least 25 mm clear of the wall. Same isolation arrangements as for timber framed walls. Take the ceiling through to the masonry of the wall and caulk or tape the junction of the ceiling with the free standing panel.
penetrations by services	Same general requirements for enclosures as for floors with concrete bases. The enclosure may be carried down to the timber floor base for platform subtype floors but in such cases the enclosure requires to be isolated from the platform. (See also Part F regarding flue-pipes which penetrate floors.)

Three basic forms of floating timber floor are available as indicated in Table H6.

Table H6
Floor type 3 : timber base with floating layer

subtype	specification	comments
A	Platform floor of 18 mm t&g flooring, joints glued, spot bonded to 19 mm plasterboard on 25 mm mineral fibre resilient layer, density in the range 60-100 kg/m^3. Floor base of 12 mm timber boarding nailed on top of timber joists, ceiling of two layers, plasterboard, staggered joints, 30 mm thick, deafening of 100 mm unfaced rock fibre, density 12-36 kg/m^3 laid within joists on ceiling.	60 kg/m^3 mineral fibre resilient layer gives best insulation but provides a soft floor
	NB. An alternative floating layer is two cement bonded particle boards, staggered joints, glued and screwed, 24 mm thick.	
B	Heavy ribbed floor with floating layer of timber or chipboard, 18 mm thick, t&g, joints glued, spot bonded to 19 mm plasterboard and nailed to 45 mm by 45 mm timber battens which bear on resilient strips of 25 mm mineral fibre of density 80-140 kg/m^3 laid on timber joists, 45 mm thick. Ceiling of two layers, plasterboard staggered joints, 30 mm thick, deafening of 100 mm unfaced mineral fibre, density 12-36 kg/m^3, laid within joists on ceiling.	

C	Ribbed floor with heavy deafening of 18 mm timber or chipboard flooring, t&g, joints glued to 45 mm by 45 mm timber battens placed either on or between joists, 45 mm thick, to bear on resilient strips of 25 mm mineral fibre, density 80-140 kg/m^3. Structural ceiling of 6 mm plywood fixed to underside of joists with two layers of plasterboard, staggered joints, 25 mm thick, deafening of mass 80 kg/m^2 laid on polyethylene liner between joists and on ceiling.	Ceiling has to support a heavy load, thus use of 6 mm plywood or metal lath for structural support

NB Alternative ceiling is 19 mm dense plaster on expanded metal lath.

Example H6

Figure H12 shows proposals for the connection of a type 3 floor (timber base with floating layer of platform type) to heavy solid masonry walls for a five storey block of flats. Check the proposals for compliance.

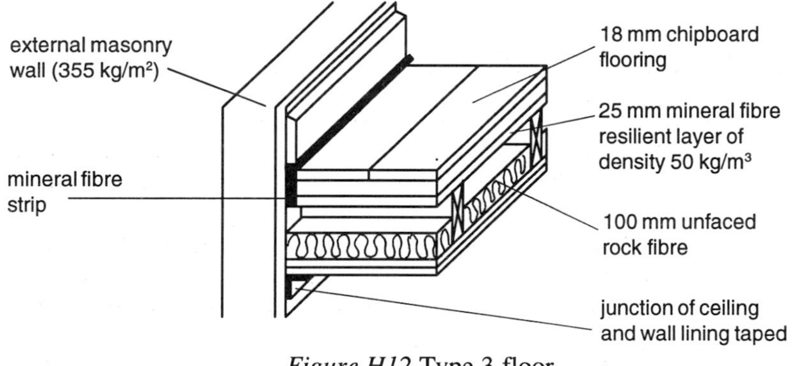

Figure H12 Type 3 floor.

Compliance check

At 355 kg/m^2 wall is heavy. All annotated information complies except for the resilient layer with a proposed density of 50 kg/m^3. This would require to be at least 60 kg/m^3 and even at that density the floor would be 'soft'.

However most important of all, the use of this specification is limited to buildings which are not more than four storeys high and is therefore inappropriate for a five storey application.

Thus the proposal fails to comply.

4. Timber base with independent ceiling

The mass of the floor base and the mass of the independent ceiling together with the isolation of the ceiling from the floor base provide the resistance to direct transmission of airborne and impact sound. In situations where concern only relates to airborne sound transmission the full construction still requires to be used.

DTS H17

Use is limited to buildings up to four storeys in height and the fire resistance must meet the criteria set out in Part D. When used with masonry walls at least three of the four bounding walls must be of heavy type.

This is a useful construction for converted properties and also for rehabilitation in tenement properties.

It is important to ensure that no bridging occurs between the floor base and the independent ceiling (this requires particular care where there are enclosures for services). To minimise weak air paths the floor base should be sealed at its perimeter and the independent ceiling should be sealed at its perimeter.

Junctions

Care must be taken at junctions to minimise indirect (flanking) transmission of airborne and impact sound as follows:

Junction of timber base with independent ceiling

Junction	Comments
with external or separating leaves	when used with masonry walls the mass of the leaf must be at least 355 kg/m^3 on at least three bounding sides. If the fourth leaf is masonry it must be at least 180 kg/m^2. Use bearers on walls to support the edges of the independent ceiling. It is useful to have a resilient strip behind the bearer to cater for any unevenness of the wall surface. Seal the junction of the wall and ceiling with tape or caulking.
with internal leaves	where masonry, these must have a mass of at least 180 kg/m^2. Where of stud partition type there is no restriction. Supporting and sealing arrangements are as external leaves.
penetrations for services	The same restrictions apply as for concrete floors with soft coverings with respect to minimising weak air paths via openings in the floor for services. Ducts for services provide additional opportunities for bridging from the base floor to the independent ceiling and care must be taken in detailing and workmanship to ensure isolation.

dry lined walls

It is important to carry the ceiling through and bring the dry lining of the wall up to meet the ceiling to avoid a weak air path behind the dry lined wall, and tape or caulk this junction.

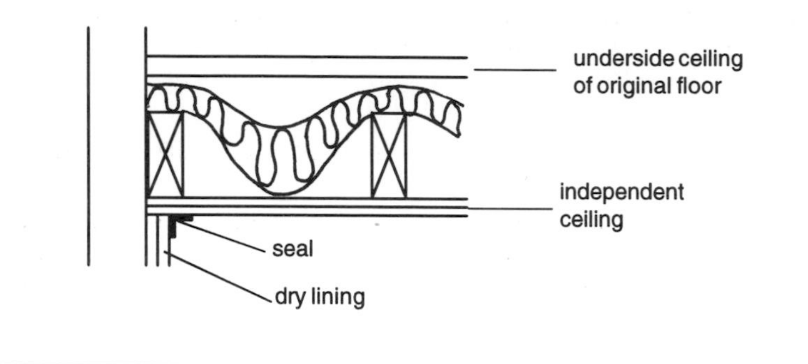

Floor base for independent ceiling

This consists of 18 mm thick, tongued and grooved, timber or chipboard on 45 mm thick joists of appropriate depth with deafening providing a mass of 80 kg/m^2 supported between the joists. The ceiling can be two layers of plasterboard, joints staggered, of thickness 30 mm, or 19 mm dense plaster on metal lath.

When upgrading existing floors the deafening may be on boards which are supported by bearers on the sides of the joists. In such cases it may be possible to utilise the existing deafening and also to utilise the existing flooring, but hardboard sheeting, 3.2 mm thick, should be laid across the top of the floor to seal air gaps.

In new floors a structural ceiling should be created by fixing 6 mm plywood to the underside of the joists, then adding the plasterboard layers of 30 mm overall thickness. The 80 kg/m^2 deafening can then be laid within the joists on top of the plywood. A polyethylene liner may be used if desired.

The independent ceiling

This will consist of joists 45 mm thick and of an appropriate depth, supported on bearers (with a resilient backing) on two opposite walls, and supporting a ceiling of two plasterboard layers, joints staggered, 30 mm thick. An absorbent blanket of 25 mm unfaced mineral fibre and of density in the range 12-36 kg/m^2 should be draped across the joists to allow maximum absorption in the cavity. The ceiling should be at least 150 mm below the underside of the floorbase and the top of the joists and the absorbent blanket should not touch the underside of the floor base.

Performance Standards

The alternative to achieving compliance via satisfactory use of Specified Constructions is to carry out acoustic tests on wall and floors. This is the mechanism used when the form of construction is sufficiently novel that it is considered to be outside the Specified Constructions. Where building control authorities are concerned about matters of workmanship in relation to the use of specified constructions, acoustic testing may also be used in these circumstances.

DTS H18

Airborne and impact tests must be carried out in accordance with the procedures set out in BS 2750 : Part 4 1980 (ISO 140/IV-1978) and Part 7 1980 (ISO 140/V11-1978). From the data collected the Weighted Standardised Level Difference (DnT,w) and the Weighted Standardised Impact Sound Pressure Level (L'nT,w) can be obtained from the procedures set out in BS 5821 : Part 1 : 1984 (ISO 717/1-1982) and Part 2 : 1984(ISO 717/2-1992).

Such airborne and impact tests may only be carried out within completed dwellings (e.g. doors and windows fitted in the rooms and in the closed position during tests), which are unfurnished.

Procedures for testing

The basic concept when testing is that a sample of four elements should be tested. For airborne tests this allows identification of a mean value and a lowest individual value of DnT,w. For impact tests this allows identification of a mean value and a highest individual value of L'nT,w.

DTS H19

The test should normally be carried out between habitable rooms of about equal volume but this may not always be possible and in worst cases even other spaces such as a hall may have to be utilised.

When measuring between rooms of unequal volume the sound source room must be the larger room. When measuring between a room and another space the sound source must be in the other space.

Walls and/or floors will comply in terms of meeting the performance standard if the tests are carried out as per the foregoing procedures and the values set out in Table H7 are met.

Table H7 Performance standards for airborne and impact sound

Airborne sound
Minimum acceptable values of weighted standardised level difference:
DnT,w

	Mean value	Individual value
Walls	53	49
Floors	52	48

Impact sound
Maximum acceptable values of weighted standardised impact sound pressure levels:
L'nT,w

	Mean value	Individual value
Floors	61	65

Example H7
The results of four airborne measurements (DnT,w) on a certain form of separating wall were as follows:

living room/living room	(both same size) 54
living room/bedroom	(bedroom smaller, source in living room) 52
bedroom to hall	(source in hall) 50
bedroom/bedroom	(both same size) 56

Does the form of construction comply with the performance standards?

Compliance check

space utilised	complies
location of sound source	complies
mean value of DnT,w = (54+52+50+56)/4 = 53	complies
lowest individual value = 50	complies

Thus this form of construction satisfies the performance standards.

The concept of a single wall or floor can be extended beyond individual rooms. For example a separating wall in a block of flats rising to several storeys and of identical construction throughout may be regarded as a single wall. In the same way a separating floor over a whole storey may be regarded as a single floor where the construction is identical throughout. There is also an underlying assumption that flat sections and plans are broadly similar on either side of the wall or floor.

Reference

Technical Standards, 1991, Part H, HMSO

CONSERVATION OF FUEL AND POWER

PART J* **Regulation 22**

BUILDING STANDARDS

Regulation

22(1) In a building to which this regulation applies, reasonable provision shall be made for the conservation of fuel and power.

(2) This regulation shall apply to all buildings other than

(a) buildings of purpose groups 2 to 7 inclusive which –

(i) are limited life buildings,

(ii) have a total floor area not exceeding 30 m^2, or

(iii) are unheated or have a space heating system which is designed to give a maximum output not exceeding 25 W/m^2 of floor area;

(b) buildings of one storey which have a fabric covering and are supported by a frame or by air pressure;

(c) circulation and service areas in buildings comprising more than one dwelling which are not part of an individual dwelling;

(d) conservatories, greenhouses, garages, stores, washhouses, water closet compartments and other accommodation which are ancillary to and form part of a building of purpose group 1 and which have external access; and

(e) buildings of purpose group 6 or 7 which have a space heating system which is designed to give a maximum output not exceeding 50 W/m^2 of floor area.

Aim

The purpose of this part is to control the rate of heat loss through the fabric by setting maximum acceptable U-values and to control the rate of heat output from space heating and hot water plant by requiring, in certain circumstances, controls to be fitted to the system. Control of heat loss rate from distribution pipework and ductwork and from storage vessels is exercised by setting minimum acceptable standards of insulation.

TECHNICAL STANDARDS

Scope

Generally the standards apply to all buildings with the exceptions indicated in regulation 22(2) above. Certain aspects of the controls for heating systems do not relate to all buildings and these will be dealt with later.

* *Amendments likely late 1996/early 1997.*

Requirements (fabric)

There are four ways in which compliance can be achieved:

J2.3 (1) *The elemental approach*
By this method every roof, external wall/floor must comply with the maximum allowable U-values and percentages of single glazing set out in the table to regulation J2.3 of the regulations and given later in this section as Table J3

J2.4 (2) *Calculated trade-offs*
To ease flexibility in design this method can be used and the U-values and areas of glazing may be varied from those of Table J3 if the total heat loss rate for the proposed design can be shown to be no greater than that which would have resulted from the elemental approach.

J2.5 (3) *Energy targets*
This method is an extension of the trade-off method which allows advantage to be taken of useful heat gains such as may arise from solar, heat recovery or artificial lighting sources. As before it must be shown that the total heat loss rate would be no greater than would arise as a result of the elemental approach.

DTS J2.2 (4) *Selection from deemed to satisfy provisions*
The designer may choose to select specifications for roofs, walls and floors from the range which are deemed to satisfy, provided the permitted areas of glazing are not exceeded.

Requirements (services)

J3.1 to 3.5 plus DTS J3.4 and 3.5 In addition to minimising heat loss rate (and thus fuel consumption) via fabric U-value and percentage glazing constraints, further measures to minimise heat loss rate are applied to the operation of services systems within the building. The measures consist of controls for certain classes of space heating and hot water storage systems and the application of thermal insulation to certain classes of storage vessels and pipes and ducts.

Background

In calculating the U-value of an element of fabric from first principles the procedure is to sum the resistances (m^2K/W) of each of the components and the U-value (W/m^2K) is then the reciprocal of this total resistance.

The summed resistances will include inside (Rsi) and outside (Rso) surface resistances, resistances of materials (obtained by dividing the thickness (m) of the material by its thermal conductivity (W/mK)) and may also include one or more cavity resistances (Rcav).

Thus $$Rtotal = Rso + Rsi + \Sigma\, {}^{t}/_{k} + Rcav \quad m^2K/W$$
$$\text{and } U = 1/Rtotal \quad W/m^2K$$

Appropriate values of surface and cavity resistances for walls, roofs and floors, and thermal conductivities for an extensive range of building materials are available in Book A of the CIBSE Guide and a sample of this information is provided here within tables J1 and J2 respectively.

Table J1 Thermal resistances (m^2K/W)

Outside surface resistance, (Rso),	walls	0.06
	roofs	0.04
Inside surface resistance, (Rsi),	walls	0.12
	roofs	0.10
	floors	0.14
Cavity resistance, (Rcav)	walls	0.18
Loft space (ventilated)		0.18
Cavity: tiles/felt		0.12

Table J2 Thermal conductivities of building materials (W/mK)

Item	**Density (kg/m^3)**	**Conductivity (W/mK)**
Brickwork (outer leaf)	1700	0.84
Brickwork (inner leaf)	1700	0.62
Concrete (outer leaf)	1700	0.84
Concrete (inner leaf)	1700	0.76
Rendering	1300	0.50
Plasterboard	950	0.16
Expanded polystyrene slab	250	0.035
Glass fibre quilt	120	0.040
Mineral fibre slab	300	0.035
Roofing felt		0.20
Sarking		0.13

Example J1
Check for compliance the following construction for outside walls of dwellings:
outer leaf, 112.5 mm brick, density 1700 kg/m³ cavity, 50 mm
inner leaf, 112.5 mm brick, density 1700 kg/m³
insulant, 50 mm glass fibre quilt, density 120 kg/m³
lining, 12.5 mm foil backed plasterboard, density 950 kg/m³

Compliance check

Item	thickness (m)	density (kg/m³)	conductivity (W/mK)	resistance (m²K/W)
Rso				0.06
outer leaf	0.1125	1700	0.84	0.1339
Rcav				0.18
inner leaf	0.1125	1700	0.62	0.1814
insulant	0.05	12	0.04	1.25
plasterboard	0.0125	950	0.16	0.0781
Rsi				0.12
			Rtotal	2.00034

U value proposed = 1/2.0034 = 0.499 W/m²K
U value allowed = 0.45 W/m²K
Fails to comply

The estimation of the U-value for most proposals, whether for walls, flat roofs or exposed floor elements, can be carried out as illustrated by Example J1. However, in the case of pitched roofs for dwellings, where ceilings are horizontal, the U-value is related to the ceiling area and not to the sloping area of the roof. This requires the use of a correction factor of $\cos\theta$ (where θ is the roof pitch angle) to allow for the difference between roof slope area and ceiling area to be applied to appropriate elements. The method is best illustrated by an example.

Example J2
Check for compliance the roof construction for a dwelling shown in Figure J1.

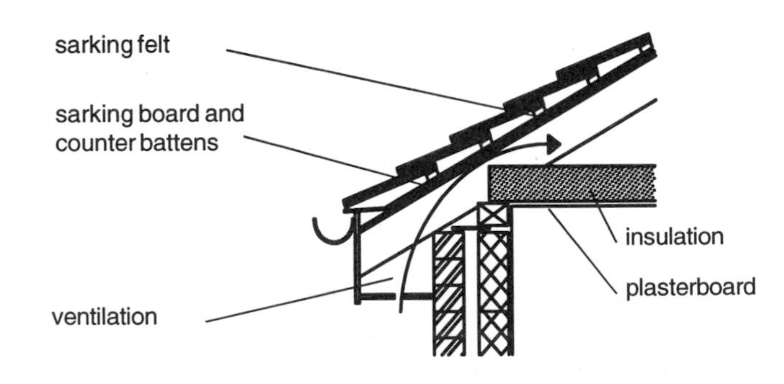

Figure J1 Details of roof and ceiling.

30° pitched roof of timber trussed rafters covered with concrete tiles on wooden battens with sarking felt underlay on timber sarking. 150 mm mineral wool slab laid between joists with ceiling of 12.5 mm foil backed plasterboard.

Compliance check

Item	thickness (m)	density (kg/m³)	conductivity (W/mK)		resistance (m²K/W)
Rso					0.04
tiles					negligible
airspace: felt/tiles					0.12
felt	0.002		0.20		0.01
sarking	0.018		0.13		0.138
				Rslope	0.309
Rslope $\cos\theta = 0.309 \times 0.866$					0.267
Rloft					0.18
insulant	0.150	30	0.035		4.286
plasterboard	0.0125	950	0.16		0.0781
Rsi					0.10
				Rtotal	4.9111

U value proposed = 1/4.9111 = 0.204 W/m²K
U value allowed = 0.25 W/m²K
Complies

COMPLIANCE
Control of fabric

In setting the standards in terms of U-values within Table J3 three categories of exposure for roofs, walls and floors are defined: J1.3

(1) exposed — this means exposed directly to the outside air, e.g. between an apartment of a dwelling and the outside air, (● U , 0.45 W/m²K applies to walls and floors)

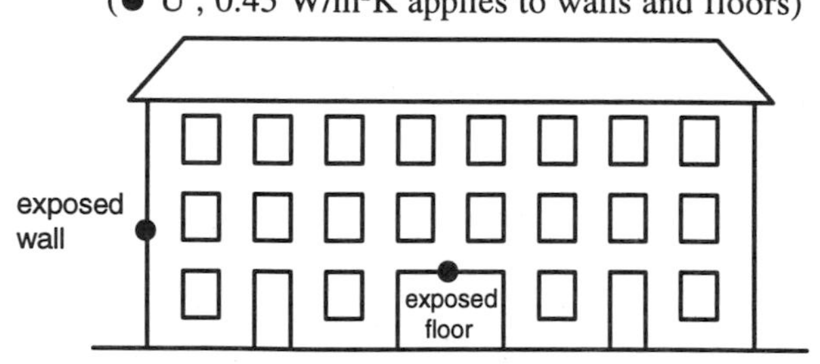

Figure J2 Exposed.

(2) semi-exposed — for a wall and/or a floor, between a building to which Part J2 applies and a building to which it does not apply, Figure J3

e.g. between an apartment of a dwelling and an outhouse such as a store openable from outside, (U,0.60 W/m²K applies)

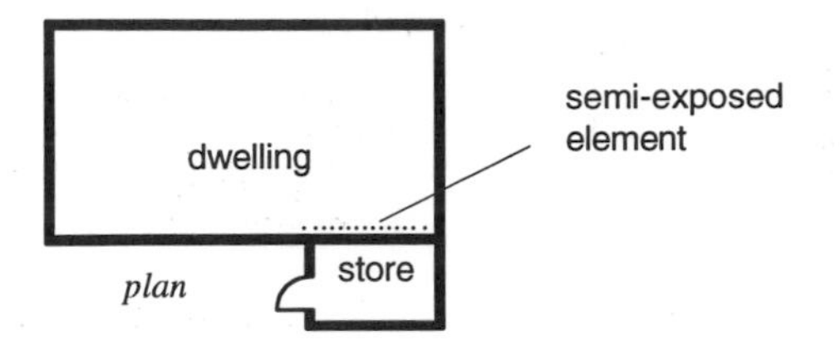

Figure J3 Semi-exposed.

(3) ventilated space — Space enclosed by structure part of which is exposed to the outside air and permanently ventilated to the outside air by openings or ducts having an aggregate area exceeding 30% of the wall area, Figure J4. Such a space is deemed to be at outside air temperature and consequently all of the internal separating elements to such a space are considered for thermal purposes to be exposed.

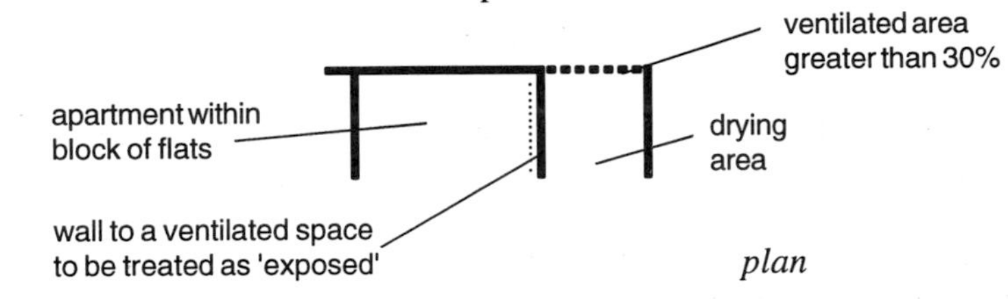

Figure J4 Ventilated space.

Rules for assessing fabric heat loss rate

J2.1 (1) No part of any wall/roof/floor may exceed a U-value of 1.2 W/m²K. This is to reduce the risk of surface condensation within buildings arising from cold bridge effects.

(2) Floors (e.g. dwellings), Figure J5.

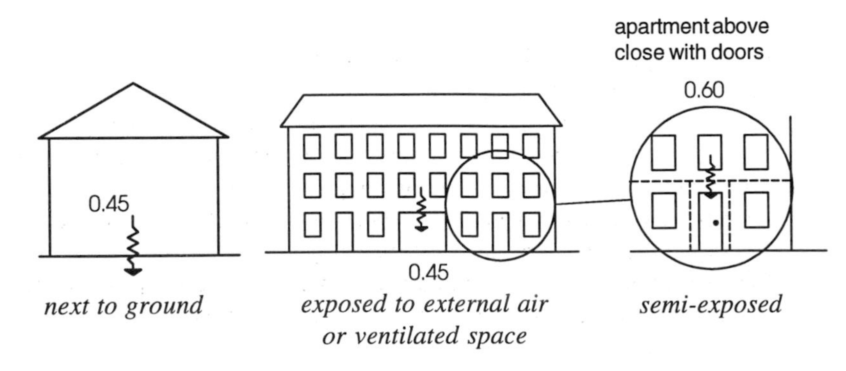

Figure J5 Floors.

(3) Walls (e.g. dwellings), Figure J6. **J2.1**

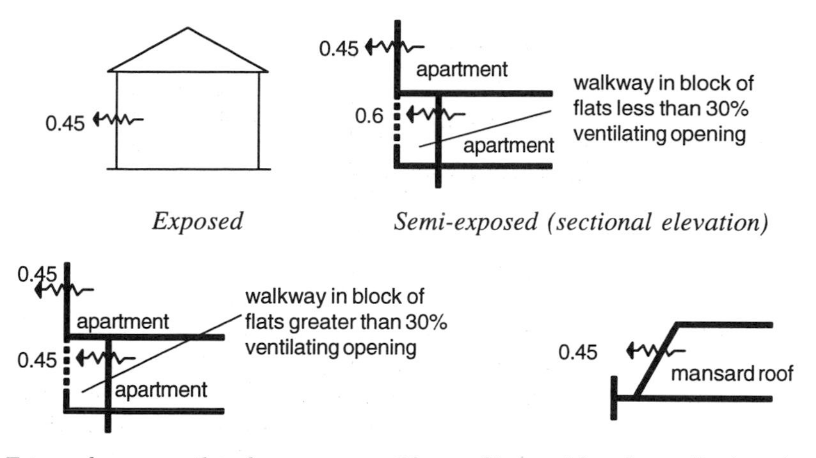

Exposed

Semi-exposed (sectional elevation)

Exposed to a ventilated space (sectional elevation)

The roof is considered a wall when the pitch of the roof is greater than 70º (e.g. mansard roof, sectional elevation)

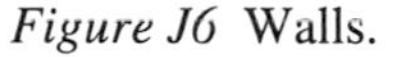

Figure J6 Walls.

(4) Doors and recessed meter boxes can be considered to have the same U-values as walls.

(5) Any part of a floor which serves as a roof is to be treated as a roof, in terms of U-value requirements, Figure J7.

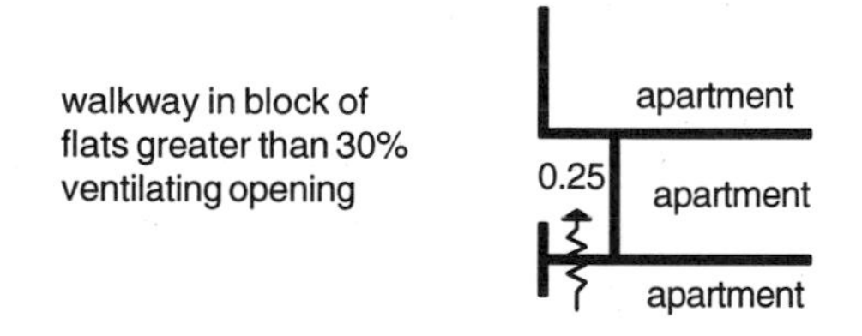

Figure J7 Section of block of flats (sectional elevation).

(6) Estimation of areas:
   (a) for rooflights or windows, measure the size of the opening to accommodate the rooflight or window; **J2.1**
   (b) where a door or a roller shutter has a glazed portion of more than 1 m^2, include this as a part of the permitted area of glazed openings, otherwise ignore this glazing when assessing the proposed area of glazing;
   (c) in estimating areas of walls/floors/roofs, use internal dimensions (in the case of a roof, in the plane of the insulant);
   (d) when dealing with dwellings, ignore any floor area and any glazed area associated with ancillary accommodation such as wash-houses, outside stores, etc.

(7) U-values for glazing (W/m^2K) to be taken as: 5.7 (single)
2.8 (double)
2.0 (triple)

(8) When assessing permitted areas of glazing, ignore:
  (a) the area of any display window (see later) at access level;
  (b) the area of wall above roof level (parapets) and below the level of the lowest floor in the building, Figure J8.

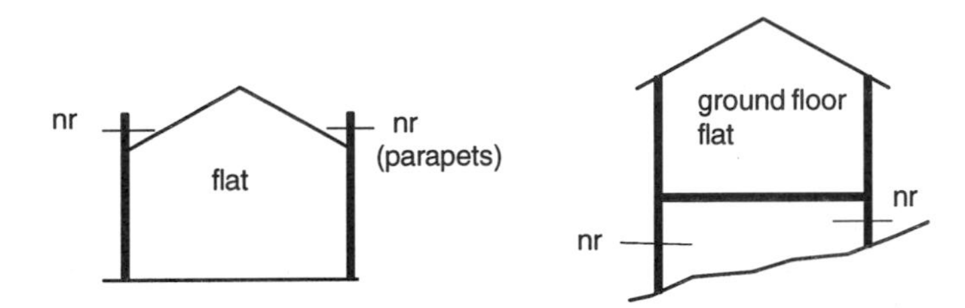

Figure J8 Areas of wall to ignore.

(9) Keep U-value calculations simple by ignoring the effects of items such as wall-ties, framing etc.

Table to J2.3

Table J3 Maximum allowable U-values and percentages of glazing

Purpose group	**U-values (W/m^2K)**				**Single glazing ($U=5.7W/m^2K$)**	
	floors next to ground	exposed floors and walls	semi-exposed floors and walls	roofs	windows	roof lights
1. Dwellings	0.45	0.45	0.60	0.25	Together 15% of floor area	
2. Other residential	0.45	0.45	0.60	0.45	25%	20%
3. Assembly, shops, offices	0.45	0.45	0.60	0.45	35%	20%
4. Industrial, storage	0.45	0.45	0.60	0.45	15%	20%

Method 1 The elemental approach

Example J3
An architect submits plans for four in a block dwellings shown in Figure J9. Do these plans comply?

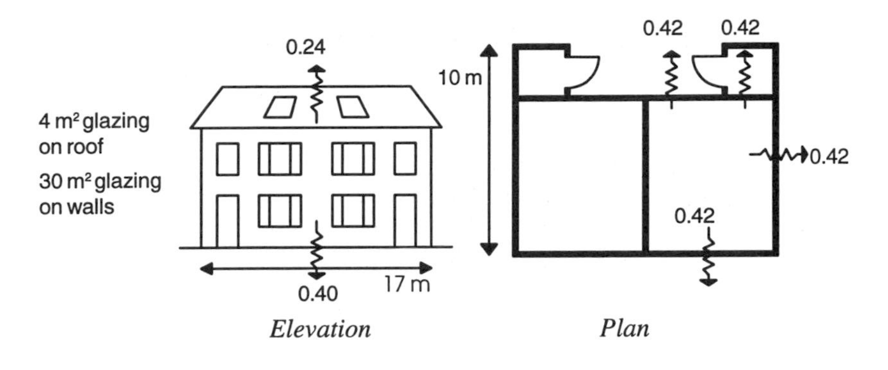

Figure J9 Plans for dwellings.

Compliance check

U-values	Walls: Proposed 0.42	Requirement not > 0.45	Complies
	NB Walls to outhouse need only match 0.60		
	Floors: Proposed 0.40	Requirement not > 0.45	Complies
	Roof: Proposed 0.24	Requirement not > 0.25	Complies
Glazing	Total floor area = 2(17 x 10) = 340 m²		
	NB Floor areas of outhouses not included		
	Proposed glazing = 4 + 30 = 34 m²		
	Requirement not > (0.15 x 340) = 51 m²		Complies

Example J4

The sketch in Figure J10 shows an architect's proposal for an office building where it is intended to use single glazed units throughout.
Check for compliance.

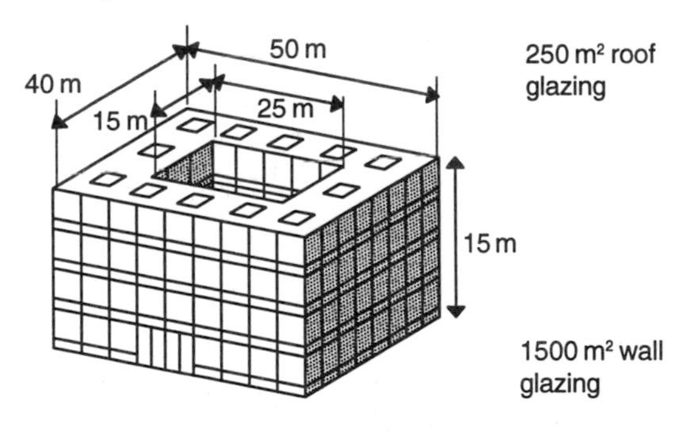

Figure J10 Proposal for office building.

Compliance check

Roof area = (50x40) – (25x15)=1625 m²
Maximum permitted roof glazing = (0.20 x 1625) = 325 m²
Proposal = 250 m² Complies
Wall area = (50+50+40+40+25+25+15+15)15 = 3900 m²
Maximum permitted wall glazing = (0.35x3900) = 1365 m²
Proposal = 1500 m² Fails to comply

J2.1 ***Display windows in shops***

It is recognised that the thermal resistance of a display window with a fixed screen behind the window (as shown in the sketch of Example J5) is very considerably higher than that offered by single glazing alone. This is dealt with by removing the area of display windows completely from the calculation of permitted areas of glazing, as illustrated by the next example.

Example J5

In revising the proposals indicated in Example J4, the architect provides for 250 m² of display windows at ground level. Does the revision now comply?

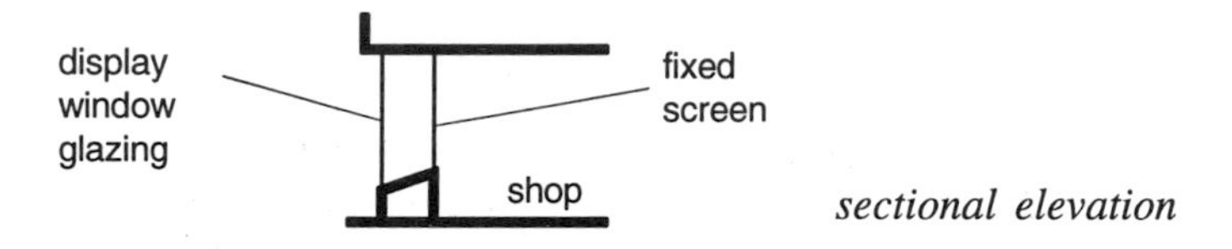

Figure J11 Revision with display windows.

Compliance check

The 250 m² of display window must be removed from the calculation.

Wall area = 3900 – 250	= 3650 m²	
Maximum permitted glazing	= (0.35 x 3650) = 1277.5 m²	
Proposal = 1500 – 250	= 1250 m²	Complies

Example J6

The sketch in Figure J12 shows an elevation of a house and garage which abuts a school building. Identify the permitted U-values of the elements labelled in the sketch.

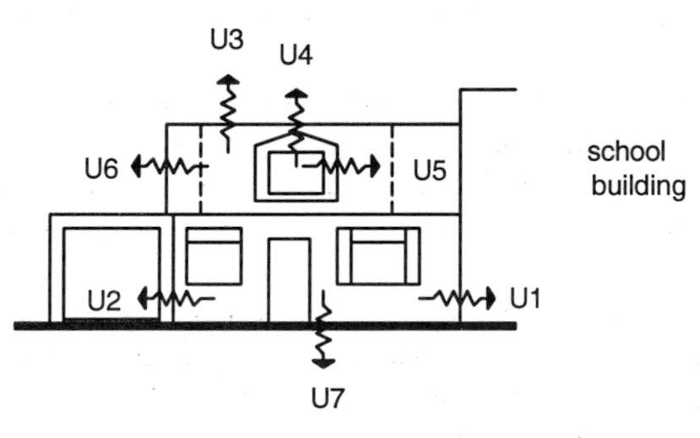

Figure J12 House and garage.

Compliance check

U1 = 0.6 U2 = 0.6 U3 = 0.25 U4 = 0.25 U5 = 0.45 U6 = 0.45 U7=0.45

Example J7

The sketch in Figure J13 shows a sectional plan of a factory building. Identify the permitted U-values of the elements labelled in the sketch.

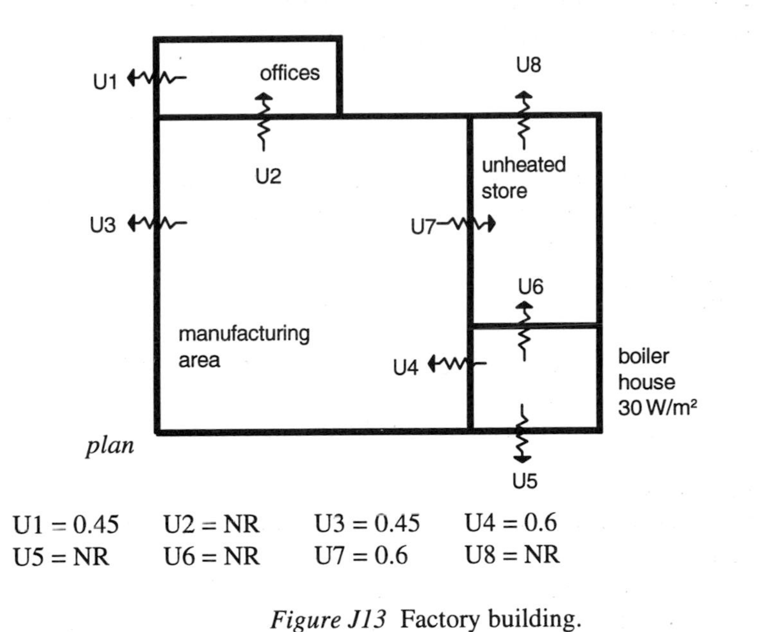

U1 = 0.45 U2 = NR U3 = 0.45 U4 = 0.6
U5 = NR U6 = NR U7 = 0.6 U8 = NR

Figure J13 Factory building.

Method 2 Calculated trade-offs

J2.4

To allow flexibility of design it is possible to exceed permitted U-values and/ or permitted percentages of glazing so long as the total heat loss rate does not exceed that which would have prevailed if the whole design had been in accordance with the U-values and percentages of glazing set out in Table J3.

When using this method, the designer's proposals are referred to as the intended construction, whereas the term notional building is taken to be related to a building complying with Table J3 in all respects together with rules 1 to 9 indicated earlier and in addition the following supplementary rules:

(10) If the total area of glazed openings (windows/rooflights) in the intended construction is less than that permitted by Table J3 then the intended area of glazing must also be used when the calculation of the total heat loss rate per degree is made for the notional building.

For example, if making such a calculation for the building of example J3, 34 m^2 of glazing (rather than the permitted glazing level of 51 m^2) would be used when dealing with both the intended and the notional constructions.

(11) If the U-value of the floor next to the ground, without insulation, is less than 0.45, then this lower value must also be used when calculating the

total heat loss rate for the notional building.

For example, if making such a calculation for the building of example J3, and the U-value of 0.4 for the floor was for an uninsulated floor, then a value of 0.4 for the floor would be used when dealing with both the intended and the notional constructions.

Calculating the total rate of heat loss per degree from fabric

J2.4 Within the Standards two tables A and B are given, A to be used for intended constructions and B for notional buildings.

Information from drawings and specifications for the intended construction is used in table A to allow calculation of the total rate of heat loss per degree through opaque fabric and glazing.

Information from Table J3 and drawings together with rules 1-11 are used in table B (assume single glazing) for the notional building to allow calculation of the total rate of heat loss per degree through the opaque fabric and glazing.

Compliance is achieved in circumstances where the total heat loss rate per degree from table A does not exceed that from table B.

The routine is best illustrated by means of a worked example.

Example J8

Check the proposed office block for compliance with the regulations, given the following data:

Length 50 m; width 40 m; height 16 m
Internal courtyard: length 25 m; width 15 m
U-values: walls 0.52; roof 0.47; floors next to ground 0.43, W/m^2K
Percentage glazing: walls 75 (double-glazed); roof nil

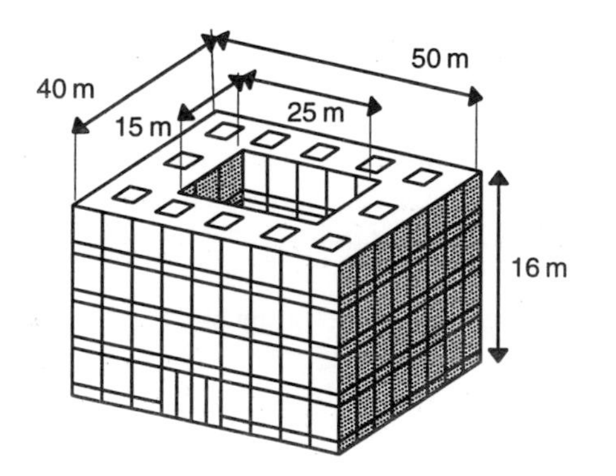

Figure J14 Proposed office block.

Compliance check

Table A Calculation for the intended construction

Element	Area (m^2)	U-value (W/m^2K)	Rate of heat loss (W/K)
External walls	1040	0.52	540.8
Glazing	3120	2.80	8736.0
Roof	1625	0.47	763.8
Ground floor	1625	0.43	698.8
		Total	10739.4

Table B Calculation for the notional building

Element	Area (m^2)	U-value (W/m^2K)	Rate of heat loss (W/K)
External walls	2704	0.45	1216.8
Glazing	1781	5.70	10151.7
Roof	1300	0.45	585.0
Ground floor	1625	0.43 (rule 11)	698.8
		Total	12652.3

Since the heat loss rate from the intended construction (10739.4 W/K) is less than that from the notional building (12652.3 W/K) then compliance has been demonstrated.

Method 3 Energy targets

This approach allows the heat loss rate of table B to be exceeded in circumstances where it can be demonstrated that advantage is taken of adventitious gains which may arise internally from people, artificial lighting, processes, or from solar flux. This provides even more flexibility of design than the trade-off method. An energy target calculation is required to demonstrate that the annual energy consumption of the intended construction is no greater than that for a building which complies with the elemental approach of method 1. **J2.5**

Related documents which can be consulted for establishing compliance or otherwise by this method are:

For dwellings:
BRE Report, BR150 1989, Building Regulations:
Conservation of Fuel and Power – The Energy Target Methods
of Compliance for Dwellings

For other buildings:
CIBSE Building Energy Code, Part 2, Section(a), 1981

DTS J2.2

Method 4 Selection from deemed-to-satisfy provisions

The designer may choose for each exposed element of the building to select an appropriate specification from the section giving provisions deemed-to-satisfy.

This section consists of a range of specifications for roofs, walls and floors which when constructed with the appropriate thicknesses of the selected insulants from the range given in table 2 to (J2.2) of the Regulations (given here as Table J4) is deemed to satisfy the standards.

Table J4 Types of insulating material

	Type/insulating material	Thermal conductivity (W/mK)
A	Wood wool slab (density not exceeding 500 kg/m^3)	0.085
B	Fibre building board	0.060
C	Cellular glass	0.050
D	Mineral fibre batt, mat or loose fill (glass or rock) urea formaldehyde foam cavity fill installed in accordance with BS 5617:1985 and BS 5618:1985	0.040
E	Mineral fibre slab (glass or rock) expanded polystyrene insulating board or loose fill; extruded polystyrene insulating board; phenol formaldehyde insulating board	0.035
F	Polyurethane insulating board	0.025

Some examples will now be given to illustrate the use of method 4.

Example J9

A designer intends to use one of the three forms of ground supported concrete floor specifications within the DTS section. The chosen arrangement is to have the damp-proof membrane immediately above the hardcore, with the insulant between this membrane and the underside of the concrete slab. There is a finishing screed on top of the concrete slab. The particular application is for an office building which measures 30 m long by 8 m wide and which has a floor with four exposed edges. Assuming that polyurethane insulating board is chosen, sketch the relevant specification indicating the appropriate thickness of insulant.

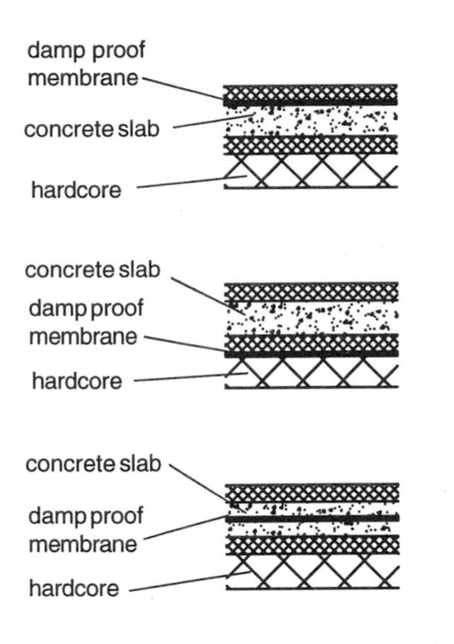

DTS J2.2 Floor type 1

1. Hardcore bed 100 mm thick of cleanbroken brick or similar inert material free from fine material and water soluble sulphates in quantities which would damage the concrete; blinded with suitable fine material and consolidated to form a level crack-free surface.

2. Concrete slab 100 mm thick with insulation, if any, laid above or below the slab; with or without a screed or floor finish.

3. Damp proof membrane above or below the slab or as a sandwich; jointed and sealed to the damp proof structure in walls, columns and other adjacent elements in accordance with the relevant clauses in Section 3 of CP 102:1973

Reference to Table J3 establishes that the relevant U-value should not exceed 0.45 W/m²K.

Next turn to the table in the DTS section (an extract from the table is shown below) which gives appropriate insulant thicknesses for floors next to the ground for various aspect ratios and edge conditions, and identify the required thickness of insulant for this specific case:

Table to floor type 1: Floors next to ground

Greater dimension of floor(m)	**Lesser dimension of floor (m)**	**Thickness (in mm) of insulating material according to type to give a U-value of 0.45 W/m²K for the following conditions**								
(a) = above (b) = up to and including		detached building (4 edges exposed)			semi detached/ end terrace (3 edges exposed)			corner building (2 adj. edges exposed)		
(a)(b)	(a)(b)	C	E	F	C	E	F	C	E	F
Floor type A: Ground supported concrete floors										
0-10	0-10	63	44	31	49	32	25	30	31	15
10-15	0-10	49	35	25	41	29	21	+	+	+
	10	30	21	15	+	+	+	+	+	+
15-20	0-10	44	31	22	38	27	19	+	+	+
	10-15	16	11	+	+	+	+	+	+	+
	15	+	+	+	+	+	+	+	+	+
20	0-10	41	29	21	37	26	18	+	+	+

Compliance check

For insulant F (polyurethane insulating board) the required thickness is 21 mm to ensure that the U-value is no greater than 0.45 W/m²K.

Realistically this will probably mean that a 25 mm board is used and the proposed specification is then:

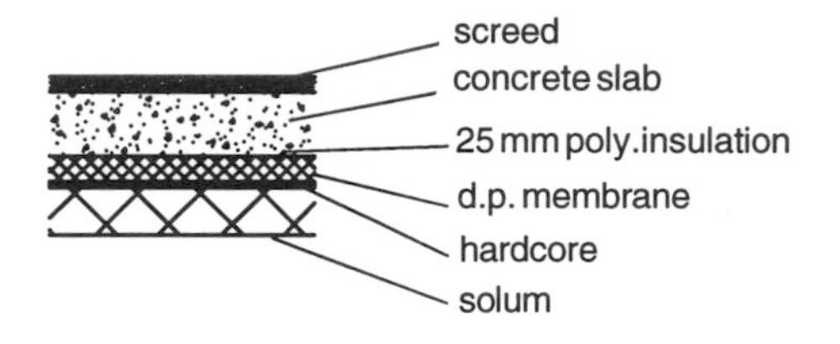

Figure J15 Specification with 25 mm board.

BRE Information Paper 3/90 provides a method for obtaining the U-value of uninsulated ground floors and/or the required insulant thickness to achieve a U-value of 0.45 W/m²K for such floors given the area and the perimeter of the floor. This method is in line with the procedure in the CIBSE Guide and is particularly suitable for use when dealing with irregularly shaped ground floors.

Example J10

A bungalow has a suspended timber ground floor as indicated on the plan shown. Determine the perimeter/area ratio and from this estimate the U-value of the uninsulated floor using the extract from table 1 of IP 3/90 given here.

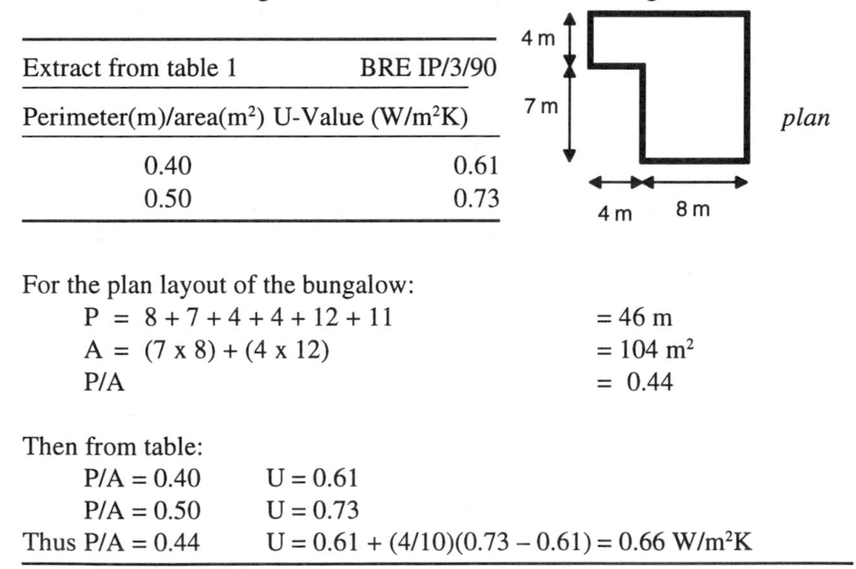

Extract from table 1	BRE IP/3/90
Perimeter(m)/area(m²)	U-Value (W/m²K)
0.40	0.61
0.50	0.73

For the plan layout of the bungalow:

P = 8 + 7 + 4 + 4 + 12 + 11 = 46 m

A = (7 x 8) + (4 x 12) = 104 m²

P/A = 0.44

Then from table:

P/A = 0.40 U = 0.61

P/A = 0.50 U = 0.73

Thus P/A = 0.44 U = 0.61 + (4/10)(0.73 – 0.61) = 0.66 W/m²K

Example J11

For the bungalow of the previous example the U-value of 0.66 W/ m²K is clearly not acceptable for an exposed floor. Estimate the appropriate thickness of insulation (of thermal conductivity 0.040 W/mK) to achieve a U-value of 0.45 W/m²K for this case using the extract from table 2 of IP 3/90 given here.

Compliance check

Extract from table 2 BRE IP 3/90

Insulation (mm) to achieve 0.45 W/m²K

Perimeter (m)/area (m²)	Thermal conductivity (W/mK) 0.035	0.040
0.40	21	24
0.50	29	34

Then from table:

P/A = 0.40 insulant thickness = 24 mm

P/A = 0.50 " " = 34 mm

P/A = 0.44 " " = 24 + (4/10)[34-24]

= 28 mm

Thus a U-value of 0.45 W/m²K can be achieved with the suspended timber ground floor, in this case by the application of 28 mm thickness of insulation of conductivity 0.040 W/mK to this floor.

Example J12

A designer of dwellings is investigating the possible use of solid masonry walls with external insulation and a rendered finish and with a plastered finish on the inside as indicated in Figure J16. The density of the masonry is 1900 kg/m³ and its thickness is 200 mm. Determine the appropriate thickness of mineral fibre slab insulation (E), for compliance.

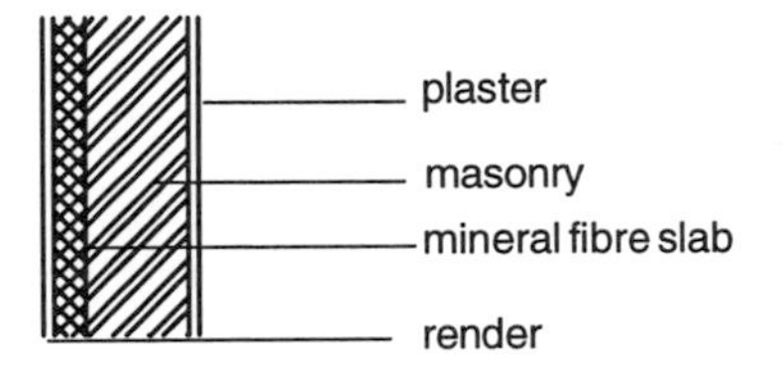

Figure J16 Solid masonry wall.

Compliance check

From Table J3, the required U-value is 0.45 W/m²K.

Table to wall type 1: Solid walls of masonry

Wall thickness (mm)	Max density (kg/m³)		Thickness (in mm) of insulating material according to type to give the U-value of: 0.6 W/m²K C	D	E	F	0.45 W/m²K C	D	E	F
200	2300	(a)	53	42	37	26	80	64	56	40
		(b)	64	51	45	32	91	73	64	46
	1900	(a)	49	39	34	24	76	61	53	38
		(b)	60	48	42	30	87	70	61	44

(a) with cladding externally
(b) rendered externally

From the extract from the regulations the required thickness of the mineral fibre slab is 53 mm. Notice also that for a situation where a wall U-value of 0.6 W/m²K applies, an insulations thickness of 34 mm would be appropriate.

Controls for space heating systems

J3.1 For dwellings, the distribution of heat from the space heating system must be controlled by a space thermostat. In the case of low pressure hot water space heating systems, thermostatic radiator valves are an acceptable means of control. For other buildings, and where the space heating medium is hot water, an additional requirement related to the control of heat distribution is the use of an external temperature sensing device (weather compensator), normally fitted on the north face of the building, which regulates either the quantity or the temperature of hot water flowing through radiators.

In addition, where the building is occupied intermittently, (schools, colleges, offices etc.) intermittent controls, involving a clock setting, must be used to ensure that the required temperature is maintained in the building only during the period of normal occupation.

For smaller systems (up to 100 kW output), a manually set clock control would suffice.

For systems of output greater than 100 kW, an optimising control is required. Such controls take account of the response of the building (rate of heat-up) to heat input, and the prevailing outside temperature, and adjust the preheating time to ensure that the building is at the operational temperature at the start of the working day.

Other controls such as a frost stat which can be used to circulate water within the space heating system without firing the boiler or can bring the boiler on to maintain minimum temperatures overnight, (such measures protect both the space heating system and the building fabric), are mentioned as possible ancillary measures which may be considered.

Where the space heating load is sufficiently large to utilise two or more gas or oil fired boilers which can each take part of the load and which together cope with the maximum load (for situations where the load is in excess of 100 kW), sequence control arrangements must be used. This allows boilers to operate close to their full load capabilities which keeps the boiler efficiencies high and minimises the consumption of fuel.

The principles of weather compensation, clock, optimising, frost and sequence controls are well covered in *Heating Services Design* by McLaughlin, McLean and Bonthron (Publishers Butterworth) while specific control application notes are available from the various manufacturers of automatic controls.

Controls for hot water storage vessels

The regulations require that every hot water storage vessel (hot water cylinders for dwellings, calorifiers for other buildings) must have a temperature detector within the water system which in conjunction with a controller (thermostat) can regulate the temperature of the stored water. **J3.3**

In addition, for larger systems (above 150 litres), a time switch must also be fitted which can start up and stop the supply of heat to the vessel. In the case of off-peak electrical systems such arrangements are built into the system. Again, *Heating Services Design* is a useful reference for understanding the control principles of such arrangements.

Insulation of hot water storage vessels

In addition to controlling the stored water temperature there is a requirement to thermally insulate the vessel to restrict the heat loss rate to not more than 90 W/m^2 of the surface area of the vessel. **J3.4**

Compliance can be by the selection of a pre-insulated cylinder which satisfies the relevant British Standard from BS 699:1984 (copper combination units). Alternatively, where insulating jackets are preferred, the relevant standard is BS 5615:1985.

Insulation of pipes and ducts

Pipes and ducts used for the distribution of warm water or air for space heating and pipes used for hot water supply systems must be adequately insulated in all circumstances other than where they make a contribution to the space heating within rooms. **J3.5**

For dwellings, compliance can be by following the relevant recommendations of BS 5422:1977. When insulating pipes, an insulant of thermal conductivity not greater than 0.045 W/mK can be used, and of thickness equal to the outside diameter of the pipe, up to a maximum thickness of 40 mm of insulant.

For all other buildings, compliance is by following the recommendations of BS 5422:1977.

Future amendments

Draft proposals for changes to Part J have been circulated for comment by the Scottish Office and amendments are currently being developed which are broadly in line with those which have been put in place in the Approved Document L for England and Wales.

The impetus for these changes is a result of thc concern of the international community about the increasing global levels of CO_2 emissions. The UK government in its commitment to reducing CO_2 emissions has sought fresh re-examination of building regulations to identify scope for improved control of such emissions.

Thus issues which could well appear in the future amendment include:

(1) new ways of calculating U-values to include the influence of cold bridges such as mortar between masonry blocks;
(2) taking account of U-values of items such as external doors;
(3) treating double-glazing rather than single glazing as the norm;
(4) means of reducing air leakage;
(5) futher improvements to heating and hot water controls and to cylinder and pipe insulation;
(6) ensuring appropriate use of mechanical ventilation and/or air conditioning in non-domestic buildings;
(7) ensuring energy efficient lighting in non-domestic buildings;
(8) provisions for an energy rating scale for dwellings.

References

Technical Standards, 1991, Part J, HMSO
CIBSE Guide, 1986, Book A
BRE Report 150, 1989
CIBSE Building Energy Code, Part 2, 1989
BRE Information paper, 3/90

VENTILATION OF BUILDINGS

PART K* — Regulation 23

BUILDING STANDARDS

Regulation

23 (1) A building to which this regulation applies shall have means of providing an adequate supply of air for users of the building.

(2) An opening in a building which is required for the purposes of paragraph (1) shall be so located as to ensure effective operation.

(3) This regulation shall apply to all buildings except buildings or any part of a building to which the Factories Act 1961 (a) applies.

(4) Paragraph (2) shall not be subject to specification in a notice served under section 11 of the Act.

Aim

The purpose of this part is to ensure that an adequate supply of air is available for people within buildings. Such ventilation may be provided naturally, mechanically or as a combination of both. In some situations, such as bathrooms and kitchens, the supply must be by a mechanical system.

TECHNICAL STANDARDS

Scope

The standards apply to all buildings except those to which the Factories Act applies. With the latter the range of situations is so diverse with regard to ventilation needs that it is a matter for the Health and Safety Executive.

Matters of ventilation impinge also on Parts E (in relation to escape routes) and F (in relation to openings for combustion).

Requirements

The standards indicate the acceptable forms of ventilation for various types of space with provisions set as areas of 1/30th of floor area for ventilators, as opening areas in mm^2 for trickle ventilators and as litre/s for mechanical systems. Other features are constraints on the location of this provision, maintenance aspects (duplicate motors in certain situations) and special requirements with regard to large garages.

** Amendment likely late 1996/early 1997.*

Definitions

A permanent ventilator is a means of providing a continuous supply of air e.g. an air brick in an outside wall (Figure K1).

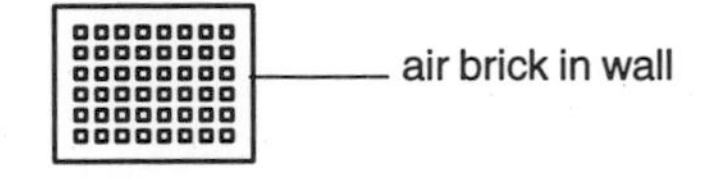

Figure K1 Permanent ventilator.

A trickle ventilator is a closeable small ventilator which is capable of providing minimum ventilation e.g. a hit and miss ventilator fitted to a window frame (Figure K2).

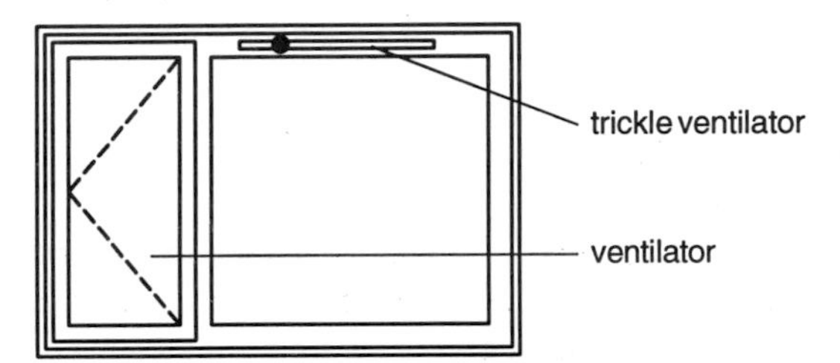

Figure K2 Ventilator and trickle ventilator.

A ventilator is a building component which is capable of being opened to provide ventilation, e.g. window, rooflight, grille, and in certain cases, a door.

Background

When assessing ventilation provision for larger buildings there is a need to assess the number of people occupying the space. This has been dealt with in earlier chapters, but as a reminder:

$$\text{Number of people in the space} = \frac{\text{Area of the space (m}^2\text{)}}{\text{Occupancy load factor}}$$

Example K1

A shop sales area measures 8 m by 10 m by 4 m high. Estimate:

(i) the number of persons occupying the space
(ii) the space per person

Compliance check

From the table of load factors in Part A the load factor is 2.0

Number of people = (8 x 10)/2.0 = 40
Space per person = (8 x 10 x 4)/40 = 8 m^3/person

As will be seen later the space per person influences the ventilation requirements for such applications.

When assessing whether an appropriate number of air changes per hour will be achieved by the use of a mechanical ventilating system with a quoted design duty of air supply or extraction rate given in m³/s or litre/s, simple calculations are required for the following relationships:

$$\text{Air change rate per hour} = \frac{\text{Fan duty (m}^3\text{/s) x 3600}}{\text{Space volume (m}^3\text{)}}$$

or where the fan duty is in litre/s and recalling that 1 m³ is equivalent to1000 litres,

$$\text{Air change rate per hour} = \frac{\text{Fan duty (litre/s) x 3.6}}{\text{Space volume (m}^3\text{)}}$$

COMPLIANCE

Introduction

The Technical Standards deal in turn with dwellings, garages (small and large), buildings other than dwellings and locations of ventilation openings subsequent to the following preliminary points:

(1) where a mechanical extract fan is fitted in a building containing an open flued appliance (e.g. extract fan in the kitchen and a gas fire in the living room of a house) it is necessary to ensure that the operation of the extract system does not have an adverse effect on the efficient operation of the appliance/flue system. The specialist recommendations of the appropr- iate fuel advisory service (for example, Scottish Gas) must be sought for the particular proposal; **K1.3**

(2) generally, the volume of any room to be ventilated is the internal cubic capacity of the room but there are exceptional cases where a portion of the room volume has to be disregarded, as illustrated in Figure K3; **K1.4**

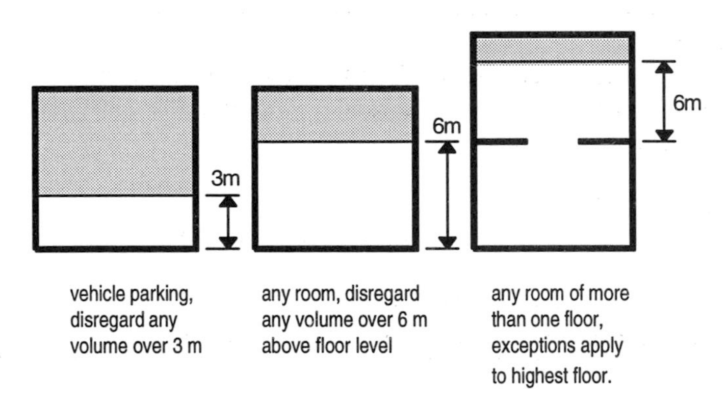

Figure K3 Exceptional cases.

(3) any requirement for an area of opening for ventilating purposes (e.g. area of openable window in a room) may be met by a single opening or by the aggregate area of two or more openings. **K1.5**

Dwellings

K2.1 The general requirement for rooms (other than kitchens) is that they must have provision for either natural ventilation or for mechanical ventilation (the exception is rooms with floor area less than 4 m^2).

Where natural ventilation is deployed there are two requirements: a ventilator with an opening area of not less than 1/30th floor area and a trickle ventilator with an opening area of not less than 4000 mm^2.

The 1/30th ventilator is normally accomplished by the use of sufficient openable area of window while the trickle ventilator is normally a hit and miss ventilating slot fitted at the top of the window frame.

Where mechanical ventilation is deployed the system will be an extract system which must be capable of operation at two speeds. At the higher speed and in relation to intermittent operation, the extraction rate must not be less than three air changes per hour. For continuous operation at the lower speed it must be capable of providing one air change per hour.

For those rooms which generate moisture, there is a basic operational philosophy that there is a need for high ventilation rates during the peak periods of moisture generation and lower or background ventilation rates at other times.

Kitchens

K2.2 A kitchen must have a mechanical extract system (except where there is an open flued appliance in the space) which is capable of an intermittent extraction rate of not less than 60 litre/s. Where there is in addition provision of natural ventilation via a 4000 mm^2 trickle ventilator this satisfies the requirements.

Otherwise, in addition to the 60 litre/s requirement the mechanical extract system must be capable of continuous operation at a lower speed to provide one air change per hour.

In kitchens where there is an open flued appliance, natural ventilation can be used; normally the ventilator is a window with a 1/30th opening area.

Example K2

A kitchen measures 4 m x 3 m and is heated by an open coal fire. The kitchen window measures 2 m x 1 m and 50% of the window is openable. The window frame has fitted to the top a hit and miss trickle ventilator which is 400 mm x 20 mm with a 60% free area.

Does this situation comply with Part K(2.2)?

Compliance check

Openable window area = 0.5 x 2 x 1 = 1.0 m^2
Floor area = 4 x 3 = 12.0 m^2
Required ventilator size = (1/30)12 = 0.4 m^2
Ventilator size complies

Total trickle ventilator area = 400 x 20 = 8000 mm^2
40% of this area constitutes metal
and 60% is free to allow air flow
Useful trickle ventilator area = 0.6 x 8000 = 4800 mm^2
Required trickle ventilator area = 4000 mm^2
Trickle ventilator area size complies

In conjunction with the open flue of the coal fire the system complies with Part K (2.2).

Baths and showers

A room containing a bath and/or shower must have a mechanical extract ventilation system capable of an intermittent extraction rate of not less than 15 litre/s. **K2.3**

The extract fan is usually activated by the light being switched on, runs during the period of occupation and over-runs for up to about 10 minutes after the light has been switched back off again at the end of occupation.

Example K3

A bathroom measures 1.8 m x 2.4 m x 2.8 m high and contains a shower cabinet and a washbasin. The bathroom has fitted a mechanical extract fan unit which can operate intermittently with an extraction rate of 0.05 m^3/s.

Calculate the number of air changes achieved during the 10 minute over-run period and check compliance with Part K (K2.3).

Compliance check

Volume of air removed in ten minutes = 0.05 x 10 x 60 = 30 m^3
Bathroom volume = 1.8 x 2.4 x 2.8 = 12.1 m^3
Number of air changes = 30/12.1 = 2.5

Since 1 m^3 = 1000 litres
Fan extraction rate = 0.05 x 1000 = 50 litre/s
Required fan extraction rate (not less than) = 15 litre/s
Thus the system complies with Part K (K2.3).

Watercloset compartments

These may be ventilated naturally or mechanically.

K2.4 Where ventilation is by natural means a ventilator (1/30th rule) must be provided. Note that there is no need here for the trickle ventilator.

Where ventilation is by a mechanical extract system, the system must be capable of providing an intermittent extraction rate of not less than three air changes per hour.

Example K4

A WC measures 1.1 m x 2.1 m x 2.7 m high.
Estimate:

(1) the necessary dimensions of a square opening window to satisfy the requirements for natural ventilation;
(2) the necessary extract rate (litre/s) for intermittent operation of a mechanical extract fan to achieve three air changes per hour.

Compliance check

Required window openable area	= (1/30) 1.1 x 2.1	= 0.077 m^2
Length of side of opening	= 0.077^{-2}	= 0.280 m

Thus an opening window, say 0.3 m x 0.3 m would comply.

Space volume	= 1.1 x 2.1 x 2.7	= 6.24 m^3
A rate of 3 air changes per hour means	3 x 6.24	= 18.72 m^3/h
	= (18.72 x 1000)/3600 = 5.2 litre/s	

K2.5 Duplication of fan motors

Where a mechanical system serves more than one dwelling it must be provided with a duplicate motor. Thus in terms of system breakdown it is acceptable for one dwelling to be discommoded but unacceptable for all those dwellings connected to say a communal extract system for bathrooms and toilets to be simultaneously discommoded.

Such systems must also be completely separate from any other ventilating plant in the building.

K2.6 Clothes drying facilities

Communal laundry areas or clothes drying rooms must have extract mechanical ventilation capable of an intermittent extraction rate of not less than ten air changes per hour. The system must also be capable of continuous operation at a lower speed to give approximately one air change per hour unless natural ventilation is provided by trickle ventilation.

Example K5

A communal laundry in a block of flats measures 6 m x 7 m x 2.8 m high and has a mechanical ventilation system, capable of extracting at its highest speed, 0.3 m^3/s of air from this drying area. Is this likely to comply with the requirements of Part K (K2.6)?

Compliance check

Volume of laundry area	= 6 x 7 x 2.8	= 117.6 m^3
Fan extract rate	= 0.3 x 3600	= 1080 m^3/h
Number of air changes per hour	= 1080/117.6	= 9.2
Required air changes per hour		= 10

Thus the fan duty narrowly fails to meet the requirements of the regulations i.e. fails to comply.

Ancillary rooms

Any room serving more than one dwelling must have provision for natural or mechanical ventilation as per the stated earlier general requirements for dwellings. **K2.7**

Example K6

A community meeting room in a block of flats measures 6 m x 7 m x 2.6 m high. There are two windows in each of two opposite walls and each window measures 1500 mm wide by 800 mm high; 25% of each window is openable. On one wall there is a ventilating grille with a sliding plate measuring 300 mm x 200 mm, with a free area of 70%. Do these provisions comply with Part K(K2.7)?

Compliance check

Actual window opening area	= 4 x 0.25 (1.5 x 0.8)	= 1.2 m^2
Floor area	= 6 x 7	= 42 m^2
Required window opening area	= (1/30)42	= 1.4 m^2
Fails to comply re ventilator		
Area of trickle ventilator	= 0.7(300 x 200)	= 42000 mm^2
Required area of trickle ventilator		= 4000 mm^2
Complies re trickle ventilator		

Common circulating areas

Where provision is by natural ventilation the ventilator requires to have an opening area of not less than 800 mm^2 per m^2 of floor area. In the case of stairs the floor area has to be considered to be the plan area of the stairwell. There is no requirement for trickle ventilation. **K3.1 and K3.2**

Where provision is by mechanical ventilation the system must be capable of continuous operation to provide approximately one air change per hour.

Example K7

A stairwell within a block of flats measures 8 m x 2.5 m in plan and the landing window (considered fully openable) measures 1.5 m by 1 m. Does this comply with Part K(K2.8)?

Compliance check

Actual window opening area	= 1.5 x 1	= 1.5 m^2
Floor area	= 2.5 x 8	= 20 m^2
Required window opening area	= 800 x 20	= 16000 mm^2

The required area is very small compared to the provision   Complies

---

Garages

K3.1 and K3.2 These regulations distinguish small garages as being within the range of 30 to 60 m^2 floor area and large garages as being in excess of 60 m^2.

A typical single garage is about 2.5 m x 5 m which is 12.5 m^2, making a typical double garage about 25 m^2. There is no requirement for the provision of ventilation for floor areas as small as this. Thus the small garage is in the range accommodating three or more cars up to about six cars for parking.

Small garages

K3.1 These, when used for parking motor vehicles, must have provision for natural or mechanical ventilation.

When natural ventilation is employed, two permanent ventilators, each with an open area of not less than 1/3000th of the floor area and so positioned as to encourage through ventilation, are required. Since ventilation is to remove carbon monoxide and other exhaust gases which are produced near floor level, one of these permanent ventilators should be not more than 600 mm above floor level.

K3.2 When mechanical ventilation is employed the extract must be capable of continuous operation and designed to give an extraction rate of not less than two air changes per hour. The system must be independent of any other ventilating system in the building and at least two-thirds of the extracted air must be from outlets not more than 600 mm above the floor.

Example K8
A garage for parking cars at ground level in an office block measures 8 m x 7 m x 3 m high. It is fitted with an extract fan and ductwork system as shown in Figure K4. Check for compliance with Part K(K3.1).

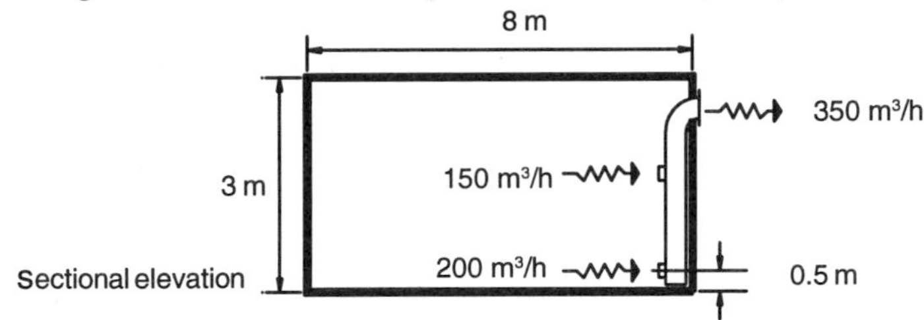

Figure K4 Proposal for ventilating a garage.

Compliance check

Area of garage	= 8 x 7	= 56 m^2	This is in the range 30-60 m^2
Volume of garage	= 8 x 7 x 3	= 168 m^3	
Two air changes		= 336 m^3	
Two air changes per hour		= 336 m^3/h	This is the requirement
Proposal		= 350 m^3/h	This complies

The system is separate from any other in the building — This complies
Location of extract grilles complies with the 600 mm rule.
Minimum amount required to be extracted at the lower level = (2/3) 336 = 224 m^3/h
Proposed amount to be extracted at the lower level = 200 m^3/h
Thus on this latter point the system fails to comply with Part K(K3.1)

Large garages

There must be provision for natural or mechanical ventilation on every storey. **K3.3** Where the system proposed is in accordance with the CIBSE Guide, Book B: 1986 (Chartered Institution of Building Services Engineers) this shall be deemed to satisfy the requirements.

CIBSE recommends for natural ventilation of above ground car parks that openings in outside walls should have an aggregate area equal to at least 5% of the floor area at each level and at least half of that area should be in opposite sides.

For mechanical ventilation CIBSE suggests between six and ten air changes per hour are necessary according to building type. Location of extract points should be carefully chosen to ensure that there is no possibility of pockets of stale air where dangerous fumes could collect.

With mechanical systems it is further suggested that an automatic carbon monoxide detector should be installed which would increase the extract ventilation rate and raise an alarm when CO concentrations rise above 100 parts per million in the air. There is also a need to install a standby fan which is connected to a secure power supply.

Example K9

A commercial organisation is proposing a city centre eight storey car parking facility measuring 55 m x 55 m x 3 m high. At each storey the intention is to provide natural ventilation openings (permanent ventilators) on one pair of opposite walls as shown in Figure K5. Would this secure compliance with Part K(K3.3)?

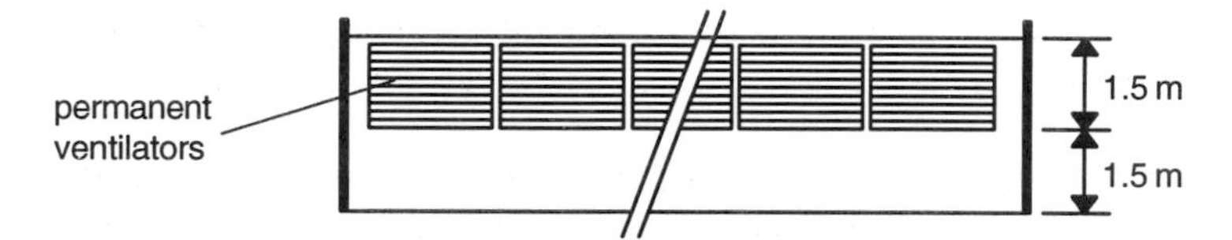

Figure K5 Natural ventilation arrangements, car park.

Compliance check

Floor area	= 55 x 55	= 3025 m^2
5% of floor area	= 0.05 x 3025	= 151.3 m^2
Required area of permanent ventilators per side		= 75.7 m^2
*Proposed area of permanent ventilators per side	= 1.5 x 55	= 82.5 m^2

*On opposite sides at each storey ignoring the area taken up by columns
Therefore the proposals comply with Part K(K3.3)

Where mechanical ventilation is proposed for larger garages it is often the practice to provide extract grilles on a grid layout on plan with a two-thirds extract at floor level (600 mm) and a one-third extract at a higher level (Figure K6).

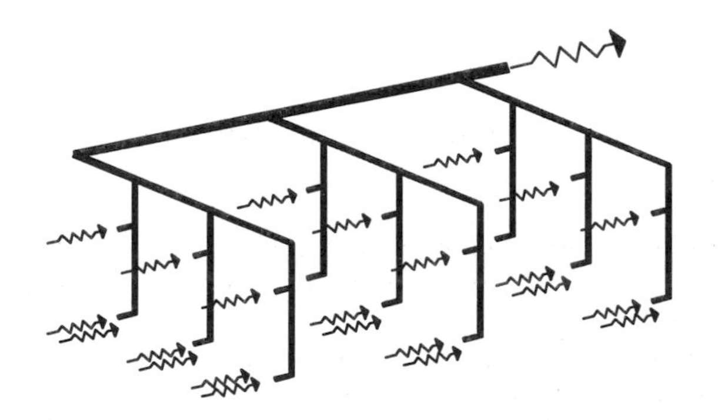

Figure K6 Grid layout.

It may be useful to have two systems, each supplying at least half the appropriate number of air changes per hour.

Other buildings

For buildings other than dwellings or garages the variety of situations which may arise will be very diverse. Where the system proposed is in accordance with the CIBSE Guide, Book B: 1986, this shall be deemed to satisfy the requirements. **K4**

Rooms

Any room in a building must have adequate provision for natural or mechanical ventilation. **K4.1**

Where provision is by natural ventilation, a ventilator such as a window or rooflight providing an opening area of 1/30th the floor area is required.

Where the volume of the room is less than 60 m^3 there is in addition a need to provide a trickle ventilator with an opening area of not less than 4000 mm^2. Thus the ventilation requirements for small rooms is in line with those for rooms in dwellings.

Example K10

An office measures 3 m x 6 m x 3 m high and is provided with two windows each having an openable element 1000 mm x 300 mm. An external wall of the room includes an airbrick with a closeable grille with a free area of 5000 mm^2. Does this arrangement satisfy the requirements of Part K(K4.1)?

Compliance check

Area of office	= 3 x 6	= 18 m^2	
Required ventilator size	= (1/30)18	= 0.6 m^2	
Proposed ventilator size	= 2(1.1 x 0.3)	= 0.66 m^2	Complies
Required trickle ventilator size		= 4000 mm^2	
Proposed trickle ventilator size		= 5000 mm^2	Complies

There are some exceptions to the foregoing general approach to the ventilation of rooms, for instance:

(1) where a room has been defined as a storage room to be used for storage which requires a controlled temperature then there are no ventilation requirements;
(2) where a room is very small, having a floor area of less than 4 m^2, then there are no ventilation requirements;
(3) where the cubic space per occupant does not exceed 3 m^3 there is a requirement that the form of ventilation must be mechanical.

Example K11

A public bar measures 10 m x 6 m x 3 m high. Does this space require mechanical ventilation as per Part K(K4.1)?

Compliance check

From Part A, Occupant load factor		=	0.5
Number of occupants	= (10 x 6)/0.5	=	120
Space per occupant	= (10 x 6 x 3)/120	=	1.5 m^3

Less than 3 m^3 per occupant, thus a mechanical system is required

The CIBSE Guide provides a table of recommended outdoor air supply rates for different applications and an extract of this information is given here as Table K1.

Table K1
Table of outdoor air supply rates

application	rate (litres/s) recommended	per/person	min.floor area m^2
offices	8	5	1.3
supermarkets	8	5	—
theatres	8	5	—
hotel bedrooms	12	8	1.7
conference rooms	25	18	6.0
restaurant kitchens	20 air changes per hour		
corridors	1.3 air changes per hour		
hotel toilets	10 air changes per hour		

Kitchens

A kitchen must have adequate provision for mechanical ventilation. The normal solution for kitchens is to extract moisture and oil laden air at source by having a canopy directly above the stove or range to capture the extract air prior to its removal by the fan and ductwork system.

Example K12

Calculate the extract rate required to be provided by a fan and ductwork system for a restaurant kitchen measuring 10 m x 8 m x 3.5 m high.

Compliance check

From Table K1, number of air changes per hour = 20

Space volume	= 10 x 8 x 3.5	= 280 m^3
Required extract rate	= 20 x 280	= 5600 m^3/h

Baths and showers
A room containing a bath and/or shower must have adequate provision for mechanical ventilation. Thus, as is the case with dwellings, ventilation by mechanical means is mandatory.

Washroom and WCs
As was the case with dwellings, ventilation can be either by natural or by mechanical means. Where natural ventilation is used, the ventilator must have an opening area of not less than 1/30th of the floor area.

Example K13
A gentlemen's room in a large hotel includes a range of washbasins and a range of WCs and the room measures 4 m x 10 m. There are four sash and case windows on one long wall, each being 800 mm long by 1700 mm deep with 50% opening area. Does the situation comply with Part K (K4.4)?

Compliance check

Floor area	= 4 x 10	= 40 m^2
Ventilator opening area required	= (1/30)40	= 1.34 m^2
Ventilator opening area installed	= 4 x 0.5(0.8 x 1.7)	= 2.72 m^2

Thus arrangements comply with Part K (K4.4)
Notice that there is no requirement here for trickle ventilation

---

Location of ventilation openings
Generally, a ventilator (e.g. window, rooflight) must open directly to the outside air. There are two exceptions, one, where within certain restrictions trickle ventilators or permanent ventilators may open into a duct, and the other where within certain restrictions, ventilators may open into a conservatory. **K5.1**

In the case of openings to a duct, the duct may not be greater than 6 m in length if vertical or within 20° of vertical or 2 m long in any other case (Figure K7).

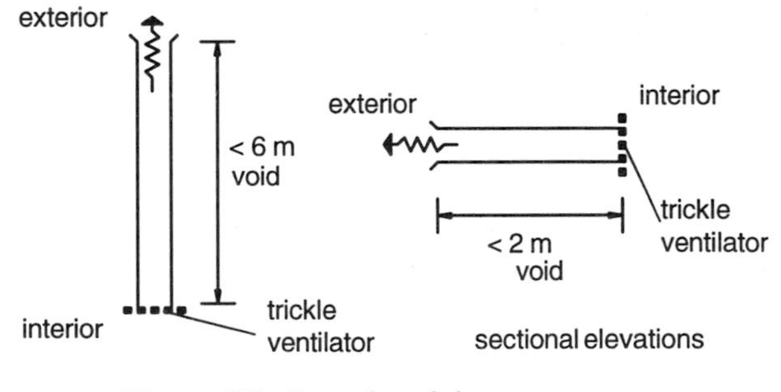

Figure K7 Lengths of duct.

In the case of a ventilator which serves a room in a dwelling and which opens into a conservatory, this may only be allowed if the conservatory is ventilated directly to the outside air by:

(1) a ventilator with an opening area of not less than 1/30th of the combined floor areas of the room plus the conservatory; and
(2) a trickle ventilator with an opening area of not less than 4000 mm².

Example K14
Check the proposal in Figure K8 for compliance with Part K (K5.1b).

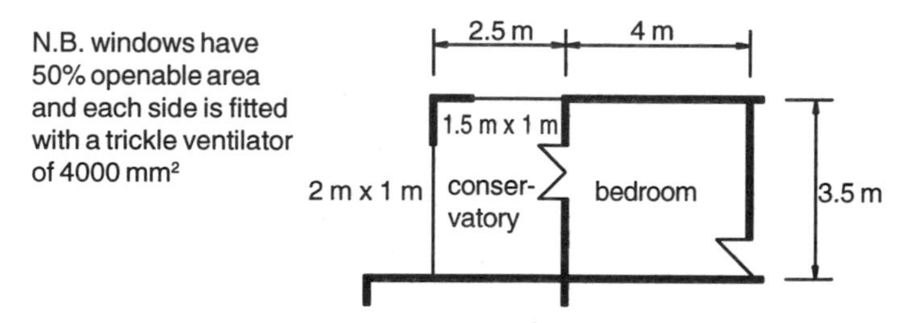

Figure K8 Bedroom with conservatory.

Compliance check

Total floor area	= (2.5 x 3.5) + (4 x 3.5)	= 22.8 m²
Ventilator opening area required	= 22.8(1/30)	= 0.76 m²
Proposed window opening area	= 0.5(1.5 + 2)	= 1.75 m²
		Complies
Required trickle ventilator		= 4000 mm²
Proposed trickle ventilator	= 2 x 4000	= 8000 mm²
		Complies

K5.2 Height of ventilator openings

Some part of the open or opening area of a ventilator must be not less than 1.75 m above the floor level of the space being ventilated. This helps to ensure reasonable fresh air distribution within the space in addition to aiding the stack effect aspect of natural ventilation.

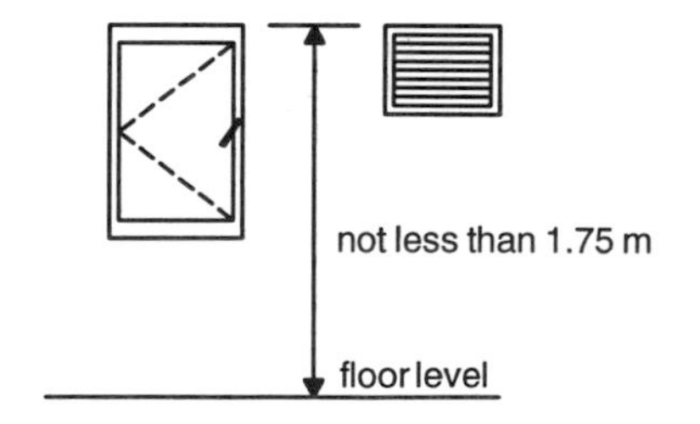

Figure K9 Position of ventilator.

Space outside ventilator openings K5.3

There must be clear air space outside the building and directly in front of any ventilators.

The clear air space, when measured at the level of the head of the ventilator, must be large enough to contain a horizontal square with a side of either 2 m or the relevant lengths derived from the rules for various situations which follow these general comments.

The clear air space must have one side in the plane of the ventilator and the whole length of the ventilator must be included in the side of the square of the required size as shown (plan) in Figure K10.

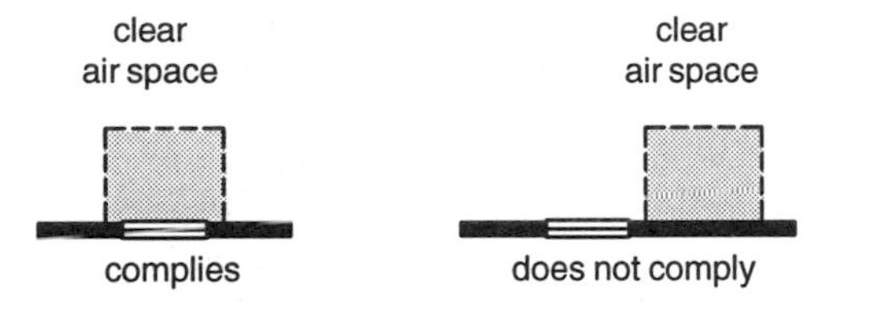

K5.4

Figure K10 Clear air space.

The square must in addition be open to the sky.

Where a roof, balcony, or other projection overhangs the ventilator at a height of less than 2 m above the head of the ventilator, the square open to the sky must be measured from the outer edge of the projection as shown (elevation) in Figure K11.

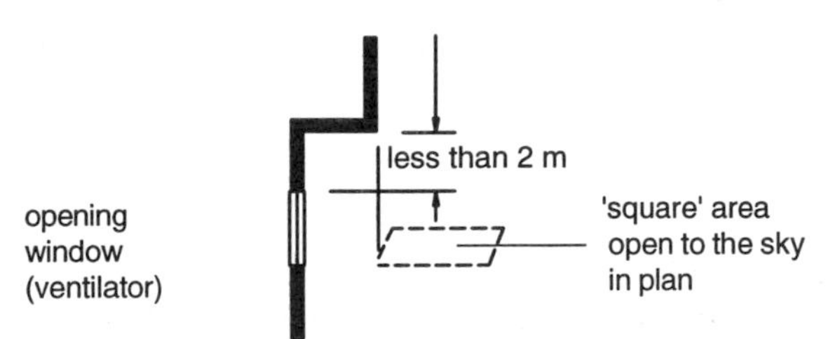

Figure K11 Projection overhanging ventilator.

The relevant length where the ventilator opens into a closed court is one-third of the height (H) of the lowest enclosing wall above the top of the ventilator as shown in plan and elevation in Figure K12 overleaf.

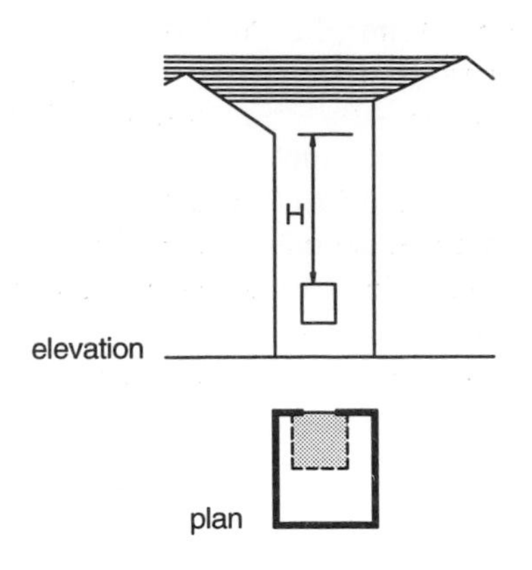

Figure K12 Ventilator in closed court.

Thus for the situation shown, given that H is 6.6 m, then the relevant length is 2.2 m and a clear space of 2.2 m x 2.2 m is required in front of the ventilator (window, in this case).

The relevant length where the ventilator opens into an open court is one quarter of either:

(1) the height (H) of the lowest enclosing wall above the top of the ventilator; or

(2) the distance (D) from the side of the ventilator to the open side of the court, whichever is the lesser.

Example K15

Calculate the relevant lengths required for the openable windows shown in Figure K13.

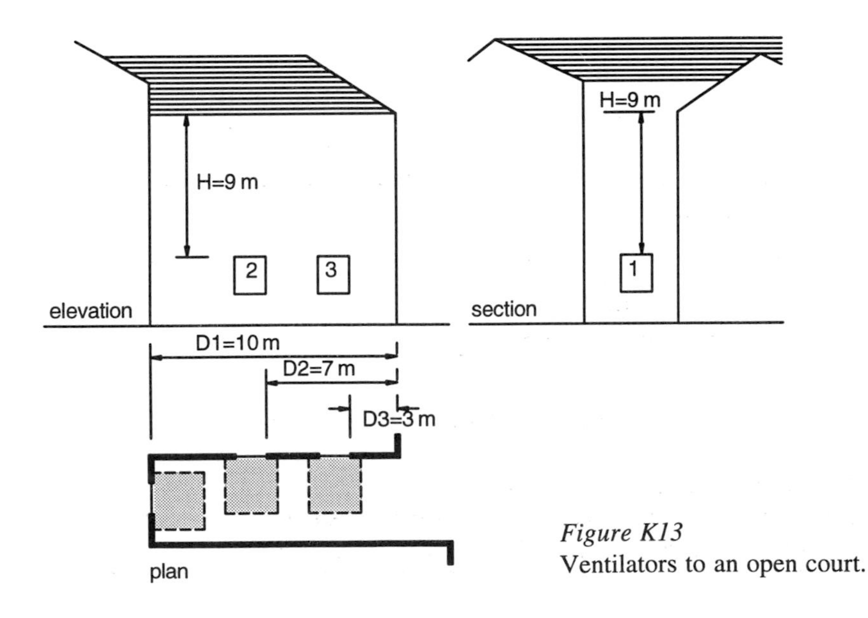

Figure K13
Ventilators to an open court.

Compliance check

Window 1 H = 9 m D1 = 10 m H is the lesser
Thus clear space required is 9/4 = 2.25 m
Clear space dimensions are 2.25 m x 2.25 m

Window 2 H = 9 m D2 = 7 m D2 is the lesser
Thus clear space required is 7/4 = 1.75 m
There is an overriding requirement that the minimum acceptable length of clear space is 2 m.
Thus clear space dimensions are 2 m x 2 m

Window 3 This will require a clear space dimension of 2 m x 2 m.

The relevant length where the ventilator faces a wall within 15 m in a passage is one sixth of either:

(1) the height (H) of the facing wall above the top of the ventilator; or
(2) the distance from the side of the ventilator to the end of the facing wall nearest the ventilator, whichever is the lesser.

Example K16

Calculate the clear space dimensions required for the openable windows shown in the proposals of Figure K14.

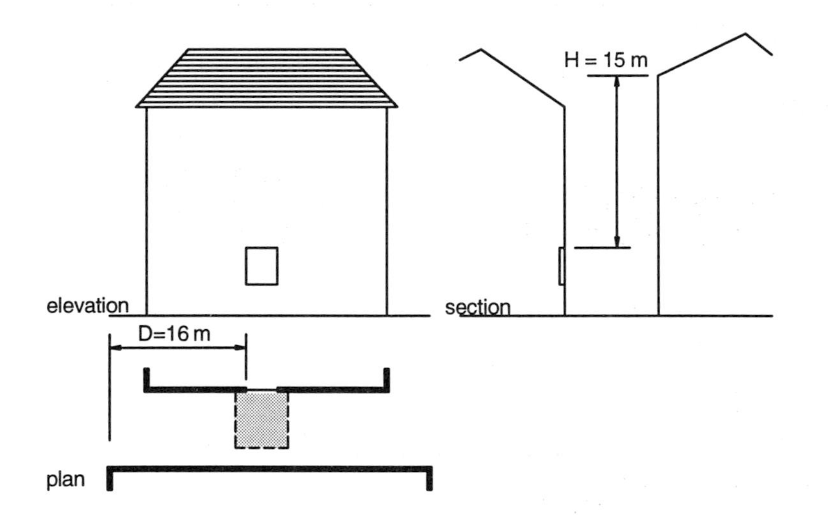

Figure K14 Ventilator to a passage.

Compliance check

H = 15 m D = 16 m H is the lesser. Relevant length = 15/6 = 2.5 m
Required dimensions of clear air space are 2.5 m x 2.5 m.

Recesses fall somewhere between the situation for open courts and for facades with flat faces. In dealing with recesses:

(1) where the ventilator is on the back wall of the recess – where the depth of the recess is greater than its length treat as an open court, otherwise treat as being equivalent to a flat face of a building (Figure K15);

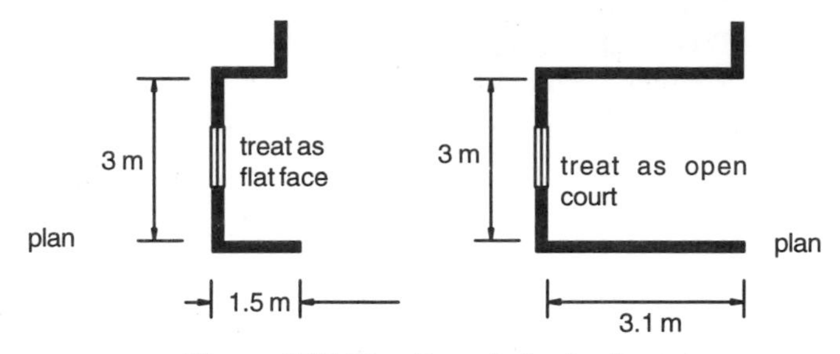

Figure K15 Ventilator in back of recess.

(2) where the ventilator is on the side wall of the recess and where the distance from the centre line of the ventilator to the outer corner of the recess is greater than the length of the recess treat as an open court, otherwise treat as being equivalent to a flat face of a building. (Figure K16.)

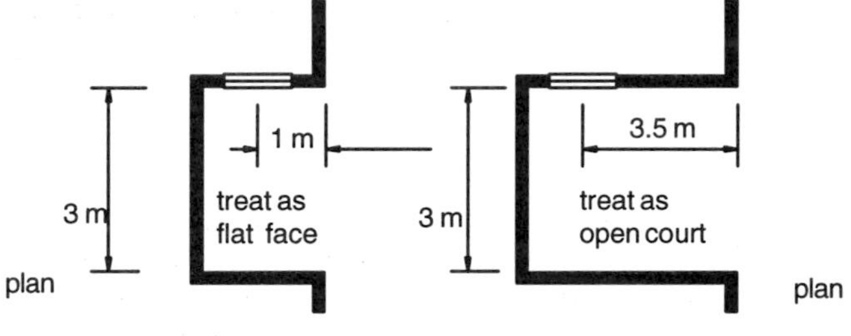

Figure K16 Ventilator in side of recess.

Openings for mechanical ventilation
Inlets and outlets to mechanical ventilation systems must be to the outside air either directly (as is the case with a window mounted fan, or a fan mounted on an external wall) or via ductwork (which may or may not include a heat exchanger) as illustrated in Figure K17

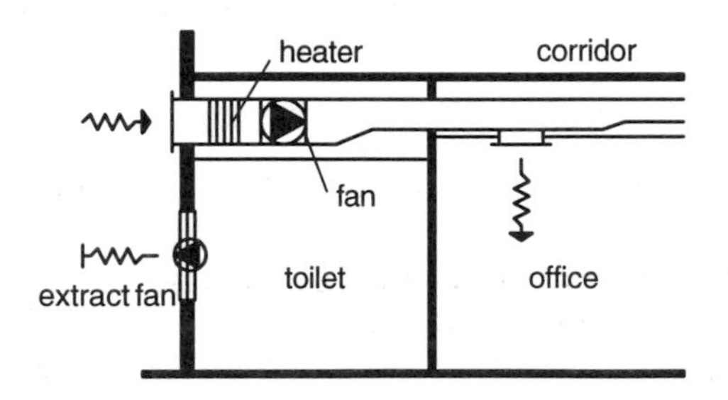

Figure K17 Ventilation intakes and outlets.

In situations where a mechanical ventilating system gathers extracted air from various spaces into a common duct system prior to discharge to an outlet, no connection to the system is permitted between the exhaust fan and the outlet (Figure K18). The reason for this is that this section of the system is under positive pressure and there is a risk of spill back of contaminated air.

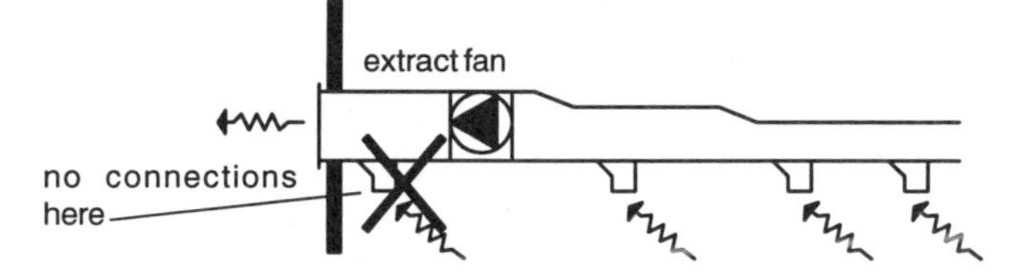

Figure K18 Common duct system.

Considerable thought must be given to the suitability of the location of outlets and inlets to mechanical ventilation systems in order to avoid:

(1) nuisance from the discharge of stale air from outlets, such as may arise, for example, where the outlet from a commercial kitchen discharges across a pavement;
(2) contamination of the air supply at the inlet of the building either by its being too close to outlets from its own or other systems or by its being too close to other sources of contamination such as smoke, steam, noxious gases or vapours such as are prevalent in city streets from vehicle emissions.

Future amendments

Draft proposals for change to Part K have been circulated for comment by the Scottish Office and amendments are currently being developed which are broadly in line with those which have been put in place in the Approved Document F for England and Wales.

Thus issues which could well appear in the future amendment include:

(1) the introduction of passive stack ventilation as an alternative to mechanical extract in dwellings or similar spaces;
(2) a further refining of the requirements relating to the concern to preclude the potentially lethal combination of extract fan and open flued appliance;
(3) increased potential for trickle ventilation (increasing the opening from 4000 mm^2 to 6000 mm^2;
(4) removal of the current requirements for ventilation openings into courtyards;
(5) requirements for evidence of commissioning and testing of air conditioning and mechanical ventilation systems;
(6) increasing design flexibility with options to provide a greater range of ventilation possibilities for buildings.

References
Technical Standards, 1991, Part K, HMSO
CIBSE Guide, Book B, 1986

DRAINAGE AND SANITARY FACILITIES

PART M — Regulations 24 & 25

BUILDING STANDARDS

Regulations

24 (1) A building shall be provided with a drainage system sufficient to ensure hygienic disposal of discharges from the building.

(2) In this regulation 'discharges' includes effluents, used water and run-off of rainwater from roofs and other exposed surfaces of the building.

25 (1) A building to which this regulation applies shall be provided with adequate sanitary facilities.

(2) This regulation shall apply to all buildings other than buildings or any part of a building to which -

(a) the Workplace (Health, Safety and Welfare) Regulations 1992 apply;

(b) s. 7 of the Factories Act 1961 applies; or

(c) the School Premises (General Requirements and Standards) (Scotland) Regulations 1967 to 1979 apply.

(3) This regulation shall not be subject to specification in a notice served under section 11 of the Act.

Aim

The drainage system for the building must ensure that all foul water and rainwater is effectively and hygienically transported through a network of sealed pipes to an acceptable point of discharge.

TECHNICAL STANDARDS

Scope

Applies is to all buildings with the exception of:

(1) buildings to which section 7 of the Factories Act 1961 applies;

(2) buildings to which the School Premises (General Requirements and Standards) (Scotland) Regulations applies;

(3) buildings administered by the Health and Safety Executive under the Workplace (Health, Safety and Welfare) Regulations 1992 with an Approved Code of Practice.

Requirements

The drainage system must be designed to accommodate the maximum discharge of foul, waste and rainwater efficiently and safely. In practice design will require to address the following criteria among others:

M2.1-2.9

(1) appropriate sizing of gutters, pipes and stacks;
(2) water and air tight jointing of pipework;
(3) adequately supported pipework and gutters;
(4) lengths and gradients;
(5) access points for clearing and maintenance;
(6) constant or increasing bore in direction of flow;
(7) ventilation of pipework.

The requirements will be met by designing and constructing the drainage system in accordance with the relevant British Standards.

Background
Until June 1994 the deemed-to-satisfy section of Part M offered a prescriptive option culled from the relevant British Standards for securing compliance. As a result of a body of user opinion expressing that this option has proved to be inadequate and restrictive, the 1994 amendment to Part M has removed this option and the deemed-to-satisfy section now simply states that the requirements will be met by systems designed and constructed in accordance with the relevant British Standards.

COMPLIANCE

Drainage system of a building

M2.1 The standards deal with the drainage system (M2.1) in terms of the need to provide safe and hygienic disposal of foul water and rainwater and also (M2.2) in terms of adequate ventilation of the drainage system.

M2.3 There must be means (M2.3) of dealing with discharges of oil, fat and grease (e.g. from restaurant kitchens, slaughterhouses etc.) and volatile substances (interceptors on the forecourts of petrol filling stations) as well as the provision of silt traps where there is the possibility of silt being washed into the below ground drainage system.

M2.4 Discharges from the drainage system are covered by M2.4 and in terms of foul water this normally requires discharge to a local authority sewer or to a public treatment works as per the Sewerage (Scotland) Act 1968. An alternative is discharge to an acceptable private sewage treatment works (which might, where appropriate, be a septic tank with suitable outfall, e.g. a remote dwelling).

Rainwater is normally taken to a local authority sewer but in certain circumstances (more likely in rural or remote situations) discharge may be to a suitable watercourse, soakaway or storage tank with an overflow to a suitable outfall.

The relevant British Standards for the design and construction of drainage systems for buildings are:

(1) BS 6367:1983 (rainwater pipes and gutters);

(2) BS 5572:1978, revised in 1994 (sanitary pipework);
(3) BS 8301:1985 (underground drainage).

Compliance with these standards would be the way of securing compliance with the Technical Standards in terms of M2. Note that any private sewage treatment works must be designed and constructed in accordance with BS 6297:1983.

DTS

Rainwater drainage (BS 6367:1983)

BS 6367:1983 includes sections dealing with materials and components, meteorological aspects (normal design intensity taken as 75 mm/h for roofs and 50 mm/h for ground level paved areas), design run-off assessment, hydraulic design of roof drainage which includes eaves, valley and parapet gutters, sizing of box and internal receivers, overflow weirs and rainwater pipes.

M2.1
DTS

The rate of run-off from roofs and paved areas and vertical surfaces may be calculated as Q, in litres/s as per the formula:

$$Q = (Ae \times I)/3600$$ where:

Ae is the effective catchment area (m^2)
I is the design rainfall intensity (mm/h)

It is assumed that the surfaces are impermeable (e.g. slates, concrete tiles for roofs, paved surfaces at ground level) with no capacity to hold back water from the drainage system.

Flat roofs — Ae is the same as the plan area of the roof.

Pitched roofs — Ae = [(b + c/2) x length] where b is the plan projection of the sloping surface and c is the eaves to ridge height.

Vertical surfaces — Normally only considered where run-off is to a lower roof or where ponding cannot be tolerated. Ae = exposed surface/2.

Design consists of choosing gutters and rainwater pipes which can carry the flow loads estimated from Ae and Q above. Table M1 provides data on flow capacities for standard (level) eaves gutters as follows:

Table M1 Eaves rainwater gutters

Gutter size (mm)	Flow capacity (litres/s)	
	True half round	Nominal half round
75	0.38	0.27
100	0.78	0.55
115	1.11	0.78
125	1.37	0.96
150	2.16	1.52

Two factors which can reduce the flow capacity are the length of the gutter and whether corners are present. Reduction factors to be applied to the flow capacities of Table M1 may be obtained from Tables M2 and M3.

Table M2 Reduction factors for capacities of long half round eaves gutters

Lg/yg	Reduction factor
50	1.00
100	0.93
150	0.86
200	0.80

where Lg is the gutter length and yg the gutter depth and modified capacity = basic capacity x RF

Table M3 Reduction factors for capacities of eaves gutters with angles

Type of angle	Reduction factor	
	Angle < 2 m from outlet	Angle between 2 m and 4m from outlet
Sharp corner	0.80	0.90
Round corner	0.90	0.95

Angles in the guttering system impede the flow (sharp corners are worse than round corners)

Once the gutter size has been determined the size of outlet (and thus size of rainwater stack) can be obtained from Table M4.

Table M4 Minimum outlet sizes for eaves gutters

Gutter size (mm)	Sharp (S) or round (R) edge	Throat diameter of outlet at one end of gutter (mm)	Throat diameter of outlet not at one end of gutter (mm)
75	S/R	50/50	50/50
100	S/R	63/50	63/50
115	S/R	63/50	75/63
125	S/R	75/63	89/75
150	S/R	89/75	100/100

Example M1

A building of plan dimensions 12 m long by 7 m wide incorporates two semi-detached dwellings. The roof pitch is 35° and eaves gutters run along the frontage and the rear with the four rainwater stacks on the gables at 0.5 m from the main line of the gutters. The corners can be assumed round.

The proposal is that 115 mm true half round gutters with 63 mm stacks will be used. Check the sizes with respect to compliance.

Compliance check

Ae = (b + c/2) x length for the front roof and the rear roof.
For front, b = 3.5 m, c = 3.5 tan 35 = 2.45 m. Length = 6 m since half the rainwater reaching the gutter flows to the right and half flows to the left to reach the two rainwater stacks.

Thus Ae = (3.5 + 1.225)6 = 28.35 m^2

Flow load Q = (75 x 28.35)/3600 = 0.59 litres/s

For a 115 mm THR gutter, Lg/yg = (6 x 1000)/(115/2) = 104 RF = 0.93

For a round corner within 2 m, RF = 0.90

Basic capacity of a 115 mm THR gutter = 1.11 litres/s
Modified capacity = (1.11 x 0.93 x 0.90) = 0.93 litres/s
This easily complies since the flow load is 0.59 litres/s

The relevant outlet (at one end) and stack for a 115 mm THR gutter is 50 mm diameter. 63 mm complies but 50 mm would do.

While the principles of sizing are the same for valley etc. gutters, the geometry means that calculation is a little less straightforward and the reader is referred to BS 6367.

Foul and waste drainage (BS 5572:1978: revised to 1994)

M2.1
M2.2
DTS

BS 5572:1994 includes a section which deals with performance covering such matters as flow and usage data for sanitary appliances, exclusion of foul air, sealing, reducing risk of blockage, access for maintenance, the phenomena of self siphonage and induced siphonage in branches, back pressure at the base of the stack, influence of bends and offsets in stacks, surcharging of the below ground system, influence of interceptor traps, influence of wind across top of stacks.

Three categories of pipework discharge systems are identified as:

(1) ventilated systems – where both the stack and the branches have ventilating pipework to minimise pressure fluctuations in the system;

(2) ventilated stack system – where relatively close grouping of appliances near the stack allows dispensation of branch vents and stack ventilation is accomplished by the use of a connected vent stack;

(3) single stack system – where lengths, sizes, gradients and connections of branches to the stack can be achieved within a set of rules and the stack itself is of sufficient diameter that all ventilating pipework can be avoided.

Table M5 taken together with Figure M1 spells out some of the rules for the design of drainage systems but for more detail see BS 5572:1994.

Table M5 Branch connections to stacks

Appliance type	Maximum length of branch pipe (m)	Minimum bore of pipe (mm)	Gradient limits Min (mm/m)	Max (mm/m)
Watercloset:				
Single – syphonic only	6	75	18	90
Single – any type	6	100	18	90
2 to 9 (max)	15	100	9	90
Urinal:				
Single bowl	3	40	18	90
2 to 5 (max) – bowl	6	50	18	90
1 to 6 (max) – stall	3	65 (Note 1)	18	90
Washbasin or bidet:				
single, either	1.7	32	see Figure M1	
or	3	40 (Note 2)	18	45
Washbasin 2 to 4	4 (Note 3)	50 (Note 2)	18	45
Sink, bath or shower	3	40	18	90
Sanitary towel macerator	3	40 (Note 1)	54	
Food waste disposal unit:				
domestic	3	50 (Note 1)	135	
industrial	3	50 (Note 1)	135	

Notes:
(1) A tubular trap must be used, not a bottle trap.
(2) Connect trap to branch pipe with a 32 mm tail 50 mm long.
(3) The branch pipe must not have bends.

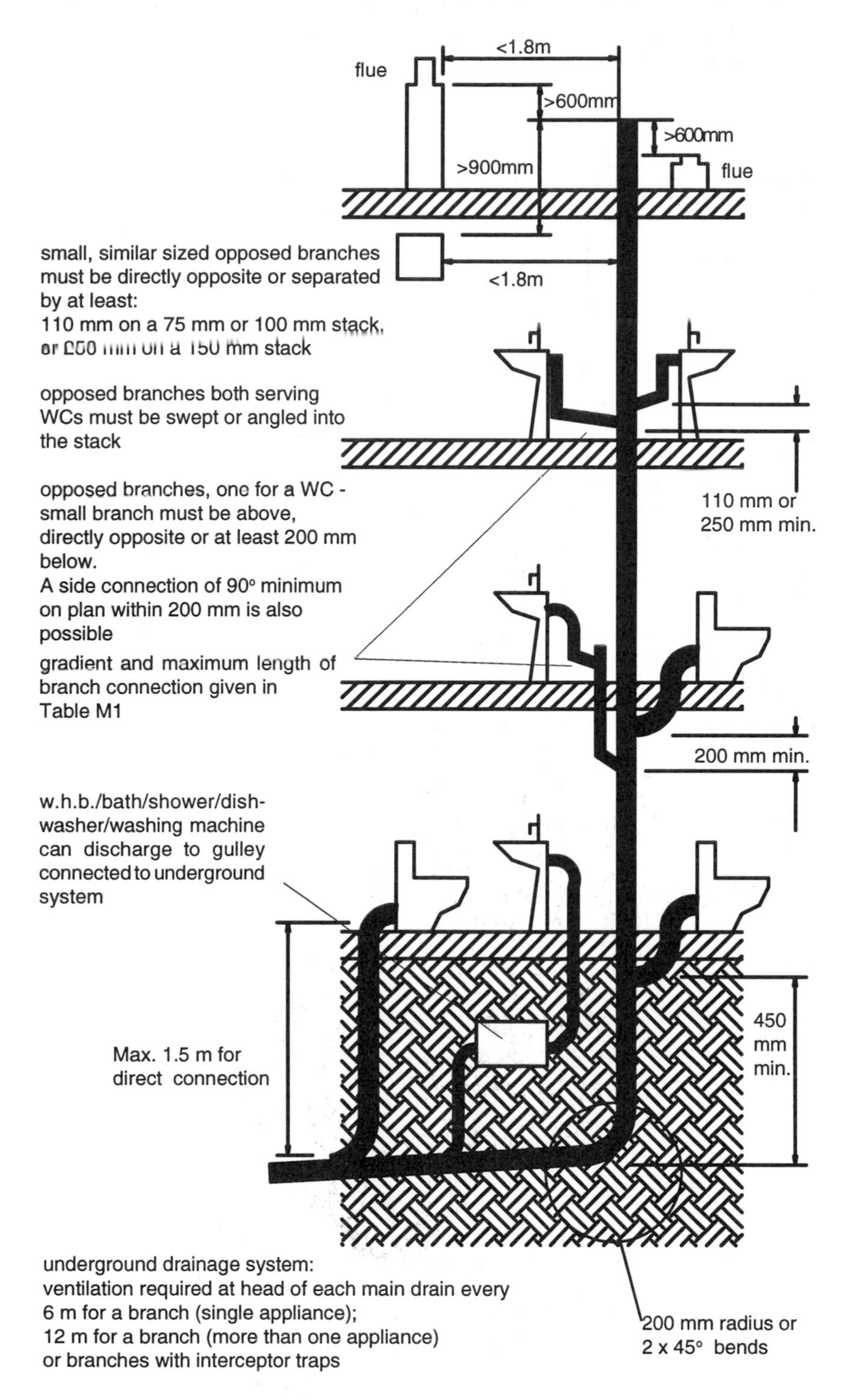

Figure M1 Drainage system requirements/branch connections to stacks.

Exclusion of foul air from the building involves the use of traps either integral with the sanitary appliance (such as with WCs) or fitted directly beneath the sanitary appliance. Table M6 provides some basic information for traps for single appliances with branch connections to the stack.

Table M6 Basic information for traps for single appliances

Appliance type	Diameter of pipe and trap (mm)	Depth of trap seal (mm)
sink	40	75
bath	40	75 (50)
WC	100 (75 min)	50
washbasin	32	75
urinal	40	75

In situations where sanitary appliances require branch lengths in excess of those given in Table M5 (e.g. washbasin branches), a possible solution might be the use of a resealing (anti-syphon) trap which, however, is noisy in operation and also requires regular maintenance.

Discharge stacks carrying foul water are connected directly to the below ground drainage system and the whole system requires to be ventilated to avoid excess pressure fluctuations in the stack. The top of the stack may or may not penetrate the roof but it must be at least 900 mm above any opening (such as a window or ventilator) which lies within 3 m of the stack. Normally there should be a wire cage covering to the stack.

Any rainwater entering the stack should be below the ventilated outlet of the stack to minimise the risk of blockage (see Figure M2).

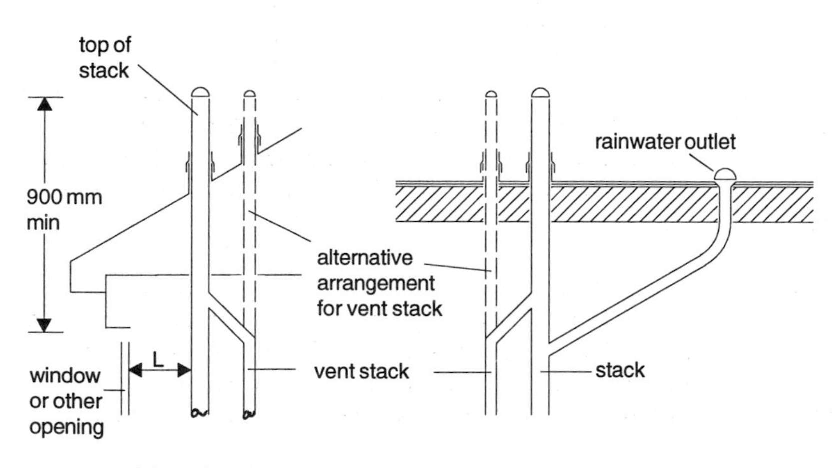

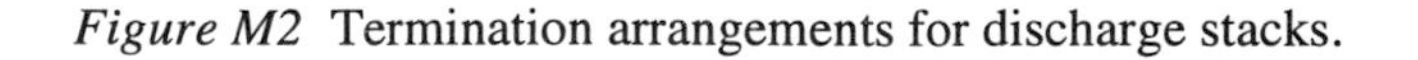
Figure M2 Termination arrangements for discharge stacks.

Stub stacks
Whilst the general rule is to ventilate all stacks to minimise any pressure fluctuations, one tightly controlled exception is the case of the stub stack which is often used to connect ground floor appliances to either a ventilated drain or to a ventilated discharge stack.

This consists of a short straight 100 mm discharge stack with its top closed and preferably fitted with an access cover. It can receive the waste discharge from one or more sanitary appliances (the number is limited – typically say a WC and a washbasin) so long as the centre line of the WC branch is not more than 1.5 m above the invert level of the ventilated drain or the connection to the ventilated discharge stack. For any other appliance connected to the stub stack the connection should not be more than 2.5 m above the invert of the ventilated drain or the connection to the ventilated discharge stack; both situations are illustrated in Figure M3.

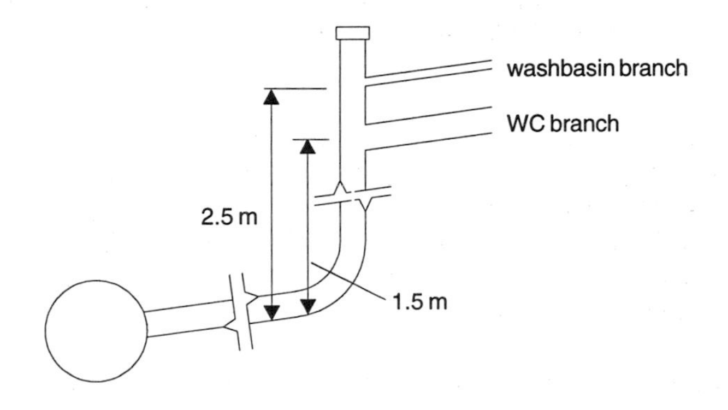

Figure M3 Rules for stub stacks.

Following these general rules for the efficient operation of the above ground system, BS 5572 then deals with the range of materials which may be used in their construction, consisting of a range of metals, plastics and for special applications such as laboratory waste discharges it may be appropriate to use borosilicate glass.

Thereafter BS 5572 provides layouts and sizing data for commonly used pipework arrangements for two discrete groups, one being bungalows and two/three storey houses (five typical arrangements given) while the other is for situations (e.g. offices, multistorey flats) where arrangements of appliances are likely to be repeated on several floors, (six typical arrangements given).

By way of example, arrangement Cb from BS 5572 which is typical of multi-storey flats is shown here as Figure M4.

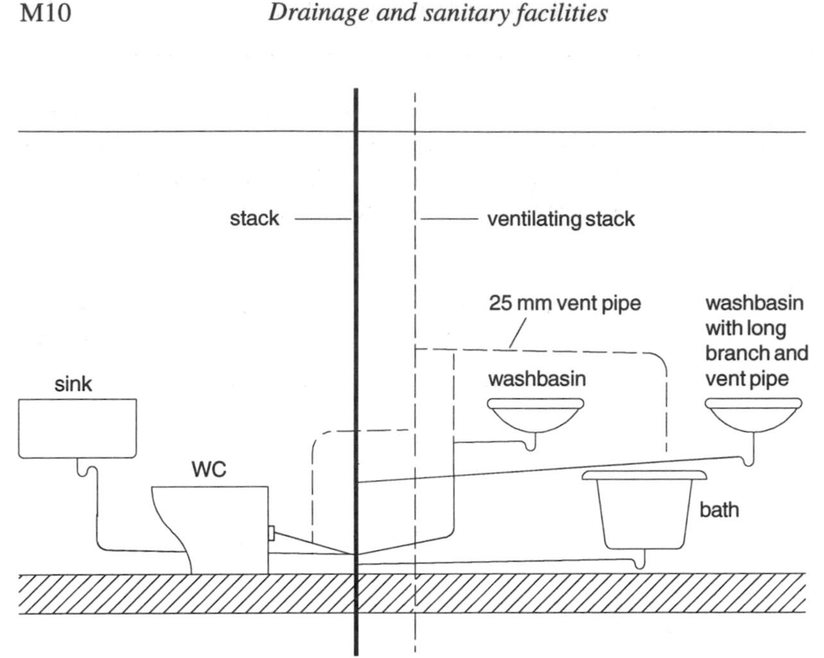

Figure M4 Ventilated pipework arrangement (multistorey flats).

The branch sizes are as given in Table M6 (e.g. bath 40 mm). The layout shows two possible arrangements for the washbasin, one close to the stack and the other at a considerable distance from the stack. It is recommended that branch vent pipes should be 25 mm diameter.

The size of the discharge stack and the ventilating stack would be influenced by the number of floors. BS 5572 includes a table to be used with arrangements such as Cb above to determine the appropriate sizes of these two stacks. The portion of that table dealing with domestic buildings is included here as Table M7.

Table M7 Discharge stack and ventilating stack sizes

Discharge stack size		**100 mm**				**150 mm**	
Usage description		**Domestic**				**Domestic**	
Number of floors		1 to 10		11 to 15		1 to 30	
Arrangements		Ca	Cb	Ca	Cb	Ca	Cb
Number of appliance groups per floor	1	0	32	50	50	0	32
	2	0	32	50	50	0	32

The arrangement in Figure M4 would be classed as one appliance group. Thus for a 10 storey building using a 100 mm discharge stack with one group per floor to the stack, a 32 mm vent stack is recommended. Even with two groups connected per floor a 32 mm vent stack would suffice.

Example M2

A designer of a 15 storey domestic building is considering the implications of using a 100 mm as against a 150 mm discharge stack with two groups per floor discharging to the stack. His proposed layout is Cb (as shown in Figure M4). What is the implication as regards size of ventilating stack?

Compliance check

From Table M7 a 100 mm stack/15 storey/2 group would require a 50 mm vent stack whereas a 150 mm stack/15 storey/2 group would require a 32 mm vent stack.

It should be noted that underlying assumptions in sizing the discharge and ventilating stacks by this method include:

(a) no offsets in wet portion of discharge stack;
(b) WC cisterns have a capacity of 9 litres;
(c) drains serving stack are not surcharged;
(d) no intercepting trap on drains serving stack.

Alternative stack sizing method (discharge units)

As an alternative to the foregoing method, discharge stacks (and in some cases branch pipes) may bc sized using discharge units. These are numerical values assigned to sanitary appliances to express their flow load producing properties. By fixing a flow limit of quarter capacity for discharge stacks (larger flows will cause plugs to form with resulting uncontrollable pressure fluctuations) and half capacity for branch discharge pipes, the maximum number of discharge units permissible for a given pipe or stack diameter can be stated. Such information is given in BS 5572 and truncated versions are presented here in Tables M8, M9 and M10.

Table M8 Discharge units for sanitary appliances

Type of appliance	Discharge units (domestic)
WC (9 litre)	7
Washbasin	1
Sink	7
Bath	7
Washing machine	3

Table M9 Discharge units for vertical stacks

Discharge stack size (mm)	Allowable number of discharge units
90	350
100	750
125	2500
150	5500

Table M10 Discharge units for branch discharge pipes

Branch pipe size (mm)	Allowable number of discharge units		
	Gradient		
	9 mm/m	22 mm/m	45 mm/m
32		1	1
40		2	8
50		10	26
65		35	95
100	230	430	1050
150	2000	3500	7500 (Note)

Note: the maximum DUs from a 150 mm discharge stack is 5500

Example M3

In a 30 storey office block there are sanitary appliances on alternate floors. These consist of ranges of eight washbasins on a single branch and eight WCs on a single branch connected to the stack. All branches are set at a gradient of 22 mm/m and the branch diameters are 50 mm (washbasin) and 100 mm (WC). The stack diameter is 150 mm. Check these three sizes for compliance.

Compliance check

(1) Washbasin branches	Discharge units = 8 x 3 = 24 DUs	
	Requirement is 65 mm (can support 35 DUs)	
	Proposal is 50 mm diameter	Fails to comply
(2) WC branches	Discharge units = 8 x 14 = 112 DUs	
	Requirement is 90 mm (can support 230 DUs)	
	Proposal is 100 mm diameter	Complies
(3) Discharge stack	DUs = 15 x (24 + 112) = 2040 DUs	
	Requirement is 125 mm (can support 2500 DUs)	
	Proposal is 150 mm diameter	Complies

Below ground drainage (BS 8301: 1985)

M2.4 DTS

BS 8301 includes a section covering the range of materials and components used, a section on design dealing among other things with connections to sanitary pipework, access for cleaning and removing blockages, sizing the below ground pipework and structural matters. Sections on work on site and on inspection testing and maintenance are also included.

The distinction between a drain and a sewer is a matter of ownership. If the below ground pipe belongs to one owner it is a drain, whereas joint ownership makes it a sewer. This is an important distinction when it comes to a question of the cost of maintenance or of removing blockages.

In many modern systems rainwater is carried in a separate drain or sewer from the foul/waste from the building, but combined systems carrying all the foul/waste/rainwater and surface water are still common. In some cases the system is partially separate with the rainwater from the building being collected in the same drain or sewer as the foul/waste while surface water from the curtilage is collected in a separate surface water sewer or drain.

The drainage system should be designed, installed and maintained so as to convey and discharge its contents without causing nuisance or danger to health, arising from leakage, blockage or surcharge.

Drainage layout

This should be as simple as possible with pipes laid in straight lines and with even gradients avoiding any unnecessary bends. Access should be provided at all bends and junctions and junctions should be oblique and in the direction of flow.

Foul drainage

The discharge is normally to a public sewer or public treatment works. The alternative acceptable is to discharge to a private sewage treatment works located and constructed in accordance with BS 6297:1983.

BS 8301 also indicates that where treatment is not practicable (e.g. single developments in remote areas), provision for storage in a cesspool which is located and constructed in accordance with BS 6297:1983 is a possible alternative provided that satisfactory arrangements can be made for emptying of the cesspool.

Gradients and pipe sizes

The key issue is to ensure self-cleansing velocities in the pipe and this is influenced by flow rate, diameter, gradient and workmanship.

Flow in foul drains is intermittent. This is especially so for drains from individual dwellings and less so for drains from blocks of flats or from high rise developments. Thus when establishing size and gradient it may be on the basis of the typical flow from one dwelling at one end of the spectrum and on simultaneous discharge (probable usage criteria) at the other end of the spectrum.

For instance for a drain from a single dwelling the design would be based on the discharge from a WC which would be 2.3 litres/s as it leaves the WC, but then it would flow as a wave in the below ground system (non-steady flow) at something less than this rate.

The discharge from multi-storey flats or from factory toilets at peak times would be assessed using a probability approach. Two approaches are possible, one involving probability data, the other using discharge units. Consider the first of these approaches.

Example M4

A factory toilet has 16 WCs connected to a below ground drain. Estimate the simultaneous discharge to the drain given the following data:

9 litre WCs,	flow rate	2.3 litres/s
	duration of discharge, t	5.0 seconds
	time between discharges, T	300.0 seconds

Solution:

Probability of discharge, p = t/T = 5/300 = 0.017

BS 8301 has a probability graph which is entered at 16 appliances read against a probability of 0.017 and yields the probable number of appliances in simultaneous use as follows:

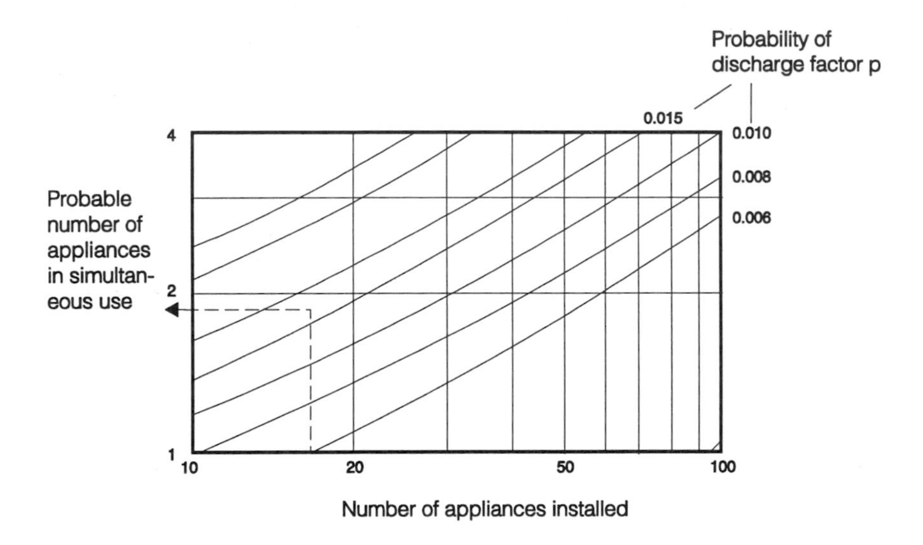

i.e. the likelihood is that two appliances are discharging simultaneously and discharge to drain = 2 x 2.3 = 4.6 litres/s.

A word of caution – this approach cannot be used for mixed appliances such as WCs and washbasins connected to the same drain since it would lead to oversizing.

The second approach is that same concept of discharge units mentioned earlier in relation to sizing the above ground pipework (see Table M8) and those same values for discharge units can be applied to the sizing of the below ground drainage system. The following example illustrates the approach.

Example M5

For a multistorey block of flats 40 flats are connected to a common below ground drain. BS 8301 advises that for domestic installations the peak flow occurs in the morning from the use of WCs, washbasins and sinks and that 14 discharge units should be allotted per flat on the basis of these appliances.

Estimate the flow rate in the drain on this basis.

Solution

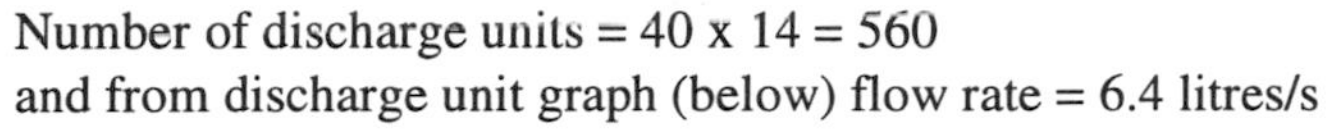

Number of discharge units = 40 x 14 = 560
and from discharge unit graph (below) flow rate = 6.4 litres/s

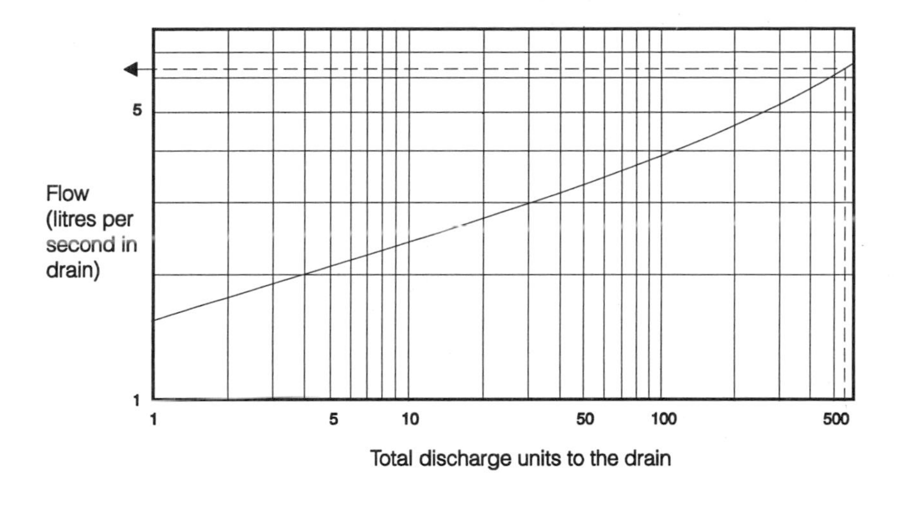

Thus by these various means the design flow rate in the pipework can be established.

Turning next to the question of suitable gradients BS 8301 gives the following advice:

(a) flows less than 1 litre/s in pipes not exceeding 100 mm bore (e.g. individual dwelling) – gradients not flatter than 1/40;
(b) flows more than 1 litre/s and with at least one WC connected a 100 mm bore pipe may be laid at a gradient not flatter than 1/80;
(c) with 150 mm bore pipes where at least five WCs are connected the pipe may be laid at a gradient not flatter than 1/150.

Where the available fall is less than that necessary to achieve these recommended gradients, avoid a solution which involves increasing the diameter of the pipe (particularly at low flows) since this will lead to reduced flow velocity in the pipe and increased likelihood for deposits to accumulate in the pipe.

Figure M5 shows a pipe sizing chart from BS 8301 which relates flow rate (litres/s), gradient, (and if necessary, discharge units) and pipe size for conditions of full bore flow, three quarters depth of flow and two thirds depth of flow.

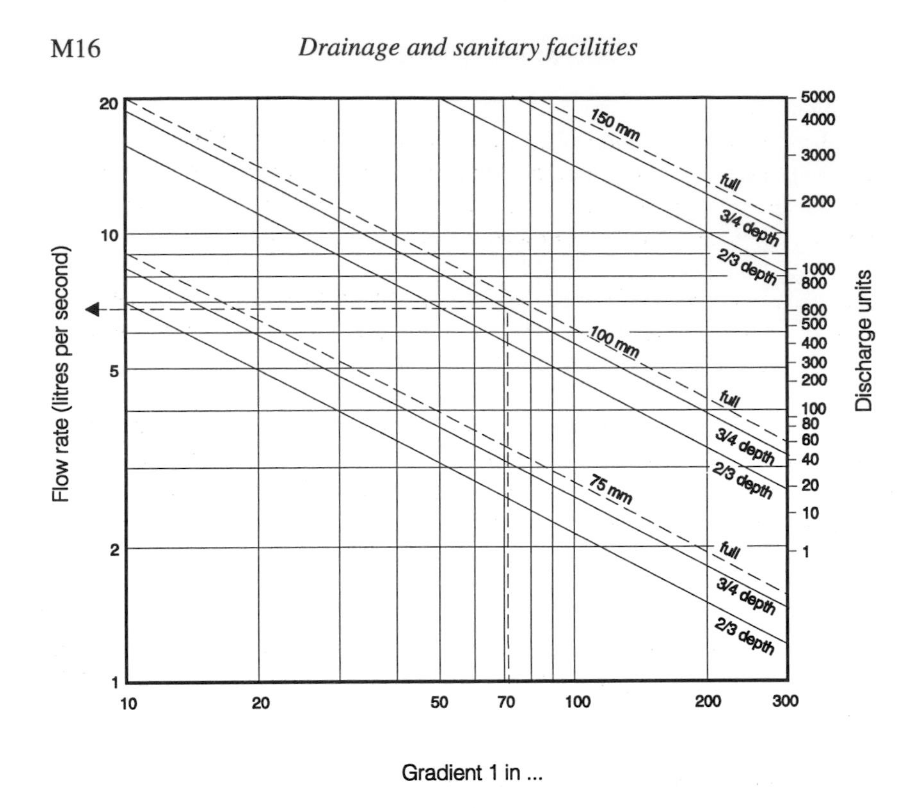

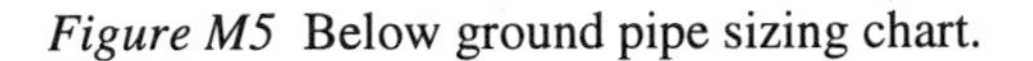
Figure M5 Below ground pipe sizing chart.

For foul drains three quarters depth of flow should be chosen under peak flow conditions to ensure adequate air movement within the below ground system. A further example will illustrate the use of this chart.

Example M6

For the multistorey flat of the previous example it is possible to have a gradient of 1/70 for the drain. From the chart obtain an appropriate size for the drain and check that the proposed gradient is suitable.

Solution

From the chart, for three quarters depth of flow, a gradient of 1/70 for a 100 mm diameter pipe would carry a flow rate of 6.9 litres/s. In this case the peak flow has been estimated as 6.4 litres/s (from example M5) which with a gradient of 1/70 would mean the depth of flow was a little less than three quarters depth, which is acceptable.

Moreover, the earlier advice regarding suitable gradients suggested that where the flow was more than 1 litre/s (with at least one WC connected) in a 100 mm bore pipe the gradient should not be flatter than 1/80. At 1/70 the suggested gradient is a little steeper.

Thus the proposal is suitable.

Connections

Where a connection is required to a foul drain from a sanitary appliance discharge stack, gully or branch drain this should be made, where convenient and practicable, at an inspection chamber or manhole. The length of a connection from an appliance should not exceed 6 m and from a group, 15 m.

Connection of a foul drain to a public sewer should be agreed with the local drainage authority and care must be taken to ensure that the minimum length of drain is subjected to backflow should the sewer surcharge.

Connection of foul drain to a septic tank or settlement tank requires arrangements to ensure that the entry velocity should be restricted so as to cause minimum disturbance to the quiescent conditions in the tank. In the case of cesspools provisions require to be made for rodding the drain and its connection to the cesspool. Further details from BS 6297.

Ventilation

Each main drain must be provided with a ventilated pipe at or near the head of the drain. More often than not this would be a ventilated discharge stack. This condition also applies to any branch:

(1) more than 6 m long serving a single appliance;
(2) more than 15 m long serving more than one appliance;
(3) serving a drain fitted with an interceptor trap.

Access

This is required for testing, inspection, maintenance and removal of debris. Access via inspection chambers and manholes can allow rodding in both directions whereas access via a rodding eye allows only rodding in the downstream direction.

The guiding principle is that every drain length should be accessible without the need to enter the building. Access provisions should be in place at the head of each run of drain and at changes in direction, gradient or pipe diameter. Table M11 gives recommended maximum distances between rodding eyes, access fittings, inspection chambers and manholes based on standard rodding techniques.

Note that where a branch drain joins another drain without the provision of an inspection chamber or manhole at the junction, access should be provided on the branch drain within 12 m of the junction.

Table M11 Maximum spacing of access points

Distance to:	From access fitting small (m)	From access fitting large (m)	From junction or branch (m)	From inspection chambers (m)	From manhole (m)
Start of external drain [Note]	12	12		22	45
Rodding eye or access bowl	22	22	22	45	45
Access fitting: small			12	22	22
Access fitting: large			22	45	45
Inspection chamber	22	45	22	45	45
Manhole	22	45	45	45	90

Note: Start is either foot of stack or outlet of ground floor appliance.

Surface water
With surface water drains or sewers, under peak flow conditions, it is assumed that the drains run full bore, that is, under full depth of flow. BS 8301 provides several pipe sizing charts after the style of Figure M5 but for different classes of surface roughness of the pipe. For more detail refer to BS 8301.

Structural considerations
The ability of a pipe to carry an imposed load can be increased by the provision of suitable bedding. A rigid pipe (materials such as clay, concrete, asbestos, cement or grey iron) has inherent strength but by the provision of a uniform support from the bedding higher loads may be carried. Flexible pipes (materials such as ductile iron, GRP, steel, pitch fibre and UPVC) deform under load and require uniform support.

BS 8301 provides information on different classes of bedding for both rigid and flexible pipes as well as information on suitable depths of cover for rigid pipes of different crushing strengths and beddings in relation to laying these under main roads, other roads or in fields or gardens. Tables M12 and M13 show selective information taken from BS 8301 and in the case of Table M13 the values are for rigid pipes of crushing strength 20 kN/m.

Table M12 Classes of bedding for rigid pipes

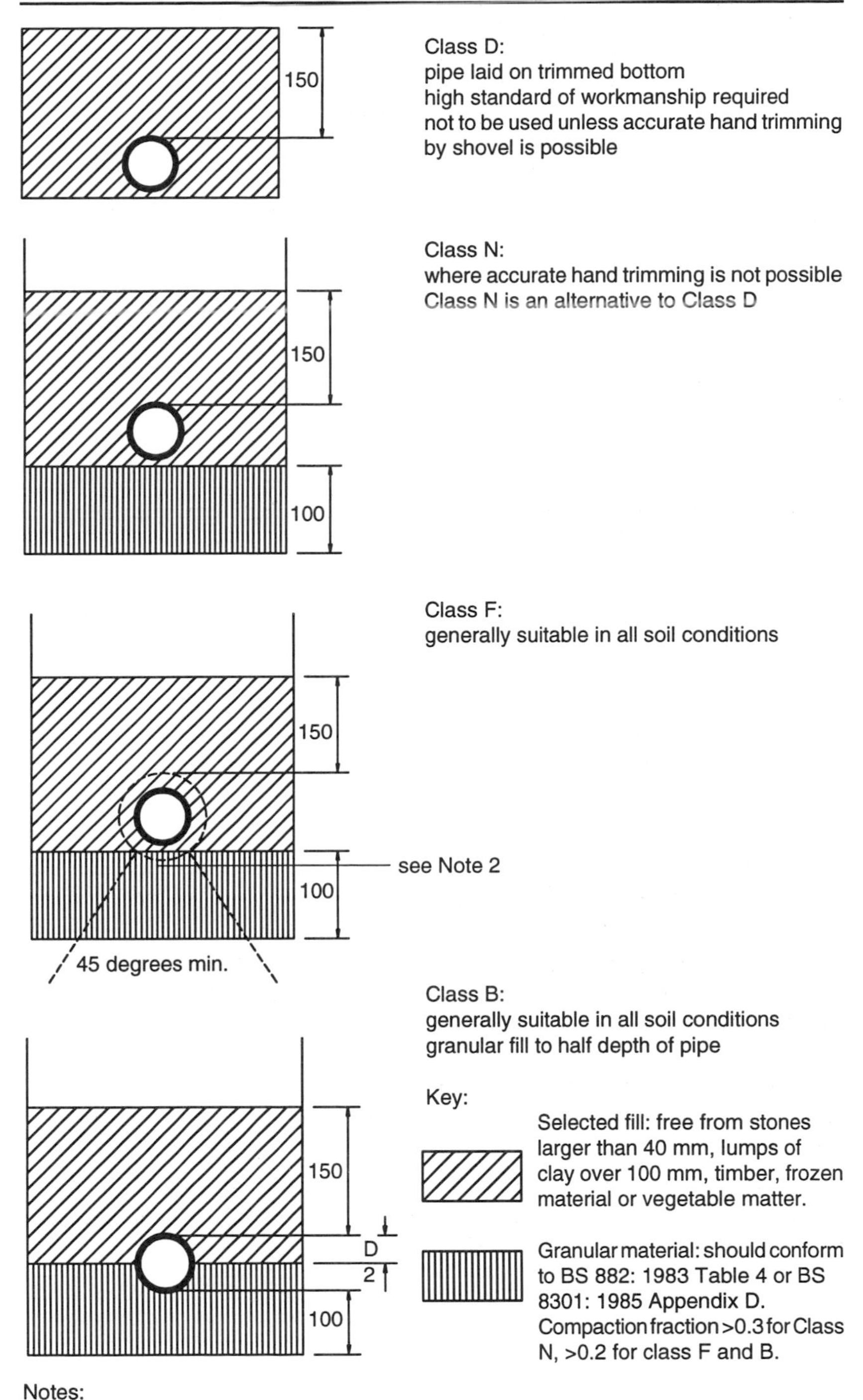

Notes:

1. Provisions may be required to prevent ground water flow in trenches with Class N, F or B type bedding.
2. Where there are sockets these should not be less than 50 mm above the floor of the trench.

Table M13 Bedding and cover for standard strength rigid pipes in any width of trench

Pipe bore (mm)	Bedding class (see Table M12)	Fields and gardens min (m)	Fields and gardens max (m)	Light traffic roads min (m)	Light traffic roads max (m)	Heavy traffic roads min (m)	Heavy traffic roads max (m)
100	D or N	0.4	4.2	0.7	4.1	0.7	3.7
	F	0.3	5.8	0.5	5.8	0.5	5.5
	B	0.3	7.4	0.4	7.4	0.4	7.2
150	F	0.6	3.9	0.7	3.8	0.7	3.3
	B	0.6	5.0	0.6	5.0	0.6	4.6

For instance, where a 150 mm concrete pipe of crushing strength 20 kN/m is laid on a granular bed (class B) partly along a road within a housing estate (light traffic road) and partly within gardens the relevant depths of cover are no less than 0.6 m and no greater than 5.0 m.

The reader is referred to BS 8301 for equivalent information associated with flexible pipes.

Drains near foundations
Special care must be taken where trenches are close to foundations. Figure M6 shows concrete filling arrangements in these circumstances.

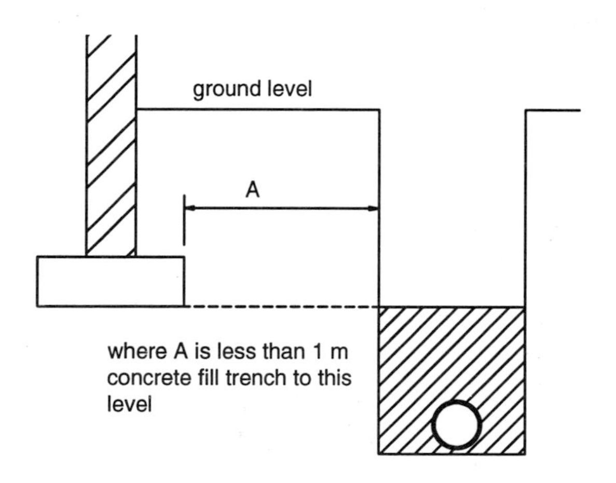

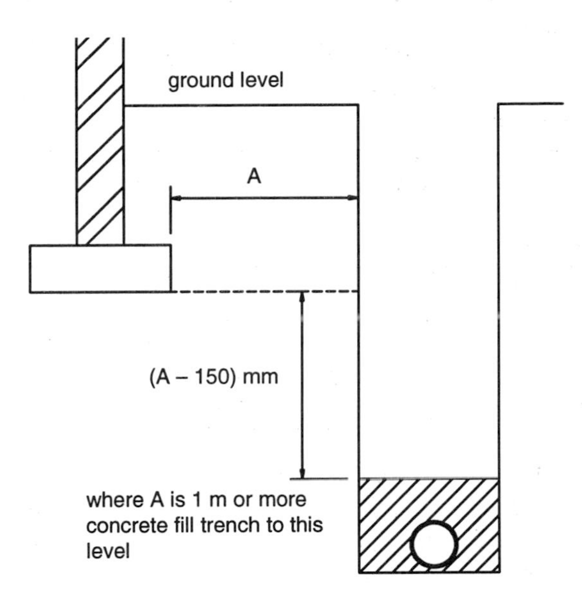

Figure M6 Concreting of drains near foundations.

Thus where A is 1300 mm, for instance, fill with concrete above the pipe to a level of 1150 mm below the foundation.

Where a drain passes through a wall or foundation it must be protected from possible movement of the wall. Figure M7 shows alternative ways of providing such protection.

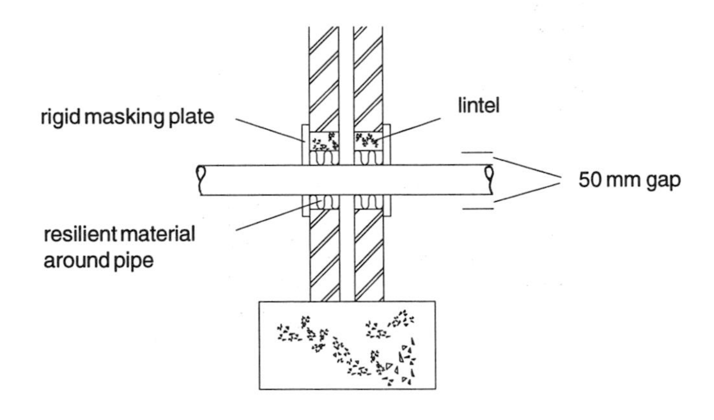

(a) pipe bridged by a lintel

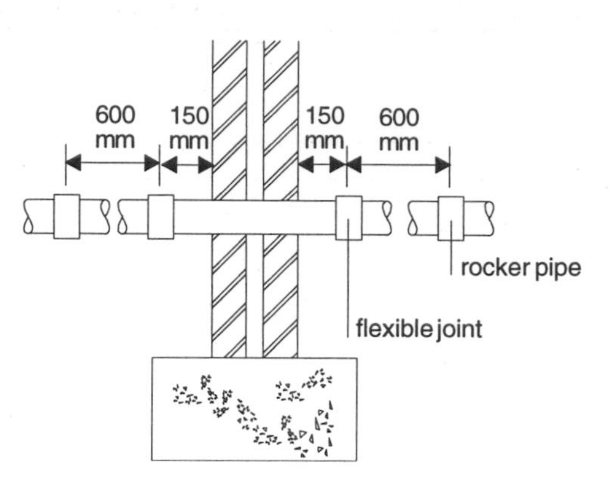

(b) pipe built into a wall

Figure M7 Drainpipe passing through a wall.

Tests

A water test may be carried out on the below ground system to check the tightness of joints. This is normally done by filling via a standpipe of the same diameter as the drain subjecting the head of the drain to 1.5 m head of water pressure and ensuring that the head is no greater than 4 m head of water pressure at the bottom of the section tested. The section to be tested should be left to stand for two hours topping up as necessary for absorption.

Over the next 30 minutes the loss of water is measured by adding top up to maintain the 1.5 m head. The leakage should not exceed 0.05 litres per metre run for a 100 mm drain (or 0.08 for a 150 mm drain).

BS 8301 provides a detailed description of the water test as well as the air test and the reader is referred there for further information.

Provision of sanitary facilities

M3.1 DTS

Any building which is not excluded by regulation 25(2), i.e. certain workplaces and school premises, must be provided with an adequate number of sanitary facilities. The requirements will be met by a provision in accordance with tables 1 to 3 (for dwellings, for the elderly and for residential homes for the elderly) or 7 to 11 (for places of public entertainment, hotels, restaurants/canteens, public houses and swimming pools) of BS 6465: Part 1:1984 as appropriate.

As an example of the type of information provided in BS 6465, Tables M14 and M15 give the recommendations for dwellings and for hotels (as per tables 1 and 8 of BS 6465).

Table M14 Minimum provision of sanitary appliances for dwellings

Type of dwelling	Appliances	Dwellings suitable for up to: 2	3	4	5	6 persons and above
On one level, e.g. bungalows and flats	WC	1*	1*	1†	1†	2‡
	Bath	1	1	1	1	1
	Washbasin	1	1	1	1	1
		and in addition one in every separate WC compartment which does not adjoin a bathroom				
	Sink and drainer	1	1	1	1	1
On two or more levels, e.g. houses or maisonettes	WC	1*	1*	1†	2‡	2‡
	Bath	1	1	1	1	1
	Washbasin	1	1	1	1	1
		and in addition one in every separate WC compartment which does not adjoin a bathroom				
	Sink and drainer	1	1	1	1	1

* May be in the bathroom
† Recommended to be in a separate compartment
‡ Of which one may be in a bathroom

There is also a general requirement, in Scotland, for water closets such that either the WC compartment must have its own washbasin or there must be a washbasin in an adjacent space which provides the sole means of access to the WC compartment. Furthermore these provisions must be separated by a door from any room or space used wholly or in part for the preparation or consumption of food.

Also in Scotland it is acceptable to provide a shower rather than a bath.

Table M15 Minimum provision of sanitary appliances for hotels

Appliances	For residential, public and staff	For public rooms: Males	Females
WCs	1 per 9 persons omitting occupants of rooms with WCs en-suite	1 per 100 persons up to 400, beyond that 1 per 250 or part thereof	2 per 100 persons up to 200, beyond that 1 per 100 or part thereof
Urinals		1 per 50 persons	
Washbasins	1 per bedroom and at least 1 per bathroom	1 per WC	1 per WC
Bathrooms	1 per 9 persons omitting occupants of rooms with baths en-suite		
Cleaners sinks	1 per 30 bedrooms, minimum 1 per floor		

Notes: 1. Some non-residential staff may be subject to regulations under the Offices, Shops and Railway Premises Act.
2. Attention is drawn to the necessity to provide facilities for the disposal of sanitary dressings.

ELECTRICAL INSTALLATIONS

PART N Regulation 26

BUILDING STANDARDS

Regulation

26 (1) Every electrical installation to which this regulation applies and every item of stationary electrical equipment connected to such an installation shall provide adequate protection against its being a source of fire or a cause of personal injury.

(2) This regulation shall not apply to an installation:

(a) serving a building or any part of a building to which the Mines and Quarries Act *1954(a) or the Factories Act 1961 applies;

(b) forming part of the works of an undertaker to which regulations for the supply and distribution of electricity made under the Electricity (Supply) Acts 1882 to 1936 or section 16 of the Energy Act**1983(b) apply; or

(c) consisting of a circuit (including a circuit for telecommunication or for transmission of sound, vision or data, or for alarm purposes) which operates at a voltage not normally exceeding 50 volts alternating current or 120 volts direct current, measured between any two conductors or between any conductor and earth, and which is not connected directly or indirectly to an electrical supply which operates at a voltage higher than those mentioned in this sub-paragraph.

(3) In paragraph (1) 'stationary electrical equipment' means electrical equipment which is fixed, or which has a mass exceeding 18 kg and is not provided with a carrying handle.

* 1954 c.70
** 1983 c.25

Aim

The purpose of this part is to ensure that electrical installations in buildings are safe with respect to the hazards such as fire, electrical shock and burns or other personal injuries which can arise from defective or inappropriate installations.

TECHNICAL STANDARDS

Scope

The standards apply to all buildings other than mine, quarry and factory buildings which have particular hazards (e.g. dampness or specialised processes) and are in any case subject to other legislation, and also statutory supply undertakers' works and extra low voltage installations which are not fed from 240 V or 415 V supplies.

The standards relate to the installation which consists of all the electrical cabling and associated wiring together with all components and fittings. This relates to permanently secured and large stationary equipment but excludes portable equipment and associated appliances.

Requirements

N2.1 These relate to the construction, installation and protection of the electrical installation.

In terms of the construction of the system, all elements (i.e. cables, conductors, components and fittings) must be sized appropriately to accommodate the likely maximum demand on the system.

Possible malfunctions of an installation can lead to excess or leakage current in the system and it is essential that the installation incorporates suitable automatic devices for protection against such possibilities.

The installation must also include switches or other means of isolation as are necessary for safe working, maintenance or convenience.

COMPLIANCE

DTS N2.1 Historically the provisions deemed to satisfy in terms of compliance have been design and installation in accordance with the relevant sections of the current edition of the *Regulations for Electrical Installations*, more commonly known as the IEE Regulations.

The 1990 edition of the Technical Standards within its 'Provisions deemed-to-satisfy' section made reference to the 15th edition of the Wiring Regulations published in 1981 by the Institution of Electrical Engineers. This has been overtaken by the 16th edition which in 1992 became a British Standard, BS 761, and the 1994 amendment to Part N establishes BS 7671:1992 as the relevant provisions deemed to satisfy the standards.

It is worth noting that certain provisions within the IEE Regulations are outside the scope of these regulations. These provisions relate to safeguard of livestock, requirements for caravans, IEE completion and inspection certificates and periodic inspection and testing.

A useful companion document to the IEE Regulations is the Handbook on the 16th Edition of the IEE Regulations published by Blackwell Science. This handbook takes situations and relates all the appropriate regulations from the IEE Regulations which apply to each situation. An example from the handbook is shown in Figure N1 which relates to rooms containing a bath tub or shower basin, reproduced by kind permission of the Electrical Contractors Association, the Electrical Contractors Association of Scotland and the National Inspection Council for Electrical Installation Contracting.

The illustration shows the towel rail to be water heated, i.e. an extraneous conductive part, thus:
547 - 03 - 02 requires the supplementary bond to be of a conductance of not less than half that of the protective conductor connected to the exposed conductive part (the infra-red heater) if the bond is mechanically protected, and 4 mm. if no protection is provided

The illustration shows the towel rail to be water heated, i.e. an extraneous conductive part, thus:
547 - 03 - 01 requires the supplementary bond, to be of a conductance of not less than half that of the protective conductor connected to the exposed conductive part if the bond is mechanically protected, and 4 mm. if no protection is provided

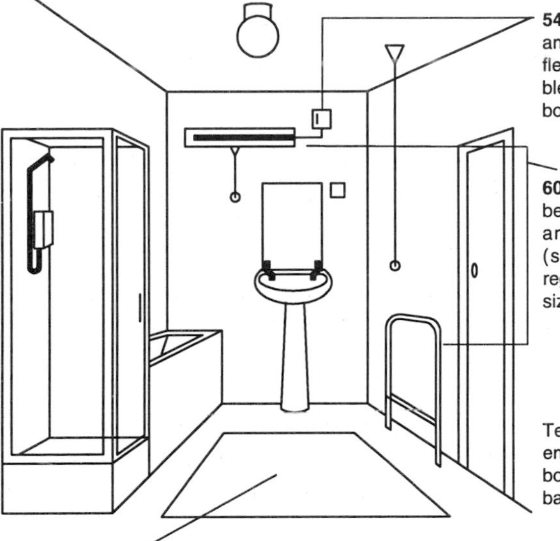

547 - 03 - 05 Where a fixed appliance is supplied by a short length of flexible cord the c.p.c. within the cable can be deemed to satisfy the bonding requirement

601 - 04 - 02 supplementary bond between exposed conductive part and extraneous conductive part (simultaneously accessible) See regulation 547-13-02 for conductor size

Tests: these must be carried out to ensure supplementary equipotential bonding is connected to the taps, bath or basin

601 - 12 - 02 Electrical heating embedded in the floor to have an earthed metallic sheath or be covered by an earthed metallic grid. Sheath or grid to be bonded to the local supplementary equipotential bonding (601-04-02)

601 - 04 - 02 Irrespective of impedance test results, this regulation requires supplementary equipotential bonding as follows:
(1) between simultaneously accessible exposed conductive parts of equipment, eg., heaters, electric towel rails, electric showers, etc.;
(2) between exposed conductive parts and simultaneously accessible extraneous conductive parts;
(3) between simultaneously accessible extraneous conductive parts.

Figure N1 Typical requirements for bonding in locations containing a bath tub or shower basin.

Note that 601 numbered references relate to relevant IEE Regulations. NB See also Part Q, Standards Q2.8,Q2.9 and Q2.10 relating to lighting points and socket outlets.

References

Technical Standards, 1991, Part N, HMSO.
Handbook on the 16th Edition of the IEE Regulations, ECA *et al.*
BS 7671:1992.

MISCELLANEOUS HAZARDS

PART P   Regulations 27 & 28

BUILDING STANDARDS

Regulation

27 (1) Subject to paragraph (2), every building shall be so constructed as to provide adequate protection for users of the building and persons in its vicinity from danger from accidents arising from:
(a) collision with projections on the exterior of the buildings and in circulation areas within the building;
(b) collision with glazing;
(c) cleaning of windows and rooflights;
(d) use of escalators or passenger conveyors;
(e) discharge of steam or hot water;
(f) the positioning of manual controls for operation of windows and rooflights;
(g) the means of access to a roof.

(2) Paragraph (1)(e) shall apply only, and paragraphs (1)(f) and (g) shall not apply to buildings of purpose group 1.

(3) In paragraph (1)(b) 'glazing' means any permanently secured sheet of glass or plastics.

28 (1) Subject to paragraph (2), every system in which heated water is stored in a building (irrespective of whether the water is heated in the system) and which does not incorporate a ventilating pipe open to the atmosphere shall be so constructed as to provide adequate protection from malfunction of the system.

(2) This regulation shall not apply to:
(a) any system in which the storage capacity for heated water does not exceed 15 litres;
(b) any parts of a system which are used solely for space heating;
(c) any system used for an industrial or commercial process.

Aim

The purpose of this Part is to minimise the risk of accidents within and around buildings stemming from designed features including mechanical equipment, and to include safety aspects which could not be easily placed within other parts of the Technical Standards.

Note

Structural aspects will next be considered with respect to the Technical Standards and compliance therewith. Thereafter the discharge of steam or hot water will be considered with respect to the Technical Standards and compliance therewith.

TECHNICAL STANDARDS (structure)

Scope

The standards apply to all buildings with a few exceptions. As has been indicated in Regulation 27(2), the safety concerns and the associated controls related to the discharge of steam or hot water have their scope confined to buildings of purpose group 1. The June 1994 amendment extended Regulation 27(1)(c) to include rooflights while widening the scope of this part by addition of items 27(1)(f) and (g).

This arose from the need to harmonise with Workplace Directive 89/391/EEC implemented in the UK by the Workplace (Health, Safety and Welfare) Regulations 1992. The intention is to ensure that harmonisation occurs at the design and build stage and that the requirements of the Workplace Regulations will be met at an early stage rather than by retrospective enforcement.

Requirements

The basic requirement is to reduce the likelihood of accidents within the built environment, ranging from collision with projections or glazing to the problems of safe cleaning of windows; accidental injury from steam and hot exhausts; reduction of explosions from unvented hot water storage systems; and emergency stopping arrangements for passenger conveyors.

Background

This section was introduced to reduce the number of accidents at home and in the work place. One of the sections with far reaching implications is the safe cleaning of windows, especially at upper floor levels in dwellings. With the introduction of double glazing an aspect of safety has been overlooked, namely breaking the window to escape in the event of fire. If the window has been designed to be cleaned from inside then opening the window also allows escape. Fixed windows with hoppers prevent such an action, therefore care must be exercised in the selection of window design.

COMPLIANCE

Danger from accidents

Compliance with this section is based on anticipating and designing out a dangerous situation. It could be argued that this is open to interpretation and dependent on personal knowledge of safety or having detailed records of incidents. Compliance depends to a large extent on the experience of the designer.

With regard to collision with projections one of the problems is in determining what constitutes a projection. The Technical Standard stipulates a projection could be on a building or part of a building. It further states that there should be no projections in circulation areas within a building (see Figure P1). **P2.1**

To clarify the matter the following examples illustrate projections:

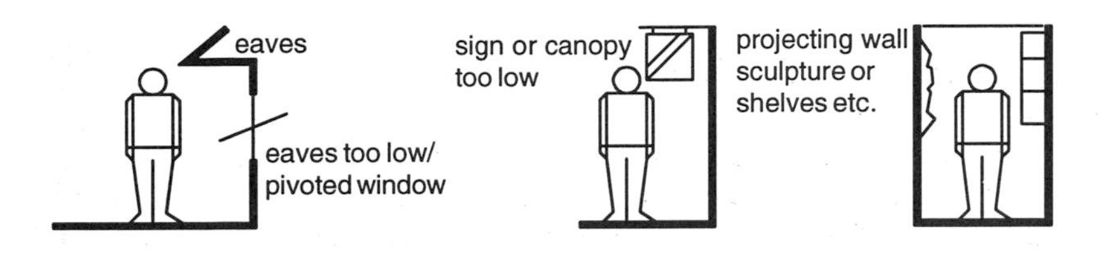

Figure P1 Obstructive projection examples.

Any door which swings both ways across a route of passage must either be glazed or have a vision panel. Other features which require safety from collision consideration in design could be windows or doors which open onto footpaths (externally) or circulation spaces (internally).

Another aspect covered under the heading of collision deals with glazing, the main requirement being that all glass used in areas where collision may be likely, for example patio doors, shop doors, balustrades etc. should be in accordance with BS 6262:1982, section 4.7.

P2.2

Requirements relating to the safe cleaning of windows are limited to dwellings. It states that any part of a window more than 4 m above adjacent ground must be easily cleaned from inside the building; for example, pivot or tilt and turn windows which should be in accordance with clauses 10.2, 10.3 and 10.4 of BS 8213: Part 1: 1991, effectively ruling out fixed windows with top hoppers; alternatively they must be easily cleaned from a load bearing surface such as a balcony, catwalk or flat roof if safe access can be gained. Such areas must have a protective barrier not less than 1.1 m in height in areas where there may be a danger of falling. The maximum height reached from such an area must not exceed 4 m (Figure P2).

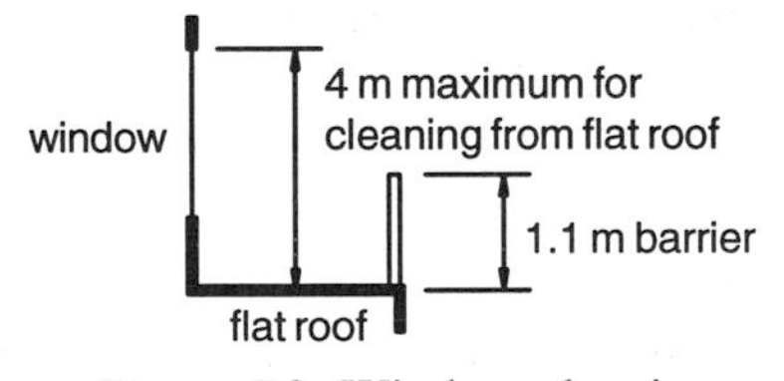

Figure P2 Window cleaning.

A further variation is where a window access system designed in accordance with clause 9 of BS 8213: Part 1: 1991 is used; this would be acceptable.

In effect by relaxing the requirement for cleaning windows from inside in certain circumstances it allows fixed windows to be installed. This does not exempt the requirements for ventilation. One exception is allowed however, that of flats and maisonettes. These are subject to P2.4b.

P2.3

The requirement for cleaning is further extended to include purpose groups 2 to 7 plus flats and maisonettes as mentioned above. The requirement stipulates that cleaning of windows should comply with P2.3 or a portable ladder can be used up to a height of 9 m above adjacent ground level or from a flat roof, balcony etc, which can take the load of the use of such a ladder. **P2.4**

Emergency stopping of escalators and passenger conveyors
Stipulates that devices to stop such equipment must be sited in accordance with BS 5656:1983 clauses 12.4 and 14.2 and be such that on activation the equipment is brought to a halt without unbalancing passengers on the escalator. **P2.5**

Roof access
Requirements here will be met by stairs, ladders and walkways complying with BS 5395: Part 3: 1985 and the requirements for the visible warning (identifying any part of the roof not capable of bearing a concentrated load of 0.9 kN per 130 mm^2) will be met by a sign A2.3 of BS 5378: Part 3: 1982. **P2.8**

The introduction of regulation 27(f) provides the local authorities with powers to prevent installation of window controls which they consider could be a danger to reach and/or operate.

TECHNICAL STANDARDS (discharge of steam or hot water)

Unvented water storage systems
These are systems which have a cold water supply fed to them directly from the rising main and therefore eliminate the need for a header tank. They find application in the domestic, commercial and industrial markets. The system operating pressure is applied via the cold water main and control of the system pressure is by means of a pressure regulating valve connected between the supply and the storage system. There must also be an isolating stop valve and a double non-return valve (to prevent back-flow). The normal expansion of the water in the system on heating, is accommodated by the expansion vessel.

Since these systems are unvented there is a need to ensure that water temperatures do not exceed 100°C. Such systems have safety devices (in addition to the thermostat) which operate sequentially, these being a non self-resetting thermal cut-out and a temperature relief valve.

The purpose of the cut-out is to shut off the energy supply to the primary heater in the event of failure of the thermostatic device. The purpose of the temperature relief valve is to spill water from the system in the unlikely event of failure of both thermostatic and thermal cut out devices. Many systems have a combined pressure/temperature relief valve which will spill water from the system in the event of excessive pressure, thereby protecting the casing of the storage vessel. In situations where there is water spillage from the system there is a need to remove the water safely. A tundish is used to collect such spillage prior to removal from the system and also to provide a visible indication that the system is operating on an emergency mode. A typical layout for a direct unvented system is shown in Figure P3.

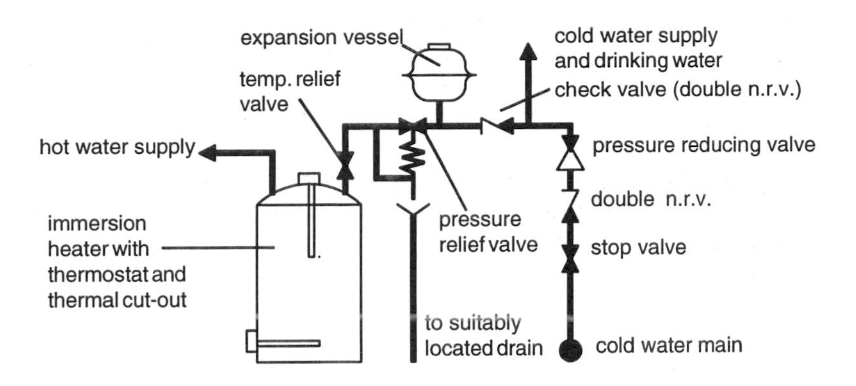

Figure P3 Direct unvented hot water storage system.

COMPLIANCE

Within the provisions which are deemed to satisfy the standards, two classes of unvented system are dealt with as follows:

P3.1 and DTS P3.1

(1) Hot water storage systems up to 500 litres/up to 45 kW power input
Such systems must be in the form of a proprietary unit or package which carries a BBA Certificate or is kitemarked to BS 7206: 1990.

A unit means a vessel for heating water or storing hot water with expansion vessel, thermostat and manually reset cut-out and incorporating within pipework, valves to stop primary flow (stop valve), prevent backflow (double non-return valve), regulate working pressure (pressure reducing valve), relieve excess pressure (pressure relief valve).

A package means a vessel for heating water or storing hot water with thermostat and manually reset cut-out together with a compatible kit of fittings as described above, which are then fitted to the system by an approved installer.

In addition to thermostatic control of heat production there is a requirement for a non self-resetting thermal cut-out and temperature relief valve as mentioned in the background notes.

In indirectly heated units/packages, the non-self resetting cut-out should be wired to a motorised valve (or some other device specified in the BBA Certificate) to shut off the flow to the primary heater. If the source of heat for the system is a boiler, the thermal cut-out may be fitted to the boiler. A possible arrangement is shown in Figure P4 where the non-resetting cut-out can interrupt the electrical supply to the thermostatic device.

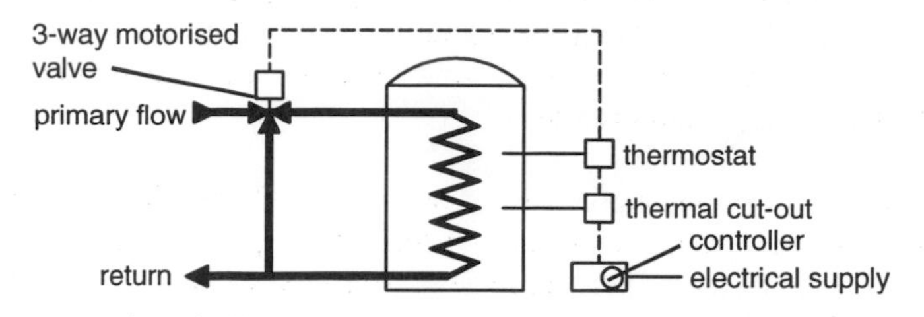

Figure P4 Location of non-self resetting cut-out.

When the thermostatic device is functioning normally it regulates the position of the ports of the three-way valve, which in turn regulates the primary water flow to the primary heating element within the storage heater, thus controlling the storage temperature of the water. When the thermal cut-out trips, the thermostatic device is de-energised and the motorised valve runs to the bypass mode thereby cutting off all primary heat supply to the storage heater. In turn the boiler thermostat cuts off the energy supply (e.g. oil or gas) to the boiler. An alternative to the use of a three-way valve is a two-way valve.

Whether a direct or indirect system is used, the temperature relief valve should be located directly on the storage vessel. The valve should be sized to give a discharge rating at least equal to the power input (BS 6283 requires that each valve is marked with its discharge rating in kW). Valves should also comply with the following measures:

- they should not be disconnected except for replacement;
- they should not be relocated in any other position;
- the valve connecting boss should not be used to connect any other devices or fittings;
- the valves should discharge via a short length of pipe of size not less than the nominal outlet size of the temperature relief valve;
- the discharge should be either direct or by way of a manifold which is large enough to take the total discharge of all the pipes connected to it;
- it should then continue via an air break over a tundish located vertically as near as possible to the valve.

The discharge collected by the tundish should be transported by a suitable pipe (normally copper) to a safe point of discharge, where it can be seen, such as a gully. The contents of the discharge will be scalding water and steam and it is important to ensure that there is no risk of contact of people with this discharge. The pipe should be laid to a continuous fall and should be no longer than 9 m in length if straight (or the equivalent in hydraulic resistance). The pipe should be of adequate size, which is at least one pipe size greater than the outlet pipe from the temperature relief valve.

(2) Hot water storage systems >500 litres/or >45 kW power input

Generally these larger systems will be for projects specified to unique designs. However they should still conform to the same general safety requirements as the small systems and be the subject of design by a qualified engineer and be installed by an approved installer.

P3.1 and DTS P3.1

The safety devices should conform to BS 6700: 1987 Specification for design, installation, testing and maintenance of services supplying water for domestic use within buildings and their curtilages, (section 2, clause 7). The system also requires to have the appropriate number of temperature relief valves as per one or other of two criteria:

either as BS 6283 Part 2:1991 or BS 6283 Part 3: 1991 giving a combined discharge rating at least the equivalent of the power input rating;

or

equally suitable temperature relief valves marked with the set temperature in °C, with a discharge rating marked in kW, measured in accordance with Appendix F of BS 6283 Part 2:1991 or Appendix G of BS 6283 Part 3: 1991, at least equivalent of the power input rating.

The temperature relief valves should be factory fitted to the storage vessel, discharge pipes to transport the discharge from the tundish to the disposal point and the non self-resetting thermal cut-outs appropriate to the heat source should be installed, all in the manner described earlier when dealing with smaller systems.

FACILITIES FOR DWELLING

PART Q — Regulation 29

BUILDING STANDARDS

Regulation

29 (1) A building of purpose group 1 shall be provided with:

(a) adequate sleeping accommodation;
(b) adequate kitchen facilities;
(c) adequate windows;
(d) adequate space heating;
(e) adequate access between its storeys; and
(f) safe and convenient access from a suitable road.

(2) Every building of purpose group 1 to which it is reasonably practicable to make available a public supply of electricity shall be provided with sufficient electricity lighting points and socket outlets.

(3) This regulation shall not be subject to specification in a notice served under section 11 of the Act.

Aim

The purpose of this part is to ensure that every dwelling will be provided with basic accommodation and necessary facilities and suitable access from the road.

TECHNICAL STANDARDS

Scope

Applies to dwellings of purpose group 1. This includes private dwellings plus flats and maisonettes; it does not include flats within residential accommodation.

Requirements

To provide the basic and necessary facilities for dwellings, flats and maisonettes for a designer to utilise his talents to the full and not to restrict design.

Background

This section has seen dramatic changes from the previous regulations. It is not the intention of this standard that the basic requirements should be adhered to rigorously. If this was the case it could result in extremely small and unhealthy accommodation. A flexible approach has therefore been taken to allow designers some freedom in design. Certain requirements originally within the Housing Standards have been relocated to other Parts in the new Standards: for example: the ventilation requirements for apartments are under (K2); sanitary requirements (M3.1–M3.2); provision of drinking water (M3.3).

The June 1994 amendments produced only one significant change to Part Q, this being with reference to access to a dwelling. This removes the onerous requirement (where there are more than two dwellings) for a common stair clear width of 1200 mm within the building, the width of which will now be controlled under Part E.

The 1994 amendment requires that a dwelling shall have access from a suitable road and where a dwelling is served by a common entrance or common stair, access shall be to the common entrance door or the foot of the common stair. In the case of more than two dwellings, the 1200 mm width still applies but only to the external access arrangements.

COMPLIANCE

Application of Part Q

These regulations apply only to purpose group 1 buildings. These include houses and flats and maisonettes for single dwelling use.

Facilities

Accommodation

Q2.1 This standard refers to apartments. An apartment means a room in a dwelling but does not include a kitchen, store or utility room. A designer does not require to stipulate the activities of each apartment, for example, lounge, dining room, bedroom etc.

The only requirement is that each apartment can accommodate a bed, a wardrobe and a chest of drawers to comply with the sizes and activity areas set out in the diagram to Q2.1 within the Standards and reproduced here as Figure Q1.

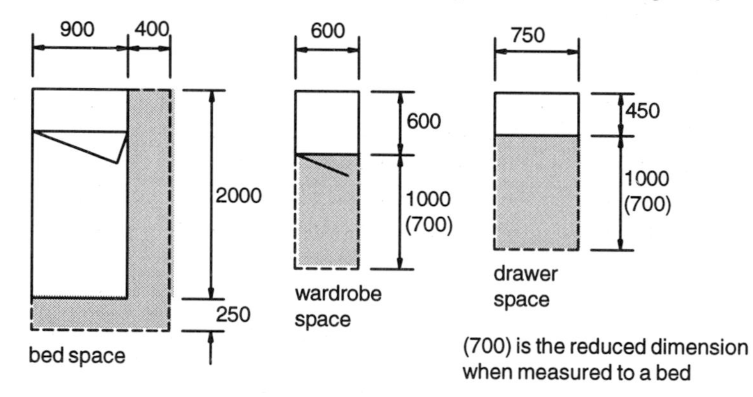

Figure Q1 Activity spaces.

The layout of an apartment would be worked out as follows:

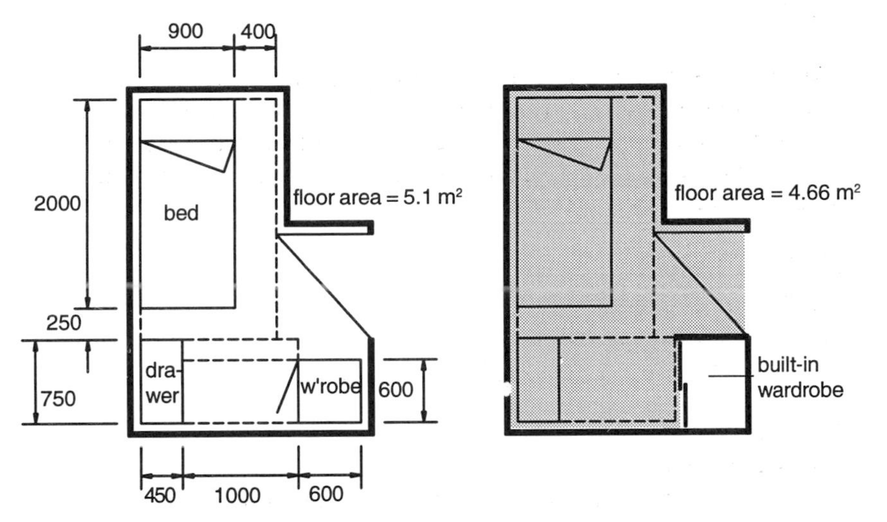

Figure Q2 Bedroom floor space requirement.

Bedrooms Q2.2

The bedroom area can be even smaller if a built-in wardrobe of equal dimension to the space requested (600 x 600 mm) is fitted, relaxing the requirement for that floor space. Refer to Figure Q2. Another stipulation is that in order to reach a bathroom, watercloset or circulation space this can only be achieved directly and not via another bedroom. Q2

Kitchens Q2.3

A dwelling must have an area designated a kitchen with either a fitted cooker which is designed for continuous burning of solid fuel (e.g. Aga cooker) or a space and the relevant cables or piping that would allow either gas, electricity or oil to be used. The relevant activity spaces for the cooker set out in the standards and reproduced here as Figure Q3, must also be provided.

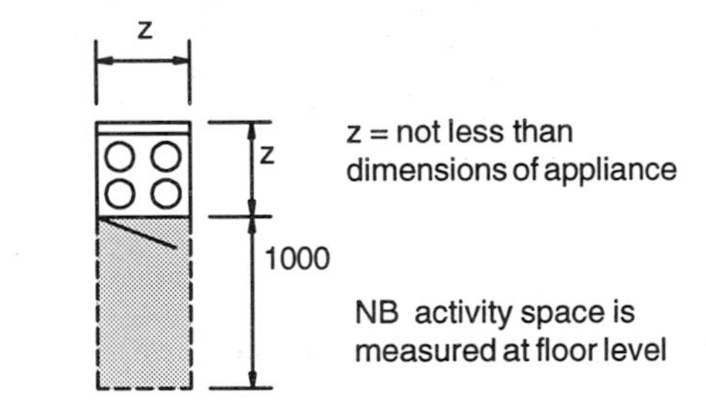

Figure Q3 Activity space for cooker.

The last requirement for the kitchen is for storage. The requirement is for a minimum of 1 m³ to be provided either within the kitchen or adjacent to the kitchen.

Q2.5 ***Windows and glazed doors***

Windows are only required for apartments. Kitchens, stores or utility rooms do not require a window, but mechanical ventilation (see Part K) in most cases will be required in kitchens. Windows (including glazed doors) are required to have an aggregate area of glass equal to at least 1/15th the floor area (for daylight provision) and be positioned in an external wall, roof or facing onto a conservatory. Dependent on size, a conservatory may be an exempted class of building.

Q2.6 ***Space heating***

Dwellings require a means of heating by either fixed heating, central heating, warm air or underfloor heating.

Whichever system is utilised it must provide a minimum of 18°C at least to one apartment when the outside temperature reaches minus 1°C, or have available not less than 3 kW of heating in at least one apartment.

Q2.7 ***Access between storeys***

Access between different levels within a dwelling must be provided in accordance with the standards set out in Part S. If the access to a different level is only for storage, then Part S does not apply; e.g., a loft ladder is acceptable for access to the loft areas, but if the loft is subsequently converted into apartments then the requirements of Part S will be necessary and a regulation stair inserted.

Lighting points

Q2.8 Dwellings are required to have at least one lighting point to the following areas: circulation space, kitchen, bathroom, watercloset compartment and any other space having a floor area greater than 2 m^2. This requirement stipulates electric supply where it is 'reasonably practicable' to have a public supply of electricity. This allows dwellings to be built in remote areas isolated from the national grid.

As a safety requirement the standard requires that for a lighting point to a stair the controlling switches be at each storey.

Socket outlets

Q2.10 The definition of 'socket outlet' is 'a fixed device containing contacts which is connected to a supply of electricity, the corresponding contacts of a plug attached to any current-using appliance'. This differs from the previous description of a power point, which had various interpretations. The new definition is quite clear; for example, within a dwelling the following minimum number of sockets are required:

(1) kitchen — 6 sockets
(2) each apartment — 4 sockets
(3) anywhere in the building — 4 additional sockets

Access to dwellings

Dwellings must have access from a suitable road to an entrance door to the dwelling as shown in Figure Q4.

Q2.11 to Q2.16

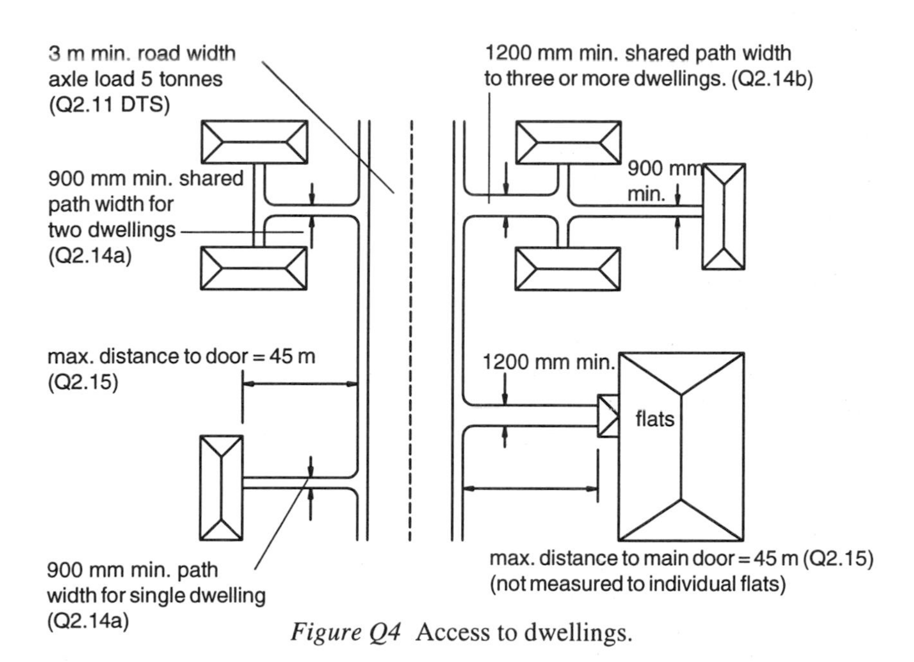

Figure Q4 Access to dwellings.

Access to dwellings must be capable of supporting pedestrian traffic, and the deemed-to-satisfy provisions give the following alternative means of satisfying Q2.16:

(1) 50 mm concrete slabs bedded on granular material; or
(2) 30 mm tarmacadam to BS 4987; Parts 1 and 2: 1993 laid on 100 mm of consolidated hardcore bottoming; or
(3) 50 mm clay or calcium silicate pavers to BS 6677: Part 1: 1986, laid in accordance with BS 6677: Part 2: 1986; or
(4) 60 mm concrete paving blocks to BS 6717: Part 1: 1993, laid in accordance with BS 6677: Part 2: 1986; or
(5) compacted and well drained gravel material.

Lifts

Figure Q5 illustrates the requirements for termination of a lift.

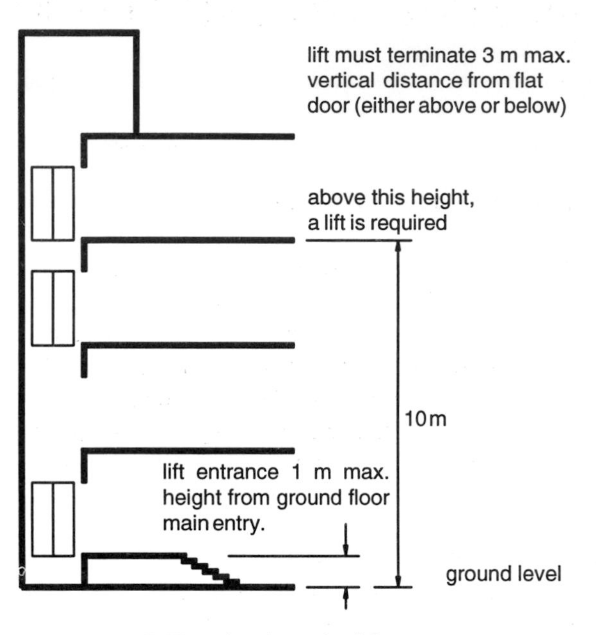

Figure Q5 Termination of a lift.

SOLID WASTE STORAGE, DUNGSTEADS AND FARM EFFLUENT TANKS

PART R **Regulations 30 & 31**

BUILDING STANDARDS

Regulations

30 (1) A building of purpose group 1 shall be provided with adequate accommodation for solid waste storage.

(2) Such accommodation shall be so constructed as to:

a) facilitate access for storage and for removal of its contents;
b) minimise risks to health and safety; and
c) prevent contamination of any water supply or watercourse.

(3) This regulation shall not be subject to specification in a notice served under section 11 of the Act.

31 A dungstead or farm effluent tank shall be so constructed, positioned and protected as to minimise risks to health and safety and prevent contamination of any water supply or watercourse.

Aim

To ensure that facilities for the storage and removal of solid waste from all buildings do not present any danger to health and safety and that dungsteads and farm effluent tanks likewise do not present any danger to health and safety.

TECHNICAL STANDARDS

Scope

These provisions apply to all buildings. In the case of factories it is helpful to discuss proposals with the Health and Safety Executive while in the case of clinical waste a useful document is DOE Waste Management Paper No. 25, Clinical Waste.

Background (solid waste)

Dwellings

An important factor here is that the distance people can be expected to carry solid waste should be no more than 30 m and assuming weekly collection, houses, flats and maisonettes up to four storey height should have access to an individual waste container (e.g. dustbin) of minimum capacity 0.12 m^3 or to a communal container of greater capacity up to 1.0 m^3. **R2.5** **R2.1** **R2.2**

For dwellings in buildings more than four storeys high, bulk containers fed from chute/hopper systems (see BS 5906: 1980 [1987]) are required. Figure R1 shows a section of a typical encased chute installation. **DTS** **R2.1** **R2.3**

R2.3
R2.7-
R2.11

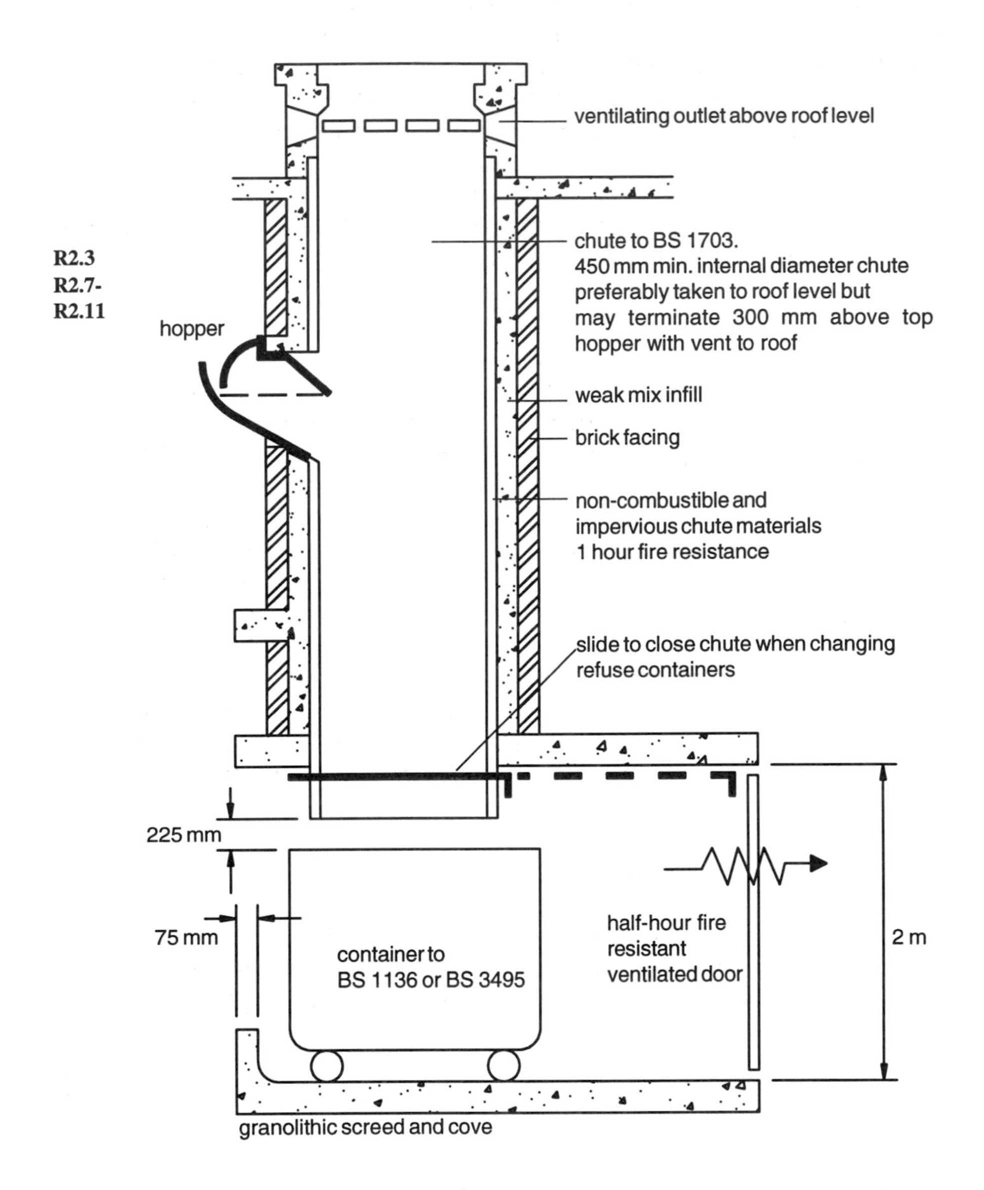

Figure R1 Refuse chute/storage facilities.

R2.7-R2.11 BS 5906 gives information on design principles for chute/hopper and storage chamber systems, with consideration of matters such as avoiding blockages within the chute, sizing, fire safety and hygiene around the hopper, within the chute and within the storage chamber. **R2.6** Thought must also be given to access for cleansing vehicles. The standard also considers other matters such as compaction, shredding and baling of refuse.

Other buildings
As indicated earlier, in addition to normal domestic refuse, other buildings such as factories, commercial, storage and hospital buildings have special problems. Consultation will be required with the local cleansing department and other relevant agencies with reference to matters such as storage capacity, storage methods, location of storage, hygiene and prevention of fire risk.

Background (dungsteads and farm effluent tanks)
Provisions relate to the construction, location, drainage, fencing and covering of dungsteads and farm effluent tanks. **R3.1-R3.4**

COMPLIANCE

Solid waste storage
Within the standards, R1 and R2 deal with the matter of solid waste storage, and within the deemed-to-satisfy provisions it is indicated that compliance can be secured by designing the provisions in accordance with BS 5906:1980 [1987].

With regard to R2.6 which deals with access provisions, paths are required to have a suitable surface for pedestrian traffic and to be at least 900 mm wide where they serve one or two dwellings and at least 1200 mm wide where they serve more than two dwellings.

R2.10b relates to walls and/or floors of private storage accommodation separating it from any other part of the building and which must have a fire resistance of at least thirty minutes. The requirements will be met where such elements are tested in accordance with BS 476: Part 8: 1972 or Parts 20 to 23:1987 as appropriate.

Dungsteads and farm effluent tanks
These must be constructed to resist the passage of moisture and there are restrictions on location. They must not be placed any closer than 15 m from any dwelling or any premises where there are food products. They must not be placed in a location where they could endanger any water supply or any watercourse. **R3.1 R3.2**

There must be adequate drainage provisions for dungsteads. **R3.3**

BS 5502: Section 1.6:1986 at clause 14 deals with measures for covering and fencing farm effluent tanks. Access can be by manhole covers or by removable sections of the cover structure. The standard also deals with suitable materials and sizes for fencing off these provisions for the protection of children and livestock. **DTS R3.4**

Compliance with R3.4 which deals with safety can be achieved by ensuring the provisions are in accordance with clause 14 of BS 5502.

STAIRS, RAMPS AND PROTECTIVE BARRIERS

PART S — Regulation 32

BUILDING STANDARDS

Regulation

32 (1) A stair or ramp which forms part of a building or which is provided to meet a requirement of these Regulations shall provide a safe means of passage for users of the building.

(2) Except where the provision of protective barriers would obstruct the use of such areas, every floor, stair, ramp, or raised accessible area which forms part of a building or which is provided to meet a requirement of these Regulations shall have a suitable protective barrier where necessary.

(3) Paragraph (1) shall not be subject to specification in a notice served under section 11 of the Act.

Aim

The purpose of this Part is to ensure that stairs, ramps and protective barriers are designed to predetermined safety criteria.

Recent amendments

The June 1994 amendments incorporated the following changes within the Technical Standards:

(1) reinstatement of some exceptions to the minimum three riser rule within S2.9;
(2) simplification of the rules for typical treads to spiral and helical stairs within S2.10, S2.11 and S2.12;
(3) removal of requirements for stepped ramps in buildings within S2.16;
(4) restrictions in relation to door openings onto top landings (S2.18) in the interests of safety;
(5) introduction of protective barriers (S3.1 and S3.4) to protect people from falling where there is a difference of level of 600 mm or more and there is a change of direction through 90 degrees;
(6) introduction of protective barriers (S3.5) to protect people from vehicles.

TECHNICAL STANDARDS

Scope

The standards apply to all stairs that link changes in level, with an exception for stepped gangways or ramps adjacent to fixed seating.

Requirements

Stairs must meet the requirements set out within the standard. One exception is granted however, to relax the requirement for protective barriers to loading bay areas.

Background

The sizes stipulated within the standard are determined ergonomically and with safety in mind. Stairs can be dangerous and care should be taken in the design of flights and the positioning of landings.

COMPLIANCE

Application

Stairs and ramps with their associated handrails and any protective barriers must comply with the standards set out in sections 2 and 3. This applies regardless of purpose group.

S1.2 Two exceptions are:

(1) a stepped gangway or ramp designed for fixed seating;

(2) industrial stairs or fixed ladders.

S1.3 Stairs designed for industrial and agricultural buildings, and this can include fixed ladders, are treated separately within S2.25 and S2.16.

The remainder of this section (S1) deals with interpretation of measurement setting out diagrammatically the concepts of tread, rise, going and nosing as indicated in Figure S1.

S1.4

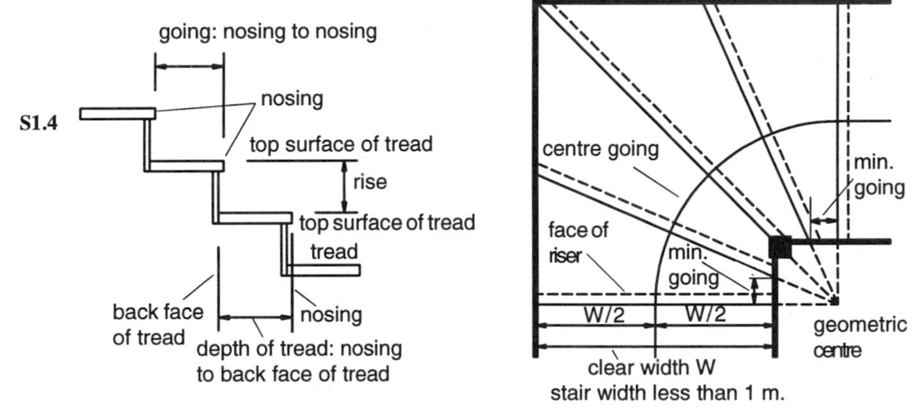

Figure S1 Stair measurement.

Stairs and ramps

General standards for all stairs

S2.2
S2.3 A straight flight must be constructed in accordance with S2.2 to S2.9. The requirements are that the flight must have uniform rises of not less than 75 mm in height with the depth of the tread not less than the going (Figure S2).

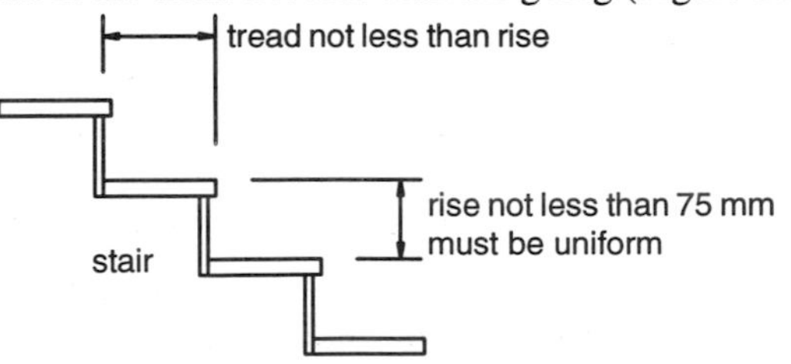

Figure S2 Stair measurements.

If the flight contains open risers the treads must overlap by not less than 16 mm. **S2.4** **S2.5**

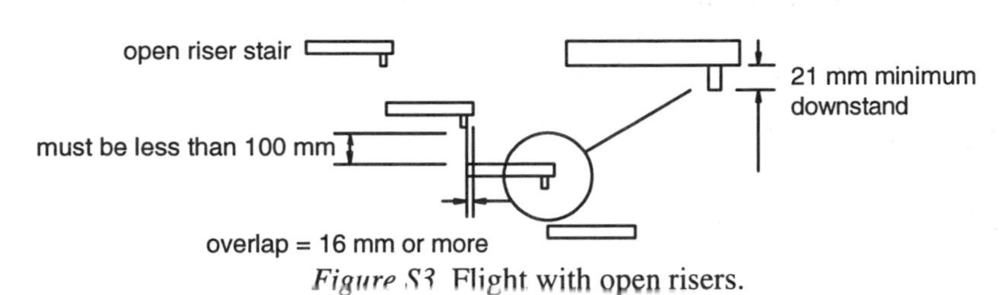

Figure S3 Flight with open risers.

In dwellings, places of assembly, recreational and residential accommodation open treads are allowed if the opening prevents the passage of a 100 mm sphere. In other words the opening must be less than 100 mm. Also the going measured along the centre line must be uniform (Figure S4).

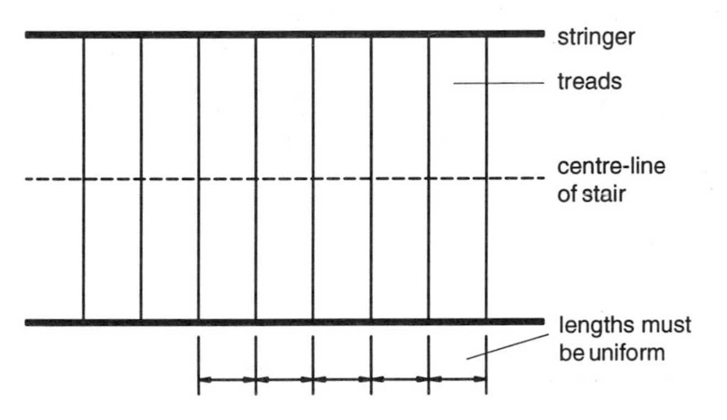

Figure S4 Going measured along centre line.

The Table S1 provides the requirements for the analysis of stairs; for example in a private stair the requirements are shown in Figure S5.

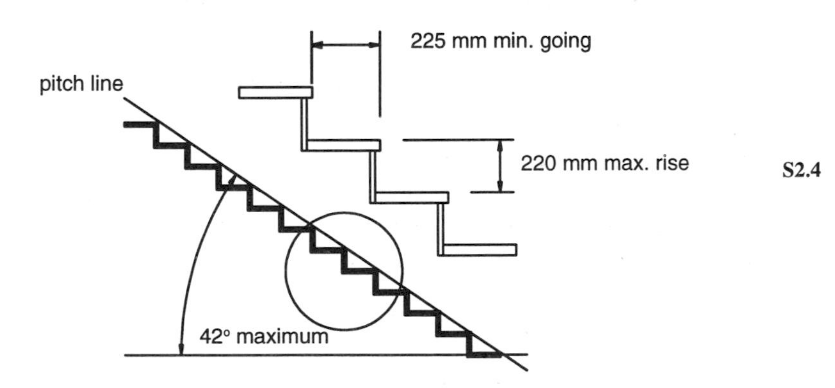

S2.4

Figure S5 Private stair requirements.

Note: minimum width excluding handrails stated in table, but stringers each may project 30 mm into the width size, for example, see Figure S6.

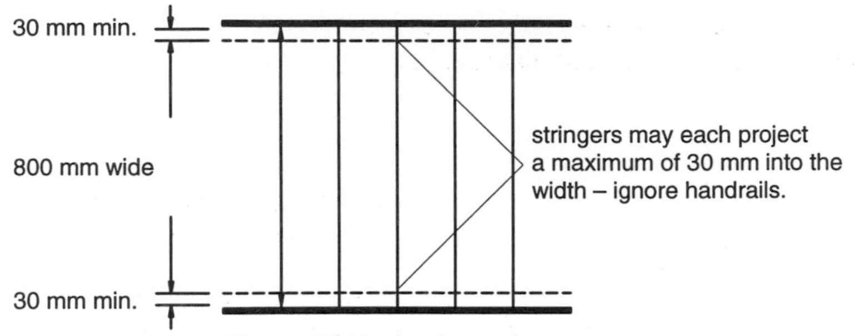

Figure S6 Projecting stringers.

Table S1 Rise, going and width of stair

Description of stair	Maximum rise (Note 1)	Minimum going (Notes 1 & 3)	Minimum width (Note 4)
1. Private stair (Note 2)	220	225	800 (Note 5)
2. Stair in building of: (a) purpose sub-group 2A used by people other than employees; or (b) purpose group 4 or 5 serving an occupancy capacity exceeding 50	180	280	1000 (Note 6)
3. Stair in building of: (a) purpose sub-group 2A used only by employees; or (b) purpose group 4 or 5 used only by employees or serving an occupancy capacity not exceeding 50; or (c) purpose sub-group 3, 6 or 7 or sub-group 2B	190	250	800 (Note 6)
4. Any other stair	190	250	900 (Note 6)

Notes:

1. The maximum rise and minimum going figures in the table do not produce a stair of a whole number of degrees pitch, but the approximate pitches are:
   stair no.2 33°
   stairs no.3 and 4 38°
2. The maximum permissible pitch of a *private stair* is 42°. The combination of rise and going must ensure that this is not exceeded.
3. In the case of *tapered treads*, subject to S2.10 to S2.12.
4. The minimum width is the width clear of handrails or other obstructions, except for stringers, each of which may project not more than 30 mm into the width or a stair lift on the straight flight of a private stair within a dwelling which may project a maximum of 350 mm into the width when in stowed or parked position.
5. 600 mm if serving only one room and/or sanitary accommodation, except a living room or kitchen.
6. In the case of an *escape stair* subject to any greater stair width required by Part E.

As a further check the going plus twice the rise must be not less than 550 mm and not more than 700 mm, for example, see Figure S7. **S2.7**

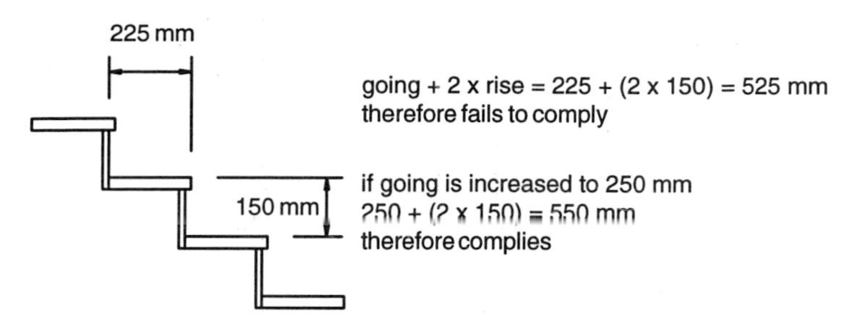

Figure S7 Further check.

A flight must have not more than 16 risers and not less than three. However, there are five exceptions to the three step rule; they are: **S2.8** **S2.9**

(1) between an external door and the ground, balcony, conservatory, porch or private garage; or
(2) in a stepped ramp; or
(3) wholly within an apartment within a dwelling; or
(4) wholly within sanitary accommodation within a dwelling; or
(5) between a landing and an adjoining level if the route of travel from the adjoining level to the next flight changes through 90°.

Rules for tapered treads and going
Flight consisting partly of straight and partly tapered treads must comply with S2.2 to S2.12 or if wholly of tapered treads with S2.2 to S2.4, S2.10 to S2.13.

For tapered treads the going must be uniform and at the inner end of the tread the going must be not less than 75 mm (Figure S8).

S2.10

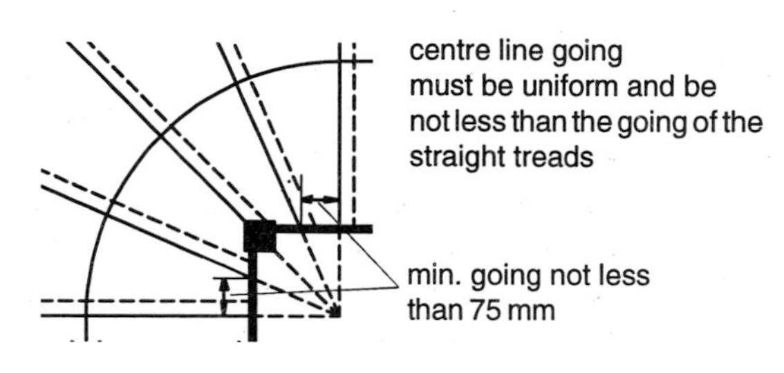

Figure S8 Tapered treads.

The centre going must conform with the following:

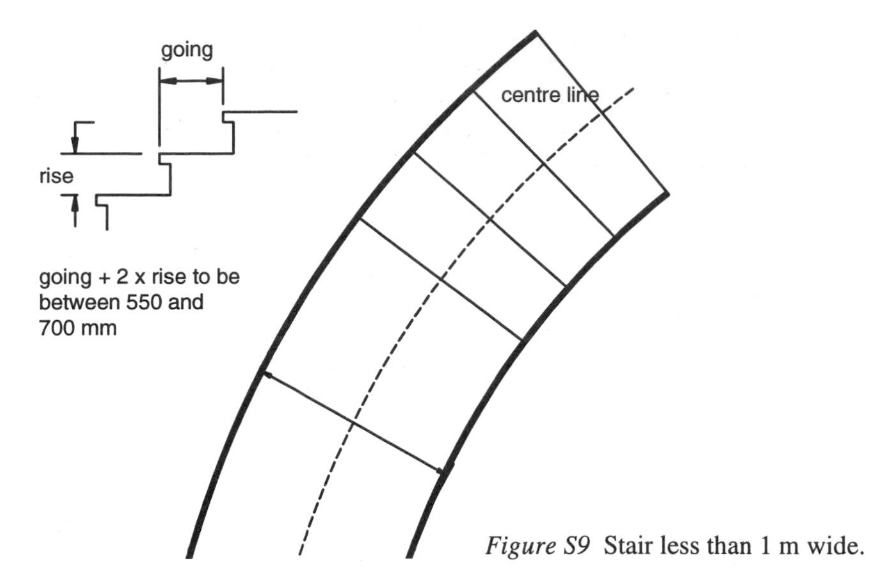

Figure S9 Stair less than 1 m wide.

S2.11 The deemed-to-satisfy provision for tapered treads is BS 585: Part 1:1989, Appendices B1 and B3. This will comply irrespective of material used, if it consists of open risers.

S2.12

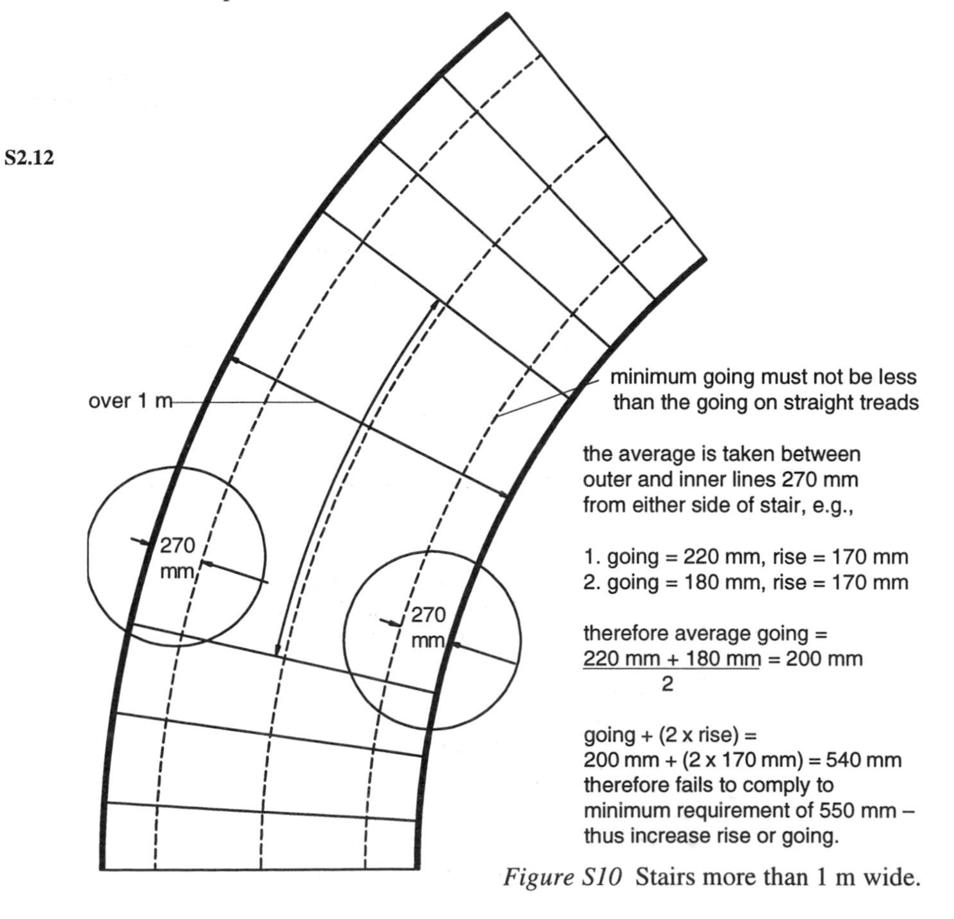

Figure S10 Stairs more than 1 m wide.

Flights consisting wholly of tapered treads

Must be designed to the deemed-to-satisfy provision which states that BS 5395: Part 2:1984 must be adhered to. It also stipulates that if such a stair forms part of an escape route then it must be designed in accordance with the table to the specification set out in the deemed-to-satisfy provision. **DTS S2.13**

Pedestrian ramps

Slopes less than 1 in 20 do not require to conform to the regulations. The gradient for other ramps must not be steeper than 1 in 12 (Figure S11). **S2.14**

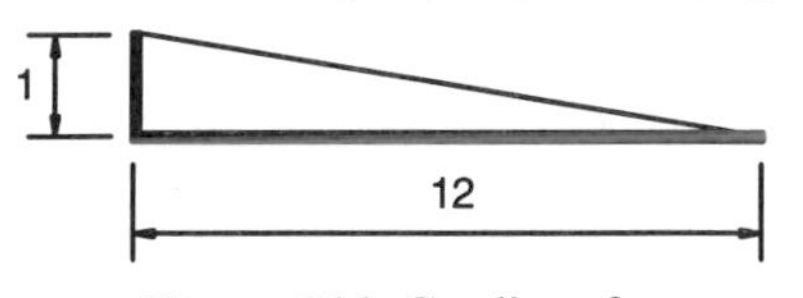

Figure S11 Gradient for ramps.

There is one exception, however, and that is to ramps giving access to a single dwelling, which must not be steeper than 1 in 8.

The length of ramps or flight must not exceed 10 m and the width of the ramp can be found in the table S1 corresponding to the purpose group. **S2.15**

Landings

Every stair or ramp requires a landing to the top and the bottom of the flight, which can be common to more than one stair. However there are two exceptions (Figure S12). **S2.17**

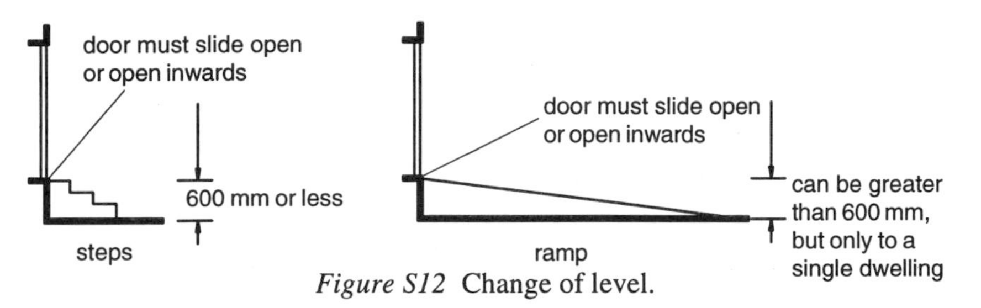

Figure S12 Change of level.

The measurement of a landing is shown in Figure S13, and must be clear of obstructions. **S2.18**

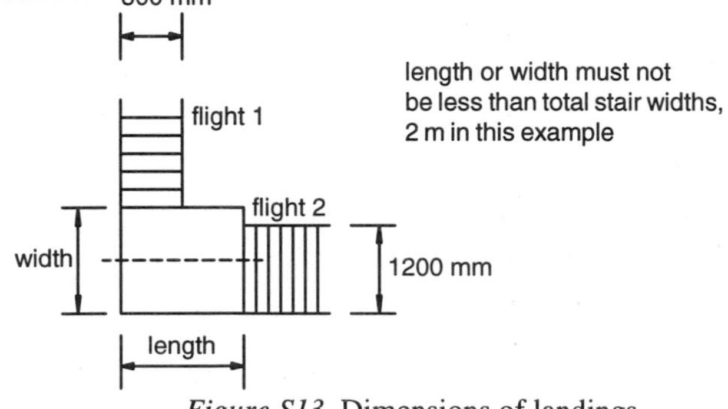

Figure S13 Dimensions of landings.

Three exceptions are allowed, as shown in Figure S14.

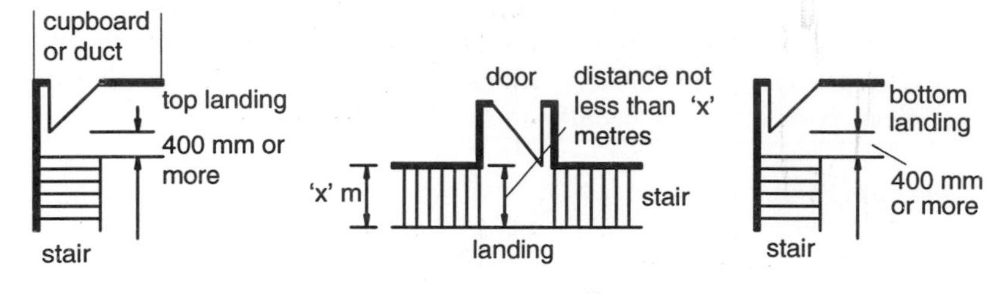

Figure S14 Exemptions.

S2.19 Landings must be level, but allowances are given for gentle slopes for drainage.

Handrails

A stair or ramp with a change of level less than 600 mm does not require handrails. Above this height the following criteria apply;

S2.20 For a stair less than 1 m wide, a handrail on one side only is required. Over 1 m in width then a handrail on both sides is required except for a flight serving a single dwelling which requires a handrail on one side.

S2.21 Stair or ramp 1.8 m or over in width, must be divided by a handrail or handrails in such a way that each section is not less than 1.1 m and not more than 1.8 m. Again an exception is allowed for stairs or a ramp to a single dwelling which does not require to be sectioned and for a stair from the entrance door and ground level which is not part of an escape route.

S2.22 The height of handrails is determined as follows:

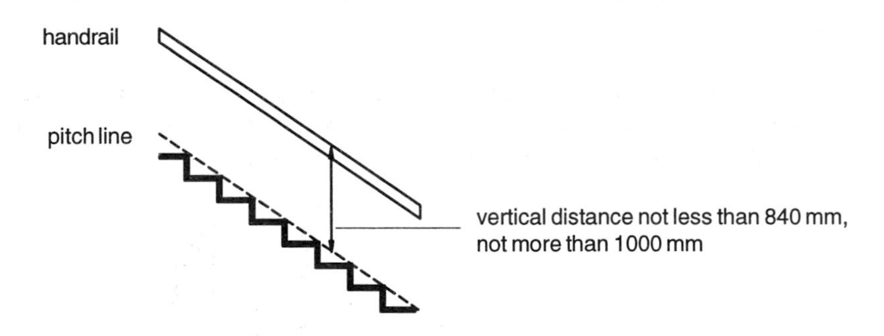

Figure S15 Height of handrails.

S2.23 Handrails should not terminate abruptly but be wreathed; if forming part of a protective barrier then this is not necessary. The exception is for handrails serving single dwellings.

Headroom

S2.24 Stairs, ramps and landings must have a clear vertical height of at least 2 m, over the entire width, for example see Figure S16.

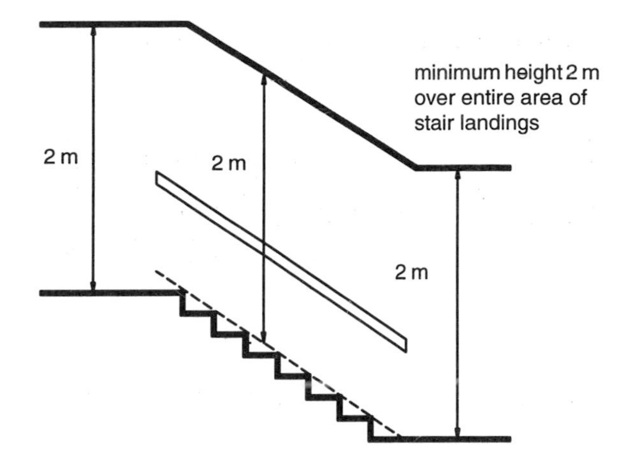

Figure S16 Clear vertical height.

Industrial stairs and fixed ladders

This refers to limited access only, for maintenance and inspection and should be in accordance with BS 5395: Part 3: 1985 or BS 4211: 1987, or by a spiral stair to BS 5395: Part 2: 1984. **DTS S2.25**

Stairs and fixed ladders in agricultural buildings

This refers to access not normally used by members of the public; these accesses must conform to BS 6180:1982. **DTS S2.26**

Protective barriers

Barriers are required at the edge of floors, stairs, ramps or other raised areas where the difference in level is 600 mm or over. An exception to this requirement is to areas such as loading bays or where a wall or partition at the edge of a floor conforms to S3.3. **S3.1**

For dwellings, assembly and recreational and residential buildings, the barriers if designed to have openings, must not allow the passage of 100 mm sphere between the openings. An exception is allowed, for example: **S3.2**

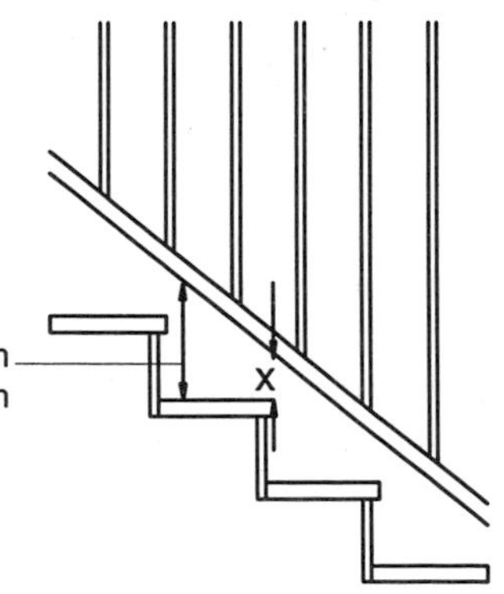

Figure S17.

S3.3 Protective barriers must be secure and the height in accordance with table S2. They must also be capable of withstanding loads in accordance with BS 6180:1982.

DTS S3.3

S3.4 The height of the protective barriers must be in accordance with the table S2 for example see Figure S18.

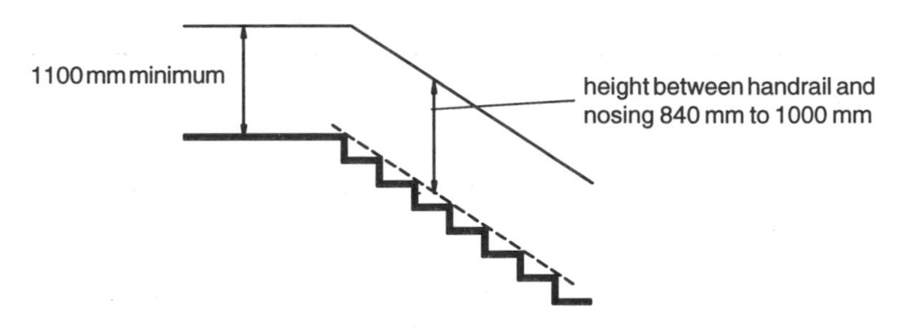

Figure S18 Height of protective barriers.

Table S2 Height for protective barriers

Location	Minimum height (mm)
At the edge of a floor, in front of walls, partitions, fixed glazing and opening windows	800
On a flight within or serving a single dwelling	840
On (a) any other flight; and (b) on a gallery or raised area within a single dwelling or a landing to a stair or ramp within or serving a single dwelling	900
In front of fixed seating	790 (Note)
Elsewhere	1100

Note: The protective barrier may be reduced to 750 mm where it has an overall width at the top of not less than 230 mm.

S3.5 S3.6 S3.7 The 1994 Amendments have introduced further measures related to protective barriers where a floor, roof or ramp is accessible to vehicles and forms part of the building. The height of such barriers must not be less than 375 mm for a floor or roof edge and 610 mm for a ramp edge and must be capable of resisting appropriate loads. Such loads to be assessed in accordance with BS 6180: 1982.

DTS S3.6

ACCESS AND FACILITIES FOR DISABLED PEOPLE

PART T Regulation 33

BUILDING STANDARDS

Regulation

33 (1) Subject to paragraph (4), a building to which this regulation applies shall be provided with adequate access and suitable aids for disabled people.

(2) In every such building in which sanitary facilities are provided, there shall be adequate provision of such facilities for disabled people.

(3) Every such building which contains fixed seating accommodation for an audience or spectators shall be provided with adequate level spaces for wheelchairs.

(4) This regulation shall apply to:

(a) surgeries with access at ground level which form part of a building of purpose group 1;

(b) building of purpose groups 2 to 6 inclusive; and

(c) car parks and parking garages of purpose sub-group 7B or 7C;

but shall not apply to a storey or part of a storey which accommodates only fixed plant or machinery and to which access is required only for maintenance purposes.

(5) In this regulation 'disabled people' means persons with a physical, hearing or sight impairment which affects their mobility or their use of buildings.

(6) This regulation shall not be subject to specification in a notice served under section 11 of the Act.

Aim

To provide accessibility for disabled people and to assist the hard of hearing and those with a sight impairment in free movement within a building and subsequent fire escape requirements from storeys accessible to disabled people.

Provision of specialist toilet facilities and wheelchair spaces in public buildings as well as providing car spaces for the disabled in car parks and garages.

TECHNICAL STANDARDS

Scope

Applies to a surgery with access at ground level which forms part of a building of purpose group 1 and to buildings of purpose groups 2, 3, 4, 5 and 6, and also to a car park or garage of purpose sub-group 7B or 7C.

Requirements

The standard has been written as functional requirements which allow the detailed aspects of design and dimensioning to be provided by other sources of information which are defined in the deemed-to-satisfy section.

Background

The title of this Part has been amended again, this time to include access as well as facilities for the disabled people. The requirements now within this section are getting closer to the definition set out in Part A in that the needs for people with a sight impairment or the hard of hearing are now being included although the requirements specified are not yet both coherent and comprehensive.

Disabled persons should now have free access throughout the building with the subsequent problems for the designer of providing unhindered movement and facilities. It follows that in the event of a fire situation there should be unhindered egress from the building. The means of escape requirements are defined in Part E5.1 which requires to provide refuge areas for disabled people to enable subsequent rescue by the fire service.

The development of a fire safety solution based on the use of refuges does not address critical behavioural issues, namely:

(1) that the disabled person may not wish to remain in the building whilst all his/her fellow colleagues have evacuated;
(2) some colleagues may remain with the wheelchair user;
(3) colleagues may aid egress by physically moving the wheelchair bound person down the escape stair.

There are additional problems of location and ease of access of refuge areas given evacuation flaws within the building.

COMPLIANCE

Application of Part T

The standard applies to a surgery with access at ground level forming part of a building of purpose group 1, and to buildings of purpose group 2, 3, 4, 5 or 6 and to a car park or parking garage of purpose sub-group 7B or 7C.

ACCESS FOR DISABLED PEOPLE

Provision of Access

The requirement for access to and within a building is split into three sections:

(1) outside to the building;
(2) access into the building;
(3) access throughout the building.

DTS T2.1 The first requirement is covered by the need to provide two car parking spaces specially designed for disabled people per 50 spaces or part thereof, providing car parking is available. The design for these spaces must be in accordance with the *Disabled Access Guide*, October 1993 edition, published by Disability Scotland.

Access into the building (requirement 2) must be directly provided from a road or from a car park within the building boundary to an entrance. The main design criteria for access are set out in T2.4. However the deemed-to-satisfy provision for T2.2 stipulates that a drop kerb is required either from the car park or the road, and should be designed in accordance with the *Disabled Access Guide* 1995.

DTS T2.2

There is also a need for the access to be level or ramped, for example as indicated in Figure T.1

DTS T2.4

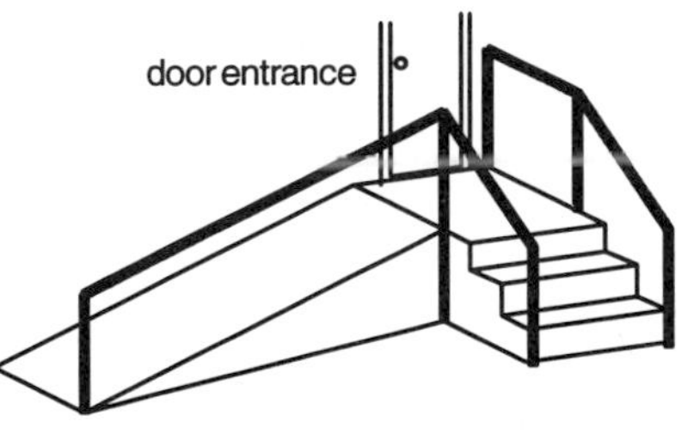

Figure T1 Disabled access.

Within the interior of the building, access must be provided to and throughout each storey conforming to the requirements set out in T2.5. Six exemptions to this provision are given:

DTS T2.3

(1) a storey containing only fixed plant or machinery where visits are intermittent for inspection and maintenance does not require access for disabled people;
(2) any catwalk, raking or openwork floor does not have to be designed to accommodate disabled people;
(3) it further states that in public areas of a restaurant or bar access to only half the area is acceptable if the area allows the disabled person access to the bar or a self-service counter in a restaurant, for example, see Figure T2;

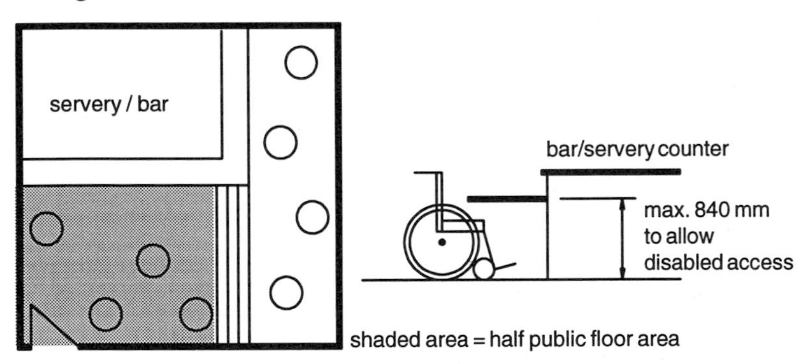

Figure T2 Access to bar/restaurant area.

(4) there is no requirement for every bedroom in buildings of purpose group 2, to have a bathroom for the disabled. This would include hotels, motels, boarding houses etc. provided toilet facilities are available that comply with T3.3, which will be discussed later;

(5) in public areas with fixed seating, the entire area is not required to have access for disabled people. However, some accommodation is required and this is covered under T3.4; again this will be discussed later;
(6) in a car park or parking garage of purpose sub-group 7B or 7C, the means of access requirements are confined to and throughout that storey or storeys where disabled car parking facilities are provided. This requirement has implications to the designer of these buildings in that the exact position of parking spaces for the disabled must be clearly indicated. However the number of such provisions is not stated.

Means of access

DTS T2.4 Access to a building must be level or ramped and be designed to be suitable for disabled people. The deemed-to-satisfy section stipulates two requirements:

(1) entrances should be designed in accordance with:
  (a) BS 5810: 1979 paragraphs 6.4 and 6.5; or
  (b) *Disabled Access Guide*, October 1993 edition, published by Disability Scotland.
(2) accessible ramps should be designed in accordance with:
  (a) BS 5810: 1979 paragraphs 6.1 to 6.2.3;
  (b) *Disabled Access Guide* (as above) with the additional design requirements of the following:
  (i) if a ramp or landing has open sides then a raised kerb 100 mm high must be provided;
  (ii) a top landing must have a tactile surface to provide people with a sight impairment of a change in level. Confusion may arise in that this requirement being for people with vision problems should indicate that tactile surfaces be placed at the top of every landing, indicating a change in level;
  (iii) a ramp requires to have a landing at every 5 m if the gradient is between 1:12 and 1:15.

DTS T2.5 Access within the building must be level or ramped throughout every storey where disabled people have access, as indicated in T2.4, or where a passenger lift is designed to accommodate disabled people it must have access to all storeys above or below the principal entrance storey. In addition further deemed-to-satisfy requirements are:

(1) access must be designed in accordance with the following:
  (a) BS 5810: 1979 paragraphs 7.4 to 7.6; or
  (b) *Disabled Access Guide*.

To complicate matters there are two exceptions:
  (i) every door across an accessible corridor or passageway must have a glazed panel not more than 900 mm to not less than 1500 mm from finished floor level; and
  (ii) where access for ambulant disabled people only is required then the width of the corridor must not be less than 1000 mm.

(2) the passenger lift must be designed in accordance with BS 5655: Part1: 1986, Part 2: 1988, Part 5: 1989 and Part 7: 1983. Again we have exceptions, in this case another five:

(i) there must be a clear landing not less than 1500 mm x 1500 mm in front of the lift entrance door or doors;
(ii) lift door or doors must have a clear width of at least 800 mm;
(iii) the lift car must be at least 1100 mm wide and 1400 mm long;
(iv) controls must be sited between 900 mm and 1200 mm above the lift floor and at least 400 mm from the front wall of the lift; and
(v) the lift must be provided with tactile call buttons on each storey, and within the lift tactile storey selector buttons, and in a lift serving more than two storeys, visual and voice indicators of the storey reached.

(3) where access is provided by a stair, it must be in accordance with BS 5810: 1979 paragraphs 7.7.1 and 7.7.2 with four exceptions:
(i) the stair must not have open risers;
(ii) handrails must be fixed at a height of 900 mm above the flight pitch and 1000 mm above the landing, be continuous around intermediate landings, and extend 300 mm beyond the first and last step nosing, for example see Figure T3.

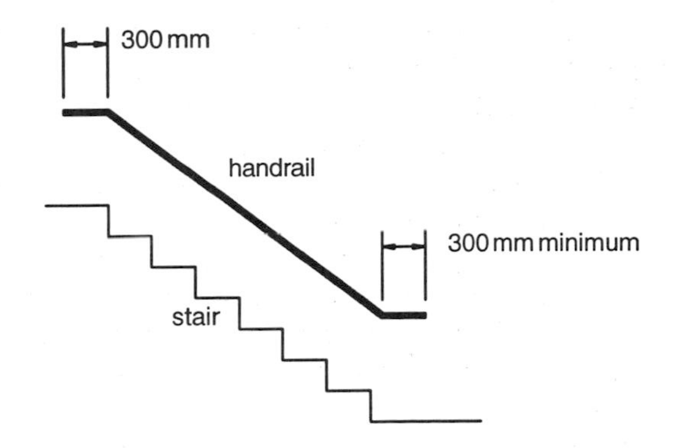

Figure T3 Handrail extension.

(iii) top landing must have a tactile surface, to warn people of a change in level, and
(iv) step nosings must be of contrasting brightness to assist people with a sight impairment.

However the exceptions to the above design requirement are as follows:
(a) to a storey or gallery in a two storey building, if the said area is less than 280 m^2 in area excluding vertical circulation, sanitary accommodation, plant rooms and a stair designed for ambulant disabled is provided, then the requirement for level/ramp access is not necessary; or
(b) to a storey or gallery of more than two storeys if the said area is less than 200 m^2 with the same conditions as (a), again the requirements for level/ramp access is not required.

If a stair is designed for the ambulant disabled and has a suitable wheelchair lift designed to BS 5776: 1979 or a suitable platform lift to BS 6440: 1983 then the requirement for level or ramped access may not be necessary. **DTS T2.6**

FACILITIES FOR DISABLED

Sanitary facilities and sanitary accommodation

The deemed-to-satisfy requirement applies to T3.1, 3.2 and 3.3 and suggests that BS 5810: 1979 or *Disabled Access Guide,* October 1993 edition, published by Disability Scotland, be consulted for design requirements.

DTS T3.1 Where sanitary facilities are provided on a storey accessible to disabled people then there must be facilities designed for wheelchair users. Five requirements are stated:

(1) at least one water closet with associated washbasin for each sex be provided; or

(2) water closet, with washbasin that is accessible to both sexes and independent of any other sanitary accommodation, is provided. The exception is a building where toilet facilities are provided for the public; at least one water closet is required to be provided for the disabled; or

(3) in a building which has a passenger lift and has toilet facilities on every storey, then toilets designed for disabled use (wheelchair) need only be provided every alternate floor. Provided that the distance travelled from any point on any floor to sanitary accommodation does not exceed 40 m, then the above requirement is acceptable. If the distance exceeds 40 m then sanitary accommodation designed for wheelchair users requires to be placed on every floor. The exception is a building classified as a car park or parking garage, in which case the toilet facilities should be positioned on the storey where car parking for the disabled has been provided or immediately above or below such a storey; or

(4) a building to which access is only available or provided for the ambulant disabled, then the toilet facilities need only be designed for that category of disabled person; or

(5) in a building where the sanitary facilities are provided for not more than 20 people, then only sanitary facilities for the ambulant disabled are required.

DTS T3.2 Further to these requirements, where changing areas have sanitary accommodation, then there must be at least one changing cubicle designed to accommodate a disabled person.

DTS T3.3 It further stipulates storeys containing bedrooms in purpose group 2B, for example hotels, motels and boarding houses which are accessible to disabled people should have not less than one bedroom in 20 providing sanitary facilities and sanitary accommodation suitable for wheelchair users. These facilities can be either en suite with a bedroom or accessible from a circulation area exclusive to the bedrooms if serving more than one. Reference to the deemed-to-satisfy section stipulates that BS 5810: 1979 will provide details of the design and dimensions for suitable sanitary facilities for the disabled.

Areas of fixed seating

DTS T3.4 Storeys accessible to disabled people which contain fixed seating for spectators or an audience, for example cinemas, theatres, opera houses etc. must provide suitable level spaces specially designed for wheelchair users. Such spaces should be in positions that are easily accessible for the wheelchair user and do not constitute an obstruction for other users of the building, and be located next to able-bodied companions. The area allocated for wheelchair users can be fitted with seating provided they can easily be removed when required for a wheelchair user.

The deemed-to-satisfy requirement states that one wheelchair space for every hundred seats should be provided, or six spaces, whichever is the greater, in other words not less than six spaces should be provided. The dimensions of an individual space should be not less than 1400 mm x 900 mm, for example see Figure T4.

DTS T3.4

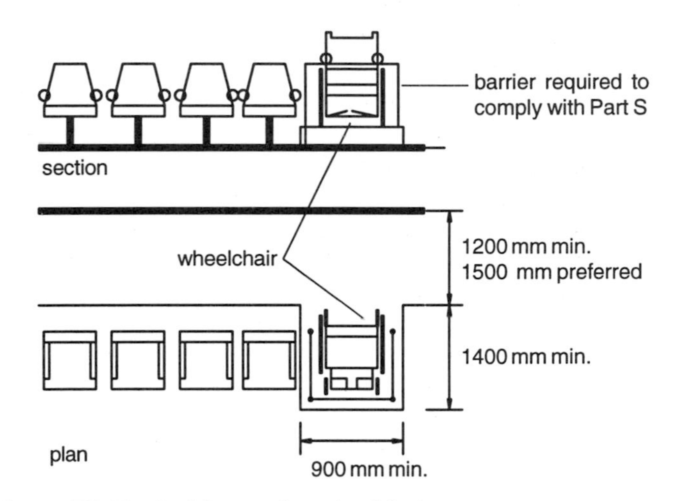

Figure T4 Typical layout for wheelchair spaces.

The exception is in sports stadia, with a seating capacity of more than 1000; the requirements then for wheelchair spaces are 10 spaces for the first 1000 plus one space for every further 2000 seats or part thereof.

AIDS TO COMMUNICATION

Aids for people with a hearing impairment must be provided within the following areas:

DTS T3.5

(a) auditoria and conference halls with a floor area exceeding 100 m^2; and
(b) booking and ticket offices where the customer is separated from the vendor by a glazed screen.

The deemed-to-satisfy section suggests that the equipment installed should provide a sound signal to a hearing aid enhanced by 20 dB and suppress any reverberation, audience or other environmental noise. This suggests that the initial sound figure cannot be estimated until the building is under occupation and after the first performance.

Recent amendments

The draft issued for consultation on 16 July 1992 proposed radical changes in Part T with the philosophical shift that handicapped people should have a right of access to all part of buildings (with the exception of dwellings and buildings used solely for storage). Similar provision has recently been introduced in England and Wales.

The proposals at that time (most of which have been consolidated in the 1994 Amendments) were:

(1) extend access for disabled people to all floors;
(2) require access for the disabled to the building from a public road and from parking provision on the site;
(3) require lift access to the upper floors;
(4) require extensive sanitary facility provision for wheelchair users and the ambulant disabled;
(5) require provisions for people with sensory impairment including wall textured surfaces, contrasting edges on nosings to steps, tactile or audible indication of floor calls in lifts, visual indication of floor calls in lifts and aids to communication for people with a hearing impairment in certain circumstances;
(6) apply to new buildings and to extensions and alterations to existing buildings.

There was also a need for complementary amendments to Part E (Means of Escape).

The 1992 proposals have been retained from the last edition to provide a link with the new February 1995 consultation document so as to provide an indication of the complexities with this section.

The 1995 draft document for further changes proposes to extend access for disabled people to dwellings and suggests that:

(1) access be provided to all ground floor dwellings including flats at ground level which have at least one apartment and a WC at entrance level;
(2) flats which are accessible by lift would require to meet the regulation;
(3) require ramped approaches;
(4) require doors to be a minimum of 800 mm wide;
(5) require doorbells/knockers to be within reach of a disabled person;
(6) within the dwelling/flat, provide the following:
    (a) access to one apartment;
    (b) provide toilet facilities;
    (c) light fittings, sockets, door handles, window controls to be accessible to disabled persons.

III Appendices

APPENDIX 1

HISTORICAL PERSPECTIVE OF BUILDING CONTROL

The development of building control in Scotland can be traced back, albeit sketchily in places, to the time of King David I in the 12th century AD.

For a long period prior to David's time, there had been a practice of granting charters to ecclesiastics. Indeed, Taylor, in *The Pictorial History of Scotland*, tells us that from the charters relating to church dealings it can be seen that 'the monks were the earliest guild brethren and possessed various exclusive privileges of trade and fisheries when scarcely any burghs were in existence'. The monks also engaged in banking and acted as a focus for excellence in various crafts (including building), which encouraged the growth of guilds.

Queen Margaret, the mother of David I, has been widely recognised as an influence for social good in Scotland and because of his mother's influence, together with the harsh economic pressures (which included defence considerations related to the cost of raising and maintaining an army), it was not a far step for David to create secular equivalents of these monastic systems.

Thus the royal burghs came into being, founded by a Royal Charter on property (royalty). Such properties were either held by the king or feuded, this latter arrangement providing a healthy source of revenue.

Indeed it might be contended that building control came into being in Scotland in AD 1119 with the foundation of the Royal Burgh of Berwick on Tweed by David I while he was known as Earl David. His vision was to make a network of towns and burghs similar to those found in Flanders (the Pas de Calais) and in Northern Belgium, but this was no easy task. In the rural landscape of Scotland at that time, because of their isolation, buildings presented little risk of creating major conflagrations or major damage from structural failure. In matters such as health and sanitation, more often than not local scattered episodes of infection could be contained, and, since in the rural areas neighbours usually gave of their time and skills to assist with traditional constructions and used traditional materials, there were not the same pressures for money's worth in materials and work standards.

The converse was the case with the increasing congestion within the new royal burghs and much of the criteria now in place in building regulations was formative at this time.

Hand in hand with the development of the burghs came the development of craft excellence within the guild and the guild brethren had a vested interest in the setting and the maintaining of standards. Thus the Dean of Guild and his Court became increasingly influential and were recognised as the source of building control. These courts were only to be found in the royal burghs.

Perhaps the archaeological remains of Roxburgh (near Kelso) could be taken as the blueprint of the earliest royal burghs. Roxburgh was founded by the same charter by which David constituted Berwick and its scarcely recognisable few streets and layout have survived, frozen in time, because this royal burgh perished early on, a victim of cross border warfare. Nevertheless along with burghs such as Perth and St Andrews it is clear that control existed in relation to what would now be considered as building control and planning matters, an example being the width of the streets.

There was originally an Association of four royal burghs in or about the time of David I, these being Berwick, Roxburgh, Edinburgh and Stirling. In the Guild Hall of Stirling today one can still see the sign of the Association, a reversed 4. The Association was known as Curia Quattor Burgorum and was the forerunner of the Convention of Scottish Local Authorities (COSLA).

An Act of 1469 put paid to democratic participation in local government (it names the Dean of Guild) with arrangements for the annual election of the burgh council by the previous council. It took nearly three and a half centuries to redress this via the Reform Act of 1832. Meanwhile the powers and jurisdiction of the Dean of Guild and his council or court were readdressed by James VI in 1593 and the concept of control via the Dean of Guild was built into the Act of Union of 1707.

Thereafter there was no further legislation relating to the Dean of Guild or his jurisdiction in matters relating to building control during the remainder of the 18th century. Neither did there appear to be any governmental department interested in overall royal burgh administration. Despite the steady growth of towns, the royal burghs continued to be restricted to their original ‘royalties’ or land grants and the jurisdiction which had been defined by custom and legislation of the two preceding centuries.

Campbell-Irons in the *Manual of the Law and Practice of the Dean of Guild Court,* tells us that ‘the period from 1785-1800 was one of electoral reform and Parliament was asked to deal with the burghs of Scotland’. A parliamentary committee was formed to deal with petitions from the royal burghs and 63 charters were produced by the burghs in the pleas. In all, or nearly all the burghs, there was a structure of guild brethren and craftsmen. Within this structure there was a Dean of Guild although the mode of election varied from burgh to burgh.

Into the 19th century it became increasingly clear that disease could be fought by improving environmental conditions. Improvements in public health through coercive action by local authorities were a development of the middle part of the 19th century.

In 1845 a Royal Commission was appointed to enquire into the causes of disease among the inhabitants of towns. The report of the Commission to Parliament tells of the conditions of those places inhabited by the labouring classes and often also by tradesmen in large towns and small towns, these being overcrowded and often in a noxious state from want of drainage, cleanliness, proper ventilation and adequate water supply. These conditions were a breeding ground for typhus, fever, cholera, consumption, scrofula and other chronic complaints. Consequently, several public health acts were enacted in the latter part of the 19th century and so far as this country was concerned this culminated in the Public Health (Scotland) Act 1897 which dealt with matters such as water supply, drainage, building structure, etc.

The Burgh Police (Scotland) Act 1892 dealt with matters of building control and the jurisdiction of the Dean of Guild Court but like much of the legislation of that period it was restrictive and inflexible as illustrated by the requirement that streets had to be 36 feet wide and that houses were not to exceed in height one and a quarter widths of the street.

Such inflexible standards were in sharp contra distinction to current building control legislation which attempts to execute control without unnecessarily inhibiting design. Another Burgh Police (Scotland) Act was passed in 1903 which also dealt with matters of building control. There had always been local variations in standards and the Housing (Scotland) Act 1950 made allowances for local authorities to make their own bye-laws on many matters which now reside within the Technical Standards 1991.

The Department of Health variously, in 1932, 1937 and 1954 published model bye-laws which were adopted in various parts of the country but nevertheless in terms of uniformity of standards the situation still remained patchy.

By this time the function of building control was generally the responsibility of the Master of Works, a post usually held by the Burgh Engineer, Burgh Surveyor or Sanitary Inspector (again we see the historical links of building control and environmental health). The ultimate control, of course, still rested with the Dean of Guild.

The need to rationalise the variation of standards which was still prevalent led to the setting up of the Committee in Building Legislation in Scotland in 1954, otherwise known as the Guest Committee. In the main the committee was charged with securing a system of building control which ensured uniformity of standards while being flexible enough to allow innovation particularly in the context of new building techniques and the development of new building materials. The outcome of the work of this committee was the Report of the Committee on Building Legislation in Scotland (1957, HMSO). An important principle established by the Guest Committee was that regulations should not be expressed in terms of methods of construction and types of materials but rather in terms of performance. This in itself was an impetus to a more flexible approach. A further impetus to flexibility was the recommendation that special cases should be treated through the vehicle of relaxations. Broadly, twin platforms of performance were to be the utilisation of specifications which were 'deemed-to-satisfy', (a link with former practice), and performance related to British Standards and Codes of Practice.

Thereafter the Building (Scotland) Act 1959 came into being giving powers and responsibilities to the Secretary of State and through him the local authorities to exercise the function of building control through unified standards (which replaced the existing bye-laws) and the making of regulations. Thus, in 1963, the Building (Scotland) Regulations 1964 were made and these came into force in 1964.

Since that time, building regulations have been re-enacted from time to time (with intervening amendments as and when necessary) to cope with changes, such as our understanding of matters of building science and health (for example, for better control of condensation, it is now mandatory to use mechanical ventilation in bathrooms and kitchens – see Part K), of economics (for example, conservation of fuel and power has become increasingly more sharply focused – see Part J) and of the needs of the individual in society (for example, the introduction of Part T which deals with the needs of handicapped people in and around buildings).

There has been an increasing awareness over this time of the need not to inhibit the creative skills which are necessary in designing buildings while at the same time maintaining reasonable minimum standards and exercising the building control function to ensure that these standards are met.

One could say that building control has come a long way since its first establishment in AD 1119 with the setting up of the Royal Burgh of Berwick by King David I. Yet, curiously enough, many of the problems of ensuring that buildings are suitable places for people required not dissimilar regulation then as today.

APPENDIX 2

TABLE OF FEES

The 1995 Amendment of the Building (Procedure) (Scotland) Regulations 1981 lists the following table of fees payable to building control departments for processing building warrant applications.

1. Application for warrant (including issue of warrant) for erection, alteration, extension or demolition of a building (whether or not combined with application for warrant for change of use)

i) where the estimated cost of the operations does not exceed £3,000	£70
ii) where the estimated cost of the operations exceeds £3,000 but does not exceed £10,000	
a) the sum of	£70
plus (b) for every £500 or part thereof exceeding £3,000	£12
iii) where the estimated cost of the operations exceeds £10,000 but does not exceed £20,000	
(a) the sum of	£238
plus (b) for every £1,000 or part thereof exceeding £10,000	£14
iv) where the estimated cost of the operations exceeds £20,000 but does not exceed £100,000	
(a) the sum of	£378
plus (b) for every £5,000 or part thereof exceeding £20,000	£28
(v) where the estimated cost of the operations exceeds £100,000 but does not exceed £500,000	
(a) the sum of	£826
plus (b) for every £50,000 or part thereof exceeding £100,000	£205
vi) where the estimated cost of the operations exceeds £500,000 but does not exceed £1,000,000	
(a) the sum of	£2466
plus (b) for every £100,000 or part thereof exceeding £500,000	£275
(vii) where the estimated cost of the operations exceeds £1,000,000	
(a) the sum of	£3841
plus (b) for every £200,000 or part thereof exceeding £1,000,000	£410

2. Application for warrant (including issue of warrant) for change of use only — £30

3. Application for amendment of warrant
   (a) for additional operations where the estimated cost of these falls within any of the sub-paragraphs detailed in paragraph 1 of this Schedule
   fee in accordance with appropriate sub-paragraph of paragraph 1 hereof
   (b) where no additional operations are involved — £30

4. Application for extension of a period of a warrant — £30

APPENDIX 3

DEFINITIONS

The following is a list of definitions of terms which are used throughout the Technical Standards.

ACCESS STAIR means a stair providing access to:

(1) one or more dwellings;

(2) shared residential accommodation; or

(3) facilities within the curtilage of such dwellings and accommodation.

ACCOMMODATION STAIR means a stair which is not an access stair, escape stair or private stair.

ACT means the Building (Scotland) Act 1959 as amended.

AGRICULTURE has the same meaning as in the Agriculture (Scotland) Act 1948(a), and

AGRICULTURAL shall be construed accordingly.

AIR SUPPORTED STRUCTURE means a structure which has a space-enclosing single-skin membrane anchored to the ground and kept in tension by internal air pressure so that it can support applied loading.

ALTERNATIVE EXIT means an exit from a dwelling which is other than through its main entrance door and is available for use at all times.

APARTMENT means a room in a dwelling not used solely as a kitchen, store or utility room.

BASEMENT STOREY means any storey which is below the level of the ground storey.

BOUNDARY means a boundary between land on which a building is situated and land in different occupation, so however that:

(1) in relation to any external side of a building it shall exclude any part of the boundary which makes an angle with that side of more than 80°;

(2) in relation to any road, whether public or private, public access way or public right of way, river, stream, canal, loch, pond, common land or public open space it shall be taken to be the centre line thereof; and

(3) the sea and its foreshore shall not be regarded as land in different occupation.

BUILDING means any structure of erection of any kind or nature whatsoever, whether temporary or permanent, and every part thereof, including any fixture affixed thereto, not being a structure of erection or part thereof consisting of, or ancillary to:

(1) any road, whether public or private, including in the case of a public road (but not in the case of a private road) any bridge on which the road is carried;

(2) any sewer or water main which is, or is to be, vested in a public authority;

(3) any aerodrome runway;

(4) any railway line;

(5) any large raised reservoir within the meaning of the Reservoirs Act 1975 (b); or

(6) wires and cables, their supports above ground and other apparatus used for telephonic or telegraphic communication;

and includes any prospective building; and in relation to the extension, alteration or change of use of a building any reference to the building shall be construed as a reference only to so much of the building as is comprised in the extension or is the subject of alteration or change of use as the case may be.

CARPORT means a roofed building for vehicle storage which is open on at least two sides except for roof supports.

CAVITY in Part D means a concealed space enclosed by elements of a building (including a suspended ceiling) or contained within an element, but not a room, cupboard, circulation space, stairway enclosure, lift well, flue or a space within a chute, duct, pipe or conduit.

CAVITY BARRIER means any construction provided to seal a cavity against the penetration of smoke or flame, or to restrict its movement within the cavity.

CHIMNEY means a structure enclosing one or more flues, not being a flue-pipe, but including any opening for the accommodation of a heat-producing appliance, but does not include a chimney-can.

CHIMNEY-STACK means that part of a chimney which rises above a roof of the building of which it forms part and includes any cope but not a chimney-can.

CLOSED COURT means an external space wholly enclosed by walls with no break more than 2 m wide in any wall.

COMPARTMENT, except in the expression watercloset compartment, means any part of a building which is divided from all other parts by one or more compartment walls or compartment floors or by both such walls and floors; and, if any part of the top storey of a building is within a compartment, that a compartment shall also include any roof space above such part of the top storey.

COMPARTMENT FLOOR and COMPARTMENT WALL mean respectively a floor and wall complying with the provisions of Part D of the Technical Standards relating to compartment floors and walls and dividing a compartment of a building or a lift well in a building from the remainder of the building.

CONSERVATORY means a building attached to and having an entrance from a dwelling and having not less than three-quarters of the area of its roof and not less than one-half of the area of its external walls made of translucent material.

CONSTRUCT includes alter, erect, extend and fit, and **CONSTRUCTION** shall be construed accordingly.

COVERED AREA means a roofed building which is open on at least two sides except for roof supports.

DEAD LOAD means the load due to the weight of all walls, permanent partitions, floors, roofs and finishes, including services and other permanent construction.

DECORATIVE FUEL-EFFECT GAS APPLIANCE means an open flued appliance designed to simulate a solid fuel open fire primarily for decorative purposes and intended to be installed so that the products of combustion pass unrestricted from the firebed to the flue. It does not include an inset live fuel effect gas fire.

DIFFERENT OCCUPATION, in relation to two adjoining buildings or parts of one building, means occupation or intended occupation of those buildings or parts by different persons.

DISABLED PEOPLE means persons with a physical, hearing or sight impairment which affects their mobility or their use of buildings.

DRAINAGE SYSTEM means the system of pipes and drains used for the drainage of a building, including all other fittings, appliances and equipment so used.

DRY FIRE MAIN means a pipe installed in a building for fire fighting purposes which is normally dry but is capable of being charged with water by pumping from a fire service appliance.

DUCT means a passage, other than a flue, used solely for conveying air, gases, or services including refuse, whether or not these are contained in separate pipes.

DWELLING means a unit of residential accommodation occupied (whether or not as a sole or main residence):

(1) by a single person or by people living together as a family; or
(2) by not more than six residents living together as a single household (including a household where care is provided for residents).

ELECTRO-MAGNETIC OR ELECTRO-MECHANICAL DEVICE SUSCEPTIBLE TO SMOKE means a device which allows a door held open by it to be operated manually at all times and to close automatically upon:

(1) the operation of an automatic smoke detector; or
(2) the manual operation of a switch fitted in a suitable position; or
(3) the failure of electricity supply to the device, apparatus or switch; or
(4) if a fire alarm system is installed in the building, operation of that system.

ELEMENT OF STRUCTURE means:

(1) a beam or column or other member forming part of a structural frame, not being a member forming part of or supporting only a roof structure, unless the roof performs the function of a floor; or
(2) an internal or external wall supporting any other structural element for which a standard of fire resistance is prescribed in the Technical Standards; or
(3) a protected area of an external wall; or
(4) a separating wall or compartment wall; or
(5) a floor other than the lowest; or
(6) a door, shutter, duct, access cover or hatch for which a standard of fire resistance is prescribed in the Technical Standards.

EMERGENCY LIGHTING means lighting, designed to come into, or remain in, operation automatically in the event of either a local or general power failure.

ESCAPE ROUTE means a route by which a person may reach a place of safety from any point of a storey. It includes protected zones and unprotected zones and, for travel distance and certain specified minimum width requirements only, the route within a room.

ESCAPE STAIR means a stair forming part of an escape route.

EXIT means a point of egress from a room or storey which forms part of, or gives access to, an escape route or place of safety.

EXPOSED in Part J in relation to a wall or floor means a wall or floor directly exposed to the outside air.
EXTERNAL WALL includes a part of a roof pitched at an angle of 70° or more to the horizontal.
FIRE DOOR means a door or shutter which, together with its frame and furniture as installed in a building, is intended, when closed, to resist the passage of fire and is capable of meeting specified performance criteria.
FIRE-STOP means a seal of fire-resisting material provided to close an imperfection of fit between elements, components or construction so as to restrict penetration of smoke and flame through that imperfection.
FLAT means a dwelling on one storey, forming part of a building from some other part of which it is divided horizontally, and includes a dwelling of which the main entrance door and associated hall are on a different storey from the remainder of the dwelling.
FLAT ROOF means a roof the slope of which does not exceed 10° from the horizontal.
FLIGHT means part of a stair or ramp uninterrupted by a landing.
FLOOR in Part G means any construction between the surface of the ground, or the surface of any hardcore laid upon the ground, and the upper surface of the floor, together with any floor finishes which are laid as part of the permanent construction.
FLUE means a passage which conveys the products of combustion to the open air.
FLUE-PIPE means a pipe forming a flue, but not a pipe fitted as a lining in a chimney.
FOUL WATER means any water contaminated by soil water, waste water or trade effluent.
FOUNDATION means that part of the structure in direct contact with, and transmitting loads to, the ground.
GALLERY means a raised floor or platform, whether level or not, which is open to the room or space in which it is situated and which:

(1) has every part of its upper surface not less than 1.8 m above the surface of the main floor of the said room or space; and
(2) occupies (or, in the case of there being more than one gallery, together occupy) not more than one-half of the area of the said room or space.

GLAZING means any permanently secured sheet of glass or plastics.
GREENHOUSE except in the expression agricultural greenhouse, means a building ancillary to a dwelling used mainly for growing plants, which is either:

(1) detached from the dwelling; or
(2) attached to, but not entered from, the dwelling.

GROUND STOREY means the storey in the building in which there is situated an entrance to the building from the level of the adjoining ground or, if there is more than one such storey, the lower or lowest of these.
HOUSE means a dwelling on one or more storeys, either detached or forming part of a building from all other parts of which it is divided only vertically.
IMPOSED LOAD means the load assumed to be produced by the intended occupancy or use, including the weight of moveable partitions, distributed, concentrated, impact, inertia and snow loads, but excluding wind loads.

INNER ROOM means a room which does not have direct access to a circulation area.

INSULATED FLUE-PIPE means a flue-pipe surrounded by a sealed air space or insulating material protected by an outer casing.

KITCHEN means any room or part of a room used primarily for the preparation or cooking of food.

LAND IN DIFFERENT OCCUPATION in relation to a building, means land occupied or to be occupied by a person other than the occupier of the land on which the building is or is to be situated.

LIMITED LIFE BUILDING means a building intended to have a life of the period specified in regulation 5.

LOW LEVEL DISCHARGE APPLIANCE means a heat-production appliance designed to discharge its products of combustion to the outside air at a height less than that required for a chimney.

MAISONETTE means a dwelling on more than one storey, forming part of a building from some other part of which it is divided horizontally.

NON-COMBUSTIBLE means that a material is resistant to combustion as determined by an appropriate test procedure as specified in Part D.

OCCUPANCY CAPACITY shall be construed in accordance with regulation 7.

OCCUPIER in relation to a dwelling, means the person inhabiting the dwelling.

OPEN COURT means an external space, other than a recess, enclosed by walls on at least three sides but with at least one break of more than 2 m in a wall. A recess becomes an open court when:

(1) a ventilator is in the back wall of the recess and the ratio of the length of the back wall to the depth of the recess is less than 1:1; or

(2) a ventilator is in a side wall of the recess and the ratio of the length of the back wall of the recess to the depth into the recess of the centre-line of the ventilator is less than 2:1.

PASSENGER CONVEYOR means a power-driven installation containing an endless moving walkway for the conveyance of persons between different levels or between different parts of the same level of a building.

PERMANENT VENTILATOR means a ventilator which provides continuous ventilation.

PLACE OF SAFETY means either:

(1) an unenclosed space in the open air at ground level; or

(2) an enclosed space in the open air at ground level leading to an unenclosed space, via an access not narrower than the total width of the exits leading from the building to that enclosed space.

PLACE OF SPECIAL FIRE RISK means any place within, or attached to, or on the roof of, a building in which there are installed one or more:

(1) solid fuel burning heating appliances, with a total installed output rating exceeding 45 kW; or

(2) oil or gas burning appliances, with total installed input rating exceeding 60 kW, other than forced air convection or radiant heaters in buildings of occupancy groups 3 to 7 inclusive; or

(3) fixed internal combustion engines, including gas turbine engines, with a total output rating exceeding 45 kW; or

(4) oil-immersed electricity transformers or switch gear apparatus with an oil capacity exceeding 250 litres and operating at a supply voltage exceeding 1000 volts.

POINT OR ORIGIN means:

(1) in the case of a flat or maisonette, the main entrance door to the flat or maisonette; or

(2) in any other case, any point within a room or storey.

PORCH means a building attached to and having an entrance from a dwelling and having a roof of opaque material.

PRIVATE STAIR OR RAMP means a stair or ramp:

(1) wholly within a dwelling; or

(2) wholly within shared residential accommodation.

PROTECTED CIRCUIT means a circuit originating at the main incoming switch or distribution board, the conductors of which are suitably protected against fire.

PROTECTED DOOR means:

(1) a self-closing fire door giving access to a protected zone; or

(2) a door leading directly to a place of safety; or

(3) a door giving access to an unenclosed external escape stair; or

(4) a door leading to an escape route across a flat roof.

PROTECTED LOBBY means a lobby within a protected zone but separated from the remainder of the zone by a self-closing fire door and a wall or screen, each having a fire resistance of at least 30 minutes.

PROTECTED ZONE means that part of an escape route which is within a building, but not within a room, and to which access is only by way of a protected door and from which exit is directly to a place of safety.

PUBLIC OPEN SPACE includes land used as a public park or for public recreation or as a burial ground.

PUBLIC ROAD has the same meaning as in the Roads (Scotland) Act 1984(a).

PURPOSE GROUP and PURPOSE SUB-GROUP mean respectively a group or sub-group of buildings specified in Schedule 3 to the Building Standards (Scotland) Regulations 1990.

REASONABLY PRACTICABLE, in relation to the carrying out of any operation, means reasonably practicable having regard to all the circumstances including the expense involved in carrying out the operation.

RELEVANT STANDARD means a standard set out in the Technical Standards which relates to the requirements of a particular provision of the Building Standards (Scotland) Regulations 1990.

ROOF in Part G means any roof, including eaves and junctions with other elements penetrating or connected to the roof.

ROOF SPACE means any space in a building between a part of the roof and the ceiling below.

ROOM means any enclosed part of a storey intended for human occupation or, where no part of any such storey is so enclosed, the whole of that storey, but excepting in either case any part used solely as a bathroom, shower room, washroom, watercloset compartment, stair or circulation area.

ROOM-SEALED APPLIANCE means a heat-producing appliance which, when in operation, has its combustion air inlet and its combustion products outlet isolated from the room or space in which it is installed.

SANITARY ACCOMMODATION includes bathrooms, watercloset compartments, and washrooms.
SANITARY FACILITIES include washbasins, baths, showers, urinals and waterclosets.
SELF-CLOSING FIRE DOOR means a fire door fitted to close automatically from any angle of swing.
SEMI-EXPOSED in relation to a wall or floor in Part J means a wall or floor between a building to which Part J applies and a building or part of a building to which it does not apply.
SEPARATING FLOOR or SEPARATING WALL mean respectively a floor or wall separating:

(1) any two adjoining buildings, or parts of one building, in different occupation; or
(2) any two adjoining buildings, or parts of one building, of different purpose groups; or
(3) any two adjoining parts of one building, where one part is in single occupation and the other in common occupation.

SERVICE OPENING means any opening to accommodate a duct, pipe, conduit or cable (including fibre optics or similar tubing).
SHARED RESIDENTIAL ACCOMMODATION means a unit of accommodation or purpose sub-group 2B having an occupancy capacity not exceeding 10, entered from the open air at ground level and having no storey at a height exceeding 7.5 m.
SITE in relation to a building, means the area of ground covered or to be covered by the building, including its foundations.
SOCKET OUTLET means a fixed device containing contacts for the purpose of connecting to a supply of electricity the corresponding contacts of a plug attached to any current-using appliance.
STOREY means that part of a building which is situated between the top of any floor and the top of the floor next above it or, if there is no floor above it, between the top of the floor and the ceiling above it or, if there is no ceiling above it, the internal surface of the roof; and for this purpose a gallery or catwalk, or an openwork floor or storage racking, shall be considered to be part of the storey in which it is situated.
SURFACE WATER means the run-off of rainwater from roofs and any paved ground surface within the curtilage of a building.
TAPERED TREAD means a stair tread in which the nosing is not parallel to the nosing of the tread or landing next above.
TECHNICAL STANDARDS means the Technical Standards for compliance with the Building Standards (Scotland) Regulations 1990 issued by the Scottish Office and dated October 1990.
TRAVEL DISTANCE has the meaning assigned to that expression by Standard E2.71.
TRICKLE VENTILATOR means a closeable small ventilator which can provide minimum ventilation.

UNPROTECTED AREA in relation to an external wall or side of a building, means:

(1) any opening, including a door or window, but not an unopenable window containing only glazing which reacts to heat to provide the required standard of fire resistance for an external wall; or

(2) any part of an external wall which has less than the prescribed standard of fire resistance; or

(3) any part of an external wall which has combustible material more than 1 mm thick attached or applied to its external face, whether for cladding or any other purpose.

UNPROTECTED ZONE means that part of an escape route which is outwith either a room or a protected zone.

UPPER STOREY means any storey which is above the level of the ground storey.

U-VALUE (or thermal transmittance coefficient) means the rate of heat transfer through 1 m^2 of a structure when the temperature at each side of the structure differs by 1°C (expressed in W/m^2K).

VENTILATED SPACE in Part J means a space which is enclosed by structure, part of which is exposed to the outside air, and permanently ventilated to the outside air by openings or ducts having an aggregate area exceeding 30% of the wall area.

VENTILATOR means a window, rooflight, grille or similar building component (and in the case of a dwelling includes a door) capable of being opened to provide ventilation.

WALL in Part G means any wall, including piers, chimneys, columns and parapets which form part of the wall.

WATERCLOSET COMPARTMENT means an enclosed part of a storey which contains a urinal or a watercloset pan and which has provision for flushing from a piped supply of water and is connected to a drainage system.

WET FIRE MAIN means a pipe installed in a building for fire fighting purposes which is permanently charged with water from a pressurised supply.

WIND LOAD means the load due to the effect of wind pressure or suction.

APPENDIX 4

BRITISH STANDARDS AND CODES OF PRACTICE

British Standards

	Number	Title	Amendment	Context within Technical Standards
BS 41	1973 (1981)	Specification for cast iron spigot and socket flue or smoke pipes and fitting		(F3.11) (F4.10)
BS 65	1988	Specification for vitrified clay pipes, fittings, joints and ducts		(F4.8-F4.9) (F4.10)
BS 449		Specification for the use of structural steel in building-		
	Part 2:1969	- Metric units	AMD416 AMD523 AMD661 AMD1135 AMD1787 AMD4576 AMD5698 AMD6255	(C2.1)
BS 476		Fire tests on building materials and structures-		
	Part 3:1958	- External fire exposure roof tests		Part D Appendix
	Part 4:1970 (1984)	- Non-combustibility test for materials	AMD2483 AMD4390	Part D Appendix
	Part 6:1968	- Method of test for fire propagation for products		Part D Appendix (E2.42)
	Part 6:1981	- Method of test for fire propagation for products		Part D Appendix (E2.42)

	Part 7:1971	- Surface spread of flame tests for materials		Part D Appendix (E2.42)
	Part 7:1987	- Method for classification of the surface spread of flame of products		Part D Appendix (E2.42)
	Part 8:1972	- Test methods and criteria for the fire resistance of elements of building construction	AMD1873 AMD3816 AMD4822	Part D Appendix (E2.45) (R2.10)
	Part 20:1987	- Method for determination of the fire resistance of elements of construction (general principles)		Part D Appendix (R210)
	Part 21:1987	- Methods for determination of the fire resistance of loadbearing elements of construction		Part D Appendix (R2.10)
	Part 22:1987	- Methods for determination of the fire resistance of non-loadbearing elements of construction		Part D Appendix (E2.45) (R2.10)
	Part 23:1987	- Methods for determination of the contribution of components to the fire resistance of a structure		Part D Appendix (R2.10)
	Part 24:1987	- Method for determination of the fire resistance of ventilation ducts		Part D Appendix
	Part 31: Section 31.1:1983	- Methods for measuring smoke penetration through doorsets and shutter assemblies - method of measurement under ambient temperature conditions		(E2.45)
BS 699:1984 (1990)		Specification for copper direct cylinders for domestic purposes	AMD5792 AMD6600	(J3.4)
BS 715:1989		Specification for metal flue pipes, fittings, terminals and accessories for gas-fire appliances with a rated input not exceeding 60 kW		(F4.10)

BS 750:1984	Specification for underground fire hydrants and surface box frames and covers		(E3.2)
BS 585	Wood stairs-		
Part 1:1981	Specification for stairs with closed risers for domestic use including straight and winder flights and quarter or half landings	AMD6510	(S2.10)
BS 952	Glass for glazing-		
Part 1:1978	- Classification		(E2.22)
BS 1181:1989	Specification for clay flue linings and flue terminals		(F4.8-F4.9)
BS 1230	Gypsum plasterboard-		
Part 1:1985	- Specification for plasterboard excluding materials submitted to secondary operations		Part D Appendix
BS 1251:1987	Specification for open-fireplace components		(F3.19)
BS 1344	Methods of testing vitreous enamel finishes-		
Part 1:1987	- Determination of resistance to thermal shock of coatings on articles other than cooking utensils		(F3.11)
Part 3:1988	- Determination of resistance to sulphuric acid at room temperature		(F3.11)
Part 5:1984	- Determination of resistance to hot detergent solutions used for washing textiles		(F3.11)
Part 7:1984	- Determination of resistance to heat		(F3.11)
BS 1449	Steel plate, sheet and strip -		
Part 1:1983	- Specification for carbon and carbon-manganese plate, sheet and strip	AMD4806	(F4.10)

	Part 2:1983	- Specification for stainless and heat-resisting steel plate, sheet and strip	AMD4807 AMD6646	(F3.11), (F4.10)
BS 1566		Copper indirect cylinders for domestic purposes -		
	Part 1:1984	- Specification for double feed indirect cylinders	AMD5790 AMD6598	(J3.4)
	Part 2:1984	- Specification for single feed indirect cylinders	AMD5791 AMD6601	(J3.4)
BS 1703 :1977		Specification for refuse chutes and hoppers		(R2.1,R2.3 R2.4)
BS 2750		Measurement of sound insulation in buildings and of building elements-		
	Part 4:1980 (ISO 140/4)	- Field measurements of airborne sound insulation between rooms		(H2.1,H2.2 H2.3)
	Part 7:1980 (ISO140/7)	- Field measurements of impact sound insulation of floors		(H2.1,H2.2 H2.3)
BS 2782		Methods of testing plastics-		
	Method 102C: 1970	- Softening point of thermoplastic moulding material (bending test)		(E2.42)
	Method 508A: 1970	Rate of burning		(D2.5), (E2.42)
	Method 508C: 1970	- Degree of flammability of thin polyvinyl chloride sheeting		(D2.5), (E2.42)
	Method 508D: 1970	- Flammability (alcohol cup test)		(D2.5), (E2.42)
	Part1:Methods 120A to 120E: 1976 (1983)	- Determination of the Vicat softening temperature of thermoplastics		(D2.5), (E2.42)
	Part1:Method 140D:1980 (1987)	- Flammability of a test piece 550mm x 35mm of thin polyvinyl chloride sheeting (laboratory method)	AMD5440	(E2.42)

Part1:Method 140E:1982 (1988)	- Flammability of a small, inclined test piece exposed to an alcohol flame (laboratory method)	AMD5439	(D2.5), (E2.42)
BS 2869	Fuel oils for non-marine use-		
Part 2:1988	- Specification for fuel oil for agricultural and industrial engines and burners (classes A2,C1,C2,D,E,F,G and H)		(F4.5)
BS 3198 :1981	Specification for copper hot water storage combination units for domestic purposes	AMD4372 AMD6599	(J3.4)
BS 4211 :1987	Specification for ladders for permanent access to chimneys, other high structures, silos and bins	AMD7064	(S2.25)
BS 4514 :1983	Specification for unplasticized PVC soil and ventilating pipes, fittings and accessories	AMD4517 AMD5584	(D2.16)
BS 4543	Factory-made insulated chimneys-		
Part 1:1976	- Methods of test for factory-made insulated chimneys		(F3.10), (F4.8-F4.9)
Part 2:1976	-Specification for chimneys for solid fuel fired appliances	AMD2794 AMD3475 AMD3878	(F3.8), (F3.10)
Part 3:1976	- Specification for chimneys for oil fired appliances	AMD2981 AMD3476	(F4.8-F4.9)
BS 4876 :1984	Specification for performance requirements for domestic flued oil burning appliances (including test procedures)		(F4.1), (F4.2)
BS 4987	Coated macadam for roads and other paved areas-		
Part 1:1993	- Specification for constituent materials and for mixtures		(Q2.16)
Part 2:1993	- Specification for transport, laying and compaction		(Q2.16)

BS 5041	Fire hydrant systems equipment-		
Part 1:1987	- Specification for landing valves for wet risers	AMD5912	(E3.6)
Part 2:1987	- Specification for landing valves for dry risers	AMD5776	(E3.6)
Part 3:1975 (1987)	- Specification for inlet breechings for dry riser inlets	AMD5504	(E3.6)
Part 4:1975 (1987)	- Specification for boxes for landing valves for dry risers	AMD5503	(E3.6)
Part 5:1974 (1987)	- Specification for boxes for foam inlets and dry riser inlets	AMD5505	(E3.6)
BS 5250 :1989	Code of practice for control of condensation in buildings		Part G Introd. (G4.1), (G4.2)
BS 5258	Safety of domestic gas appliances-		
Part 1:1986	- Specification for central heating boilers and circulators		(F2.2), (F5.6), (F5.11)
Part 2:1975	- Cooking appliances	AMD3285 AMD4925	(F2.2), (F5.11)
Part 4:1987	- Specification for fanned-circulation ducted-air heaters		(F2.2), (F5.11)
Part 5:1989	- Specification for gas fires		(F2.2), (F5.11)
Part 6:1988	-Specification for refrigerators and food freezers		(F2.2), (F5.11)
Part 7:1977	- Storage water heaters		(F2.2), (F5.11)
Part 8:1980	- Combined appliances: gas fire/back boiler		(F2.2), (F5.11)
Part 9:1989	- Specification for combined appliances:fanned-circulation ducted-air heaters/circulators		(F2.2), (F5.11)

Standard	Part	Title	Amendments	Reference
	Part 10:1980 (1983)	- Flueless space heaters (excluding catalytic combustion heaters)(3rd family gases)	AMD4411	(F2.2), (F5.11)
	Part 11:1980 (1983)	- Flueless catalytic combustion heaters (3rd family gases)	AMD4412	(F2.2), (F5.11)
	Part 12:1980	- Decorative gas log and other fuel effect appliances (2nd and 3rd family gases)	AMD5434	(F2.2), (F5.11)
	Part 13:1986	- Specification for convector heaters		(F2.2), (F5.11)
	Part 14:1984	- Specification for barbecues (3rd family gases)		(F2.2), (F5.11)
BS 5262 :1976		Code of practice. External rendered finishes	AMD2103 AMD6246	(G3.1)
BS 5266		Emergency lighting-		
	Part 1:1988	- Code of practice for the emergency lighting of premises other than cinemas and certain other specified premises used for entertainment		(E2.44)
BS 5268		Structural use of timber-		
	Part 2:1991	- Code of practice for permissible stress design, materials and workmanship		(C2.1)
	Part 3:1995	- Code of practice for trussed-rafter roofs		(C2.1)
BS 5306		Fire extinguishing installations and equipment on premises-		
	Part 0:1986	- Guide for the selection of installed systems and other fire equipment	AMD5695 AMD6653	(D2.1)
	Part 1:1976 (1988)	- Hydrant systems, hose reels and foam inlets	AMD4649 AMD5756	(E3.6)
	Part 4:1986	- Specification for carbon dioxide systems		(D2.1)

Part 5	- Halon systems -		
Section 5.1:1982	- Halon 1301 total flooding systems	AMD4667	(D2.1)
Section 5.2:1984	- Halon 1211 total flooding systems		(D2.1)
Part 6	- Foam systems-		
Section 6.1:1988	- Specification for low expansion foam systems		(D2.1)
Section 6.2:1989	- Specification for medium and high expansion foam systems		(D2.1)
Part 7:1988	- Specification for powder systems		(D2.1)

BS 5314	Specification for gas heated catering equipment-		
Part 1:1976	- Ovens	AMD3122 AMD5398	(F2.2)
Part 2:1976	- Boiling burners	AMD3137 AMD5397	(F2.2)
Part 3:1976	- Grillers and toasters	AMD3146 AMD5399	(F2.2)
Part 4:1976	- Fryers	AMD3151 AMD5400	(F2.2)
Part 5:1976	- Steaming ovens	AMD3169 AMD5401	(F2.2)
Part 6:1976	- Bulk liquid heaters	AMD3172 AMD5402	(F2.2)
Part 7:1976	- Water boilers	AMD3173 AMD5403	(F2.2)
Part 8:1979	- Griddle plates	AMD5404	(F2.2)
Part 9:1979	- Boiling pans	AMD5405	(F2.2)
Part 10:1982	- Heated rinsing sinks	AMD5406	(F2.2)
Part 11:1979	- Hot cupboards	AMD5407 AMD5872	(F2.2)
Part 12:1979	- Bains-marie	AMD5408	(F2.2)

Standard	Part	Title	Amendments	References
	Part 13:1982	- Brat pans	AMD5409	(F2.2)
BS 5378		Safety signs and colours-		
	Part 3:1982	- Specification for additional signs to those given in BS 5378: Part 1		(P2.8)
BS 5386		Gas burning appliances-		
	Part 1:1976	- Gas burning appliances for instantaneous production of hot water for domestic use	AMD2990 AMD5832	(F2.2), (F5.11)
	Part 2:1981 (1986)	- Mini water heaters (2nd and 3rd family gases)		(F2.2), (F5.11)
	Part 3:1980	- Domestic cooking appliances burning gas	AMD4162 AMD4405 AMD4878 AMD5220 AMD6642 AMD6883	(F2.2), (F5.11)
	Part 4:1983	- Built-in domestic cooking appliances		(F2.2), (F5.11)
	Part 5:1988	- Specification for gas burning instantaneous water heaters with automatic output variation (2nd and 3rd family gases)		(F2.2), (F5.11)
BS 5390 :1976(1984)		Code of practice for stone masonry	AMD4272	(G3.1)
BS 5395		Stairs, ladders and walkways-		
	Part 2:1984	- Code of practice for the design of helical and spiral stairs		(S2.13) (S2.25)
	Part 3:1985	- Code of practice for the design of industrial type stairs, permanent ladders and walkways		(S2.25) (P2.8)
BS 5410		Code of practice for oil firing-		
	Part 1:1977	- Installations up to 44 kW output capacity for space heating, hot water and steam supply purposes	AMD3637	(F4.13), (F6.1)

	Part 2:1978	- Installations of 44 kW and above output capacity for space heating, hot water and steam supply purposes		(F2.2), (F6.1)
BS 5422:1977		Specification for the use of thermal insulating materials	AMD2599 AMD5742	(J3.5)
BS 5440		Installation of flues and ventilation for gas appliances of rated input not exceeding 60kW (1st, 2nd and 3rd family gases)-		
	Part 1:1990	- Specification for installation of flues		(F5.4), (F5.5), (F5.8)
	Part 2:1989	- Specification for installation of ventilation for gas appliances		(F5.3)
BS 5446		Components of automatic fire alarm systems for residential premises-		
	Part 1:1990	- Specification for self-contained smoke alarms and point type smoke detectors		(E4.1)
BS 5502		Buildings and structures for agriculture-		
	Section 1.6:1986	- Human and animal welfare		(R3.4)
	Section 3.6:1986	- Reference data space		(S2.26)
	Part 22:1993	- Code of practice for design, construction and loading		(C2.2)
	Part 80:1990	- Code of practice for design and construction of workshops, maintenance and inspection facilities		(S2.26)
BS 5534		Slating and tiling-		
	Part 1:1978 (1985)	- Design	AMD2734 AMD3554 AMD4777 AMD5781	(G3.1)

BS 5546:1979	Code of practice for installation of gas hot water supplies for domestic purposes (2nd family gases)		(F5.1)
BS 5572:1978	Code of practice for sanitary pipework	AMD3613 AMD4202	(M2.1-M2.2)
BS 5588	Fire precautions in the design and construction of buildings -		
Part 4:1978	- Code of practice for smoke control in protected escape routes using pressurisation	AMD5377	(E2.18-E2.19), (E2.45)
Part 5:1986	- Code of practice for firefighting stairways and lifts		(E3.15), (E3.17)
Part 8:1988	- Code of practice for means of escape for disabled people		(E5.1)
BS 5615:1985	Specification for insulating jackets for domestic hot water storage cylinders		(J3.4)
BS 5617:1985	Specification for urea-formaldehyde (UF) foam systems suitable for thermal insulation of cavity walls with masonry or concrete inner and outer leaves		(G3.1) (J2.2)
BS 5618:1985	Code of practice for thermal insulation of cavity walls (with masonry or concrete inner and outer leaves) by filling with urea-formaldehyde (UF) foam systems		(G3.1), (J2.2)
BS 5628	Code of practice for use of masonry-		
Part 1:1992	- Structural use of unreinforced masonry	AMD7745	(C2.1), (C3.1)
Part 3:1985	- Materials and components, design and workmanship	AMD4974	(C2.1) (G3.1) (H2.1,H2.2 H2.3)
BS 5655	Lifts and service lifts-		
Part 1:1986	- Safety rules for the construction and installation of electric lifts	AMD5840	(E3.17) (T2.5)

Standard	Title	Amendments	Reference
Part 2:1988	- Safety rules for the construction and installation of hydraulic lifts	AMD6220	(T2.5)
Part 5:1989	- Specification for dimensions of standard lift arrangements		(T2.5)
Part 7:1983	- Specification for manual control devices, indicators and additional fittings		(T2.5)
BS 5656:1983	Safety rules for the construction and installation of escalators and passenger conveyors		(P2.5)
BS 5776:1979	Specification for powered stairlifts	AMD4027 AMD6523	(T2.6)
BS 5810:1979	Code of practice for access for the disabled to buildings		(T2.3-T2.4) (T3.1-T3.3)
BS 5821	Methods for rating the sound insulation in buildings and of building elements-		
Part 1:1984 (ISO717/1)	- Method for rating the airborne sound insulation in buildings and of interior building elements		(H2.1,H2.2, H2.3)
Part 2:1984 (ISO717/2)	- Method for rating the impact sound insulation		(H2.1,H2.2 H2.3)
BS 5839	Fire detection and alarm systems for buildings-		
Part 1:1988	- Code of practice for system design, installation and servicing	AMD6317 AMD6874	(E4.1)
BS 5906:1980 (1987)	Code of practice for storage and on-site treatment of solid waste from buildings		(R2.1,R2.3, R2.4),(R2.6) (R2.7)(R2.9)
BS 5930:1981	Code of practice for site investigations		Part G Appendix
BS 5950	Structural use of steelwork in building-		

Part 1:1990	- Code of practice for design in simple and continuous construction: hot rolled sections	AMD6972	(C2.1) (C3.1)
Part 2:1992	- Specification for materials, fabrication and erection: hot rolled sections	AMD7766	(C2.1)
Part 4:1994	- Code of practice for design of composite slabs with profiled steel sheeting		(C2.1)
Part 5:1987	- Code of practice for design of cold formed sections	AMD5957	(C2.1)
Part 8:1990	- Code of practice for fire resistant design		(C2.1) Part D Appendix
BS 6180 :1982	Code of practice for protective barriers in and about buildings	AMD4858	(S3.3) (S3.6)
BS 6229 :1982	Code of practice for flat roofs with continuously supported coverings		(G3.1)
BS 6232	Thermal insulation of cavity walls by filling with blown man-made mineral fibre-		
Part 1:1982	- Specification for the performance of installation systems	AMD5428	(G3.1)
Part 2:1982	- Code of practice for installation of blown man-made mineral fibre in cavity walls with masonry and/or concrete leaves		(G3.1)
BS 6262:1982	Code of practice for glazing buildings		(E2.22), (P2.2)
BS 6283	Safety devices for use in hot water systems-		
Part 2:1991	- Specification for temperature and pressure relief valves for pressures up to and including 10 bar		(P3.1)

Part 3:1991	- Specification for combined temperature and pressure relief valves for pressures up to and including 10 bar		(P3.1)
BS 6297:1983	Code of practice for design and installation of small sewage treatment works and cesspools	AMD6150	(M2.4)
BS 6367:1983	Code of practice for drainage of roofs and paved areas	AMD4444	(M2.1)
BS 6399	Loading for buildings-		
Part 1:1984	- Code of practice for dead and imposed loads	AMD4949 AMD5881 AMD6031	(C2.2)
Part 3:1988	- Code of practice for imposed roof loads	AMD6033	(C2.2)
BS 6440:1983	Code of practice for powered lifting platforms for use by disabled people		(T2.6)
BS 6461	Installation of chimneys and flues for domestic appliances burning solid fuel (including wood and peat)-		
Part 1:1984	- Code of practice for masonry chimneys and flue-pipes		(F3.8), (F3.10), (F4.8-F4.9)
Part 2:1984	- Code of practice for factory-made insulated chimneys for internal applications		(F3.8), (F3.10), (F3.13), (F4.8-F4.9)
BS 6465	Sanitary installations-		
Part 1:1984	- Code of practice for scale of provision, selection and installation of sanitary appliances		(M3.1)
BS 6661:1986	Guide for design, construction and maintenance of single-skin air supported structures		(E2.44), (E2.68)

BS 6676	Thermal insulation of cavity walls using man-made mineral fibre batts (slabs)-	
Part 1:1986	- Specification for man-made mineral fibre batts (slabs)	(G3.1)
Part 2:1986	- Code of practice for installation of batts (slabs) filling the cavity	(G3.1)
BS 6677	Clay and calcium silicate pavers for flexible pavements-	
Part 1:1986	- Specification for pavers	(Q2.16)
Part 2:1986	- Code of practice for design of lightly trafficked pavements	(Q2.16)
BS 6700:1987	Specification for design, installation, testing and maintenance of services supplying water for domestic use within buildings and their curtilages	(P3.1)
BS 6714:1986	Specification for installation of decorative log and other fuel effect appliances (1st, 2nd and 3rd family gases)	(F5.1)
BS 6717	Precast concrete paving blocks-	
Part 1:1993	- Specification for paving blocks	(Q2.16)
BS 6915:1988	Specification for design and construction of fully supported lead sheet roof and wall coverings	(G3.1)
BS 7206:1990	Specification for unvented hot water storage units and packages	(P3.1)
BS 7501:1989	General criteria for the operation of testing laboratories	(B2.1)
BS 7502:1989	General criteria for the assessment of testing laboratories	(B2.1)
BS 7503:1989	General criteria for laboratory accreditation bodies	(B2.1)

BS 7511:1989	General criteria for certification bodies operating product certification		(B2.1)
BS 7512:1989	General criteria for certification bodies operating quality system certification		(B2.1)
BS 7513:1989	General criteria for certification bodies operating certification of personnel		(B2.1)
BS 7514:1989	General criteria for suppliers declaration of conformity		(B2.1)
BS 7671:1992	Requirements for electrical installations, IEE Wiring Regulations, 16th edition		(N2.1)
BS 8004 :1986	Code of practice for foundations		(C2.1)
BS 8110	Structural use of concrete-		
Part 1:1985	- Code of practice for design and construction	AMD5917 AMD6276 AMD7583	(C2.1) (C3.1)
Part 2:1985	- Code of practice for special circumstances	AMD7973	(C2.1) (C3.1)
Part 3:1985	- Design charts for singly reinforced beams, doubly reinforced beams and rectangular columns	AMD5918	(C2.1)
BS 8118	Structural use of aluminium-		
Part 1:1991	- Code of practice for design		(C2.1)
Part 2:1991	- Specification for materials, workmanship and protection		(C2.1)
BS 8200 :1985	Code of practice for design of non-loadbearing external vertical enclosures of buildings		(G3.1)
BS 8208	Guide to assessment of suitability of external cavity walls for filling with thermal insulants-		
Part 1:1985	- Existing traditional cavity construction	AMD4996	(G3.1)

BS 8213	Windows, doors and roof lights-		
Part 1:1991	- Code of practice for safety in use and during cleaning of windows and doors (including guidance on cleaning materials and methods)		(P2.3)
BS 8298:1989	Code of practice for design and installation of natural stone cladding and lining		(G3.1)
BS 8301:1985	Code of practice for building drainage		(G2.2) (M2.1-M2.3)
BS 8303:1986	Code of practice for installation of domestic heating and cooking appliances burning solid mineral fuels	AMD5723	(F2.2), (F3.2)

CODES OF PRACTICE

	Number	Title	Amendment	Context within Technical Standards
CP 3		Code of basic data for the design of buildings-		
	Chapter V	- Loading		
	Part 2:1972	- Wind loads		C2.2
CP 102:1973		Code of practice for protection of buildings against water from the ground	AMD1511 AMD2196 AMD2470	(G2.3,G2.5, G2.6) (J2.2)
CP 143		Code of practice for sheet roof and wall coverings-		
	Part 1:1958	- Aluminium, corrugated and troughed	PD4346	(G3.1)
	Part 5:1964	- Zinc		(G3.1)
	Part 10:1973	- Galvanized corrugated steel. Metric units		(G3.1)
	Part 12:1970 (1988)	- Copper. Metric units	AMD863 AMD5193	(G3.1)
	Part 15:1973 (1986)	- Aluminium. Metric units	AMD4473	(G3.1)
CP 144		Roof coverings-		
	Part 3:1970	- Built-up bitumen felt. Metric units	AMD2527 AMD5229	(G3.1)
	Part 4:1970	- Metric asphalt. Metric units		(G3.1)
CP 297:1972		Precast concrete cladding (non-loadbearing)		(G3.1)
CP 1007:1955		Maintained lighting for cinemas		(E2.44)

DRAFTS FOR DEVELOPMENT

Number	**Title**	**Context within Technical Standards**
DD 93:1984	Methods for assessing exposure to wind-driven rain	(G3.1)
DD 175:1988	Code of practice for the identification of potentially contaminated land and its investigation	Part G Appendix

INTERNATIONAL STANDARDS

Number	**Title**	**Context within Technical Standards**
ISO 140/4-1978	See BS 2750:Part 4:1980	(H2.1,H2.2, H2.3)
ISO 140/7-1978	See BS 2750:Part 7:1980	(H2.1,H2.2, H2.3)
ISO 717/1-1982	See BS 5821:Part 1:1984	(H2.1,H2.2, H2.3)
ISO 717/2-1982	See BS 5821:Part 2:1984	(H2.1,H2.2, H2.3)

Note: Copies of British Standards and British Standards Codes of Practice, Drafts for Development and International Standards may be purchased from the British Standards Institution, Linford Wood, Milton Keynes, MK14 6LE (Tel: 01908 221166). Copies of British Board of Agrément Certificates may be purchased from the British Board of Agrément, PO Box 195, Bucknalls Lane, Garston, Watford, Herts, WD2 7NG (Tel: 01923 670844).

APPENDIX 5

EUROPEAN STANDARDS FOR CONSTRUCTION

The following list has been selected for this publication from a list produced by BSI from the PERINFORM database. It includes standards and draft standards already published on topics relating to the construction industry.

Document identifier	Publication date	Title
EN 1	1980-02	Flued oil stoves with vaporising burners
EN 2	1972-08	Classification of fires
EN 2 AMD 1	1983-03	Classification of fires
EN 3-1	1975-06	Portable fire extinguishers; Part 1
EN 3-1 AMD 1	1987-01	Portable fire extinguishers; Part 1
EN 3-2	1978-07	Portable fire extinguishers; Part 2
EN 3-4	1984-01	Portable fire extinguishers; Part 4
EN 3-4 AC 1	1984-07	Portable fire extinguishers; Part 4
EN 3-5	1984-01	Portable fire extinguishers; Part 5: Complementary requirements and tests
EN 3-5 AMD 1	1986-03	Portable fire extinguishers; Part 5 Complementary requirements and tests; amendment to EN 3-5:1982.08
EN 20	1979-06	Wood preservatives; Determination of the preventive action against Lyctus brunneus (Stephens) (Laboratory method)
EN 21	1988-11	Wood preservatives; Determination of toxic values against Anobium punctatum (De Geer) by larval transfer (Laboratory method)
EN 22	1974-09	Wood preservatives; Determination of eradicant action against Holyotrupes bajulus (Linnaeus) larvae (Laboratory method)
EN 24	1974-12	Doors; Measurement of defects of general flatness of door leaves
EN 25	1975-01	Doors; Measurement of dimensions and of defects of squareness of door leaves
EN 26	1977-09	Gas burning appliances for instantaneous production of hot water for domestic use

EN 26 AMD 3	1984-03	Gas burning appliances for instantaneous production of hot water for domestic use
EN 26 AMD 4	1984-03	Gas burning appliances for instantaneous production of hot water for domestic use
EN 26 AMD 5	1986-10	Gas burning appliances for instantaneous production of hot water for domestic use
EN 26 AMD 6	1988-06	Gas burning appliances for instantaneous production of hot water for domestic use
EN 30	1979-01	Domestic cooking appliances burning gas
EN 30 AMD 2	1979-10	Domestic cooking appliances burning gas
EN 30 AMD 3	1985-07	Domestic cooking appliances burning gas
EN 30 AMD 4	1985-07	Domestic cooking appliances burning gas; Amendment 4 to EN 30:1979.01
EN 31	1977-03	Pedestal wash basins; Connecting dimensions
EN 31 AMD 1	1987-10	Pedestal wash basins; Connecting dimensions
EN 32	1977-02	Wall hung wash basins; Connecting dimensions
EN 32 AMD 1	1987-10	Wall hung wash basins; Connecting dimensions
EN 33	1979-04	Pedestal W.C. pan with close coupled cistern; Connecting dimensions
EN 34	1977-02	Wall hung W.C. pan with close coupled cistern; Connecting dimensions
EN 35	1977-03	Pedestal bidets over rim supply only; Connecting dimensions
EN 35 AMD 1	1987-10	Pedestal bidets over rim supply only; Connecting dimensions
EN 36	1977-03	Wall hung bidets over rim supply only; Connecting dimensions
EN 36 AMD 1	1987-10	Wall hung bidets over rim supply only; Connecting dimensions
EN 37	1979-04	Pedestal W.C. pan with independent water supply; Connecting dimensions
EN 38	1977-03	Wall hung W.C. pan with independent water supply; Connecting dimensions
EN 39	1976-12	Steel tubes for working scaffolds; Requirements, tests
EN 40-1	1976-06	Lighting columns; Part 1:Definitions and terms
EN 40-2	1976-06	Lighting columns; Part 2:Dimensions and tolerances
EN 40-3	1983-08	Lighting columns; Part 3:Materials
EN 40-4	1982-02	Lighting columns; Part 4:Surface protection of metal lighting columns

EN 40-5	1984-02	Lighting columns; Part 5:Base compartments and cableways
EN 40-6	1982-02	Lighting columns; Part 6:Loads
EN 40-8	1983-02	Lighting columns; Part 8:Verification of structural design for testing
EN 40-9	1983-02	Lighting columns; Part 9:Special requirements for reinforced and prestressed concrete lighting columns
EN 42	1975-10	Methods of testing windows; Air permeability tests
EN 43	1985-06	Methods of testing doors; Behaviour under humidity variations of door leaves placed in successive uniform climates
EN 46	1988-11	Wood preservatives; Determination of the preventive action against recently hatched larvae of Hylotrupes bajulus (Linnaeus) (Laboratory method)
EN 47	1988-11	Wood preservatives; Determination of toxic values against recently hatched larvae of Hylotrupes bajulus (Linnaeus) (Laboratory method)
EN 48	1988-11	Wood preservatives; Determination of the eradicant action against larvae of Anobium punctatum (Dr Geer) (Laboratory method)
EN 49	1976-09	Wood preservatives; Determination of the toxic values against Anobium punctatum (Dr Geer) by egg laying and larvial survival (Laboratory method)
EN 54-1	1976-10	Components of automatic fire detection systems, Part 1:Introduction
EN 54-5	1976-10	Components of automatic fire detection systems; Part 5:Heat sensitive detectors; containing a static element
EN 54-5 AMD 1	1988-04	Components of automatic fire detection systems; Part 5:Heat sensitive detectors; point detectors containing a static element
EN 54-6	1984-11	Components of automatic fire detection systems; Part 6:Heat-sensitive detectors; rate of rise point detectors without a static element
EN 54-6 AMD 1	1988-04	Components of automatic fire detection systems; Part 6:Heat-sensitive detectors; rate of rise point detectors without a static element
EN 54-6 AC 1	1984-11	Components of automatic fire detection systems; Part 6:Heat-sensitive detectors; rate of rise point detectors without a static element

EN 54-7	1984-11	Components of automatic fire detection systems; Part 7:Point type smoke detectors, Detectors using scattered light, transmitted light or ionization
EN 54-7 AMD 1	1988-04	Components of automatic fire detection systems; Part 7:Point type smoke detectors, Detectors using scattered light, transmitted light or ionization
EN 54-7 AC 1	1984-11	Components of automatic fire detection systems; Part 7:Point type smoke detectors, Detectors using scattered light, transmitted light or ionization
EN 54-8	1982-11	Components of automatic fire detection systems; Part 8:High temperature heat detectors
EN 54-8 AMD 1	1988-04	Components of automatic fire detection systems; Part 8:High temperature heat detectors
EN 54-8 AC 1	1984-11	Components of automatic fire detection systems; Part 8:High temperature heat detectors
EN 54-9	1984-11	Components of automatic fire detection systems; Part 9:Fire sensitivity test
EN 54-9 AC 1	1985-01	Components of automatic fire detection systems; Part 9:Fire sensitivity test
EN 58	1984-03	Sampling bituminous binders
EN 58 AMD 1	1986-04	Sampling bituminous binders
EN 73	1988-11	Wood preservatives, accelerated ageing of treated wood prior to biological testing, evaporative ageing test
EN 74	1988-06	Couplers, loose pigots and base plates for use in working scaffolds and falsework made of steel tubes; Requirements and test procedures
EN 77	1977-08	Methods of testing windows; Wind resistance tests
EN 78	1977-08	Methods of testing windows; Form of test report
EN 79	1985-06	Methods of testing doors; Behaviour of door leaves placed between two different climates
EN 80	1978-11	Wall hung urinals without built-trap; Connecting dimensions
EN 80 AMD 1	1978-12	Wall hung urinals without built-trap; Connecting dimensions
EN 81-1	1985-12	Safety rules for construction and installation of lifts and service lifts; Part 1:Electric lifts
EN 81-2	1987-11	Safety rules for construction and installation of lifts and service lifts; Part 2:Hydraulic lifts

EN 84	1989-06	Wood preservatives; Accelerated ageing of treated wood prior to biological testing, leaching procedure
EN 84 AC 1	1990-04	Wood preservatives; Accelerated ageing of treated wood prior to biological testing, leaching procedure; amends 1
EN 85	1980-04	Methods of testing doors; Hard body impact test on door leaves
EN 86	1980-04	Methods of testing windows; Water tightness test under static pressure
EN 87	1984-01	Ceramic floor and wall tiles; Definitions, classification and marking
EN 98	1984-01	Ceramic tiles; Determination of dimensions and surface quality
EN 99	1984-01	Ceramic tiles; Determination of water absorption
EN 100	1984-01	Ceramic tiles; Determination of modulus rupture
EN 101	1984-12	Ceramic tiles; Determination of scratch hardness of surface according to Mohs
EN 102	1984-01	Ceramic tiles; Determination or resistance to deep abrasion; Unglazed tiles
EN 103	1984-01	Ceramic tiles; Determination of linear thermal expansion
EN 104	1984-01	Ceramic tiles; Determination of resistance to thermal shock
EN 105	1982-08	Determination of crazing resistance; Glazed tiles
EN 106	1984-01	Ceramic tiles; Determination of chemical resistance; Unglazed tiles
EN 107	1980-10	Methods of testing windows; Mechanical test
EN 108	1980-10	Methods of testing doors; Test for deformation in the plane of the leaf
EN 111	1984-09	Wall hung rinse basins; Connecting dimensions
EN 111 AMD 1	1987-01	Wall hung hand rinse basins; Connecting dimensions
EN 113	1980-04	Wood preservatives; Determination of toxic values of wood preservatives against wood destroying Basidiomycetes cultured on an agar medium

EN 113 AMD 2	1985-07	Wood preservatives; Determination of toxic values of wood preservatives against wood destroying Basidiomycetes cultured on an agar medium
EN 115	1983-11	Safety rules for the construction and installation of escalators and passenger conveyers
EN 117	1989-12	Wood preservatives; Determination of toxic values against Reticluitermes santonensis de Feytaud (laboratory method)
EN 117 AC 1	1990-04	Wood preservatives; Determination of toxic values against Reticluitermes santonensis de Feytaud (laboratory method) amends 1
EN 118	1990-03	Wood preservatives; Determination of preventive action against Reticluitermes santonensis de Feytaud (laboratory method)
EN 120	1984-01	Particleboards; Determination of formaldehyde content; Extraction method called perforator method
EN 121	1984-01	Extruded ceramic tiles with low water absorption (E<kleiner =>3%); Group A 1
EN 122	1984-01	Ceramic tiles; Determination of chemical resistance; Glazed tiles
EN 124	1986-10	Gully tops and manhole tops for vehicular and pedestrian areas; Design requirements, type testing, marking
EN 129	1984-07	Methods of testing doors; Test for deformation in torsion of the door leaves
EN 130	1984-07	Methods of testing doors; Test for the change in stiffness of the door leaves by repeated torsion
EN 152-1	1988-09	Test methods for wood preservatives; Laboratory method for determining the protective effectiveness of a preservative treatment against blue stain in service; Part 1; Brushing procedure
EN 152-2	1988-03	Test methods for wood preservatives; Laboratory method for determining the protective effectiveness of a preservative treatment against blue stain in service; Part 2; Application by methods other than brushing
EN 153	1990-05	Methods for measuring the energy consumption of electric mains operated household refrigerators, frozen food storage cabinets, food freezers and their combinations, together with associated characteristics

EN 154	1984-01	Ceramic tiles; Determination of resistance to surface abrasion; Glazed tiles
EN 155	1984-01	Ceramic tiles; Determination of moisture expansion using boiling water; Unglazed tiles
EN 159	1984-01	Dust pressed ceramic tiles with water absorption $E > 10\%$; Group Bill
EN 162	1985-06	Methods of testing doors; Soft and heavy body impact test on door leaves
EN 163	1985-01	Ceramic tiles; Sampling and basis for acceptance
EN 176	1884-01	Dust-pressed ceramic tiles with a water absorption ($E \leq 3\%$); Group BI
EN 177	1984-12	Dust-pressed ceramic tiles with a water absorption of $3\% < E \leq 6\%$ (Group Blla)
EN 178	1984-12	Dust-pressed ceramic tiles with a water absorption of $6\% < E \leq 10\%$ (Group Bllb)
EN 186-1	1985-01	Ceramic tiles: Extruded ceramic tiles with a water absorption of $3\% < E \leq 6\%$ (Group Alla) Part 1
EN 186-2	1985-01	Ceramic tiles; Extruded ceramic tiles with a water absorption of $3\% < E \leq 6\%$ (Group Alla) Part 2
EN 187-1	1985-01	Ceramic tiles; Extruded ceramic tiles with a water absorption of $6\% < E \leq 10\%$ Group Allb) Part 1
EN 187-2	1985-01	Ceramic tiles; Extruded ceramic tiles with a water absorption of $6\% < E \leq 10\%$ (Group Allb) Part 2
EN 188	1985-01	Ceramic tiles; Extruded ceramic tiles with a water absorption of $E > 10\%$ (Group Alll)
EN 196-1	1987-05	Methods of testing cement; Determination of strength
EN 196-2	1987-05	Methods of testing cement; Chemical analysis of cement
EN 196-3	1987-05	Methods of testing cement; Determination of setting time and soundness
ENV 196-4	1987-05	Methods of testing cement; Quantitive determination of constituents
EN 196-5	1987-05	Methods of testing cement; Pozzolanicity test for pozzolanic

EN 196-6	1989-12	Methods of testing cement; Determination of fineness
EN 196-7	1989-12	Methods of testing cement; Methods of taking and preparing samples of cement
EN 196-21	1989-12	Methods of testing cement; Determination of the chloride, carbon dioxide and alkali content of cement
EN 200	1989-06	Sanitary tapware; General technical specifications for single taps and mixer taps (normal size 1/2) PN 10; minimum flow pressure of 0,05 MPa 0,5) bar
EN 202	1985-01	Ceramic tiles; Determination of frost resistance
ENV 206	1990-03	Concrete; Performance, production, placing and compliance criteria
EN 212	1986-06	Wood preservatives; Guide to sampling and preparation of wood preservatives and treated timber for analysis
EN 215-1	1987-07	Thermostatic radiator valves; Part 1: Requirements and test methods
EN 215-1 AC 1	1987-07	Thermostatic radiator valves; Part 1 requirements and test methods
EN 232	1990-06	Baths; connecting dimensions
EN 233	1989-01	Wallcoverings in roll form; Specification for finished wallpapers, wall vinyls and plastic wall coverings
EN 234	1989-01	Wallcoverings in roll form; Specification for wallcoverings for subsequent design
EN 235	1989-01	Wallcoverings in roll form; Vocabulary and symbol
EN 246	1989-06	Sanitary tapware; General specifications for flow rate regulators
ENV 247	1990-06	Heat exchangers; Terminology
EN 248	1989-06	Sanitary taps; General technical specifications for electrodeposited nickel chrome coatings
EN 251	1990-06	Shower trays; Connecting dimensions
EN 252	1989-06	First test method for determining the relative protective effectiveness of a wood preservative in ground contact

EN 252 AC 1	1989-11	First test method for determining the relative protective effectiveness of a wood preservative in ground contact; amends EN 252 June 1989
EN 253	1990-09	Preinsulated bonded pipe systems for under ground hot water networks; Pipe assembly of steel service pipes, polyurethane thermal insulation and outer casing of high density polyethylene
EN 254	1990-09	Preinsulated bonded pipe systems for under ground hot water networks; Pipe assembly of steel service pipes, polyurethane thermal insulation and outer casing of high density polyethylene
EN 255-1	1988-10	Heat pumps; Heat pump units with electrically driven compressors for heating or for heating and cooling; Part 1:Terms, definitions and designation
EN 263	1987-06	Specification for cast acrylic sheet for baths and shower trays for domestic purposes
ENV 305	1990-06	Heat exchangers; Definitions of performance of heat exchangers and general test procedure for establishing performances of all heat exchangers
ENV 306	1990-06	Heat exchangers; Methods of measuring the parameters necessary for establishing the performance
ENV 307	1990-06	Heat exchangers; Guidelines to prepare installation, operating and maintenance instructions required to maintain the performance of each type of heat exchanger
EN 327	1990-11	Heat exchangers; Test procedures for establishing performance of forced convection air cooled refrigerant condensers
EN 10002-1	1989-12	Metallic materials; tensile testing; Part 1: Method of test
EN 10002-1 AC	1990-06	Metallic materials; tensile testing; Part 1: Method of test; amendment to EN 10002-1
EN 10020	1988-11	Definition and classification of grades of steel
EN 10025	1990-03	Hot rolled products of non-alloy structural steels, technical delivery conditions

EN 10036	1989-01	Chemical analysis of ferrous materials; Determination of total carbon in steels and irons; Gravimetric method after combustion in a stream of oxygen
EN 10045-1	1989-12	Metallic materials; Charpy impact test; Part 1 Test method
EN 10071	1989-01	Chemical analysis of ferrous materials; Determination of manganese in steels and irons; Electrometric titration method
EN 10136	1989-01	Chemical analysis of ferrous materials; Determination of nickel in steels and irons; Atomic absorption spectrometric method
EN 10142	1990-12	Continuously hot-dip zinc coated low carbon steel sheet and strip for cold forming; Technical delivery conditions
EN 10177	1989-01	Chemical analysis of ferrous materials; Determination of calcium in steels; Flame atomic absorption spectrometric method
EN 10178	1989-01	Chemical analysis of ferrous materials; Determination of niobium in steels; Spectrophotometric method
EN 10179	1989-01	Chemical analysis of ferrous materials; Determination of nitrogen (trace amounts) in steels; Spectrophotometric method
EN 10181	1989-01	Chemical analysis of ferrous materials; Determination of lead in steels; Flame atomic absorption spectrometric method
EN 10184	1989-05	Chemical analysis of ferrous materials; Determination of phosphorus in steels and irons; Spectrophotometric method
EN 10188	1989-05	Chemical analysis of ferrous materials; Determination of chromium in steels and irons; Flame atomic absorption spectrometric method
EN 10202	1989-04	Cold reduced electrolytic chromium/chromium oxide coated steel
EN 20216	1990-03	Writing paper and certain classes of printed matters; trimmed sizes; A and B series (ISO 216: 1975)
EN 22860	1985-12	Earth-moving machinery; Minimum access dimensions (ISO 2860-1983, edition 3, 1983-06-01) incorporating an agreed common modification

EN 23164	1985-09	Earth-moving machinery; Laboratory evaluations of roll-over and falling-object protective structures; Specifications for the deflection-limiting volume (ISO 3164-1979, edition 2 with Amendment 10 1980)
EN 23411	1988-03	Earth-moving machinery; Human physical dimensions of operators and minimum operator space envelope (ISO 3411-1982, 2nd edition)
EN 23449	1988-03	Earth-moving machinery; Falling-objective protective structures; Laboratory tests and performance requirements (ISO 3449-1984, 3rd edition)
EN 25353	1988-11	Earth-moving machinery and tractors and machinery for agriculture and forestry; Seat index point (ISO 5353:1978, AMD 1: 1981, AMD 2: 1984)
ENV 26385	1990-06	Ergonomic principles of the design of work systems (ISO 6385: 1981)
EN 26927	1990-11	Building construction; Jointing products; Sealants; Vocabulary (ISO 6927: 1981)
EN 27389	1990-11	Building construction; Jointing products; Determination of plastic recovery
EN 27390	1990-11	Building construction; Determination of resistance to flow (ISO 7390: 1987)
EN 27574-1	1988-12	Acoustics; Statistical methods for determining and verifying stated noise emission values of machinery and equipment; Part 1: General considerations and definitions (ISO 7574-1: 1985)
EN 27574-2	1988-12	Acoustics; Statistical methods for determining and verifying stated noise emission values of machinery and equipment; Part 2: Method for stated values for individual machines (ISO 7574-2: 1985)
EN 27574-3	1988-12	Acoustics; Statistical methods for determining and verifying stated noise emission values of machinery and equipment; Part 3: Simple (transition) method for stated values for batches of machines (ISO 7574-3: 1985)

EN 27574-2	1988-12	Acoustics; Statistical methods for determining and verifying stated noise emission values of machinery and equipment; Part 4: Method for stated values for batches of machines (ISO 7574-2: 1985)
EN 28233	1990-12	Thermoplastics valves: torque; Test method (ISO 8233:1988)
EN 28339	1990-11	Building construction; Jointing products; Sealants; Determination of tensile properties (ISO 8339:1984)
EN 28340	1990-11	Building construction; Jointing products; Sealants; Determination of tensile properties at mainland extension (ISO 8339:1984)
EN 28394	1990-11	Building construction; Jointing products; Determination of extrudability on one-component sealants (ISO 8394:1988)
EN 28659	1990-12	Thermoplastic valves; Fatigue strength; Test method (ISO 8659:1989)
EN 29000	1990-02	Quality management and quality assurance standards; Guidelines for selection and use
EN 29001	1990-02	Quality systems; Model for quality assurance in design/development, production, installation and servicing
EN 29002	1990-02	Quality systems; Model for quality assurance in production and installation
EN 29003	1990-02	Quality systems; Model for quality assurance in final inspection and test
EN 29004	1990-02	Quality management and quality system elements; Guidelines
EN 29046	1990-11	Building construction; Jointing products; Determination of adhesion, cohesion properties at constant temperatures (ISO 9046:1987)
EN 29048	1990-11	Building construction; Jointing products; Determination of extrudability of sealants using standardised apparatus (ISO 9048:1987)
EN 45001	1989-09	General criteria for the operation of testing laboratories
EN 45002	1989-09	General criteria for the assessment of testing laboratories

EN 45003	1989-09	General criteria for laboratory accreditation bodies
EN 45011	1989-09	General criteria for certification bodies operating product certification
EN 45012	1989-09	General criteria for certification bodies operating quality system certification
EN 45013	1989-09	General criteria for certification bodies operating certification of personnel
EN 45014	1989-09	General criteria for supplier's declaration of conformity
EN 50006-3	1975-01	Limitation of disturbances in electricity supply networks caused by domestic and similar appliances equipped with electronic devices
EN 50007	1981-01	Low voltage switchgear and controlgear for industrial use. Single hole mounted control switches and indicator lights mounting dimensions
EN 50008	1987-07	Low voltage switchgear and controlgear for industrial use. Inductive proximity switches, form A, for direct current, 3 or 4 terminals
EN 50010	1987-07	Low voltage switchgear and controlgear for industrial use; Inductive proximity switches; Methods for measuring the operating distance and operating frequency
EN 50011	1978-01	Low voltage switchgear and controlgear for industrial use; Terminal marking, distinctive number and distinctive letter for particular contractor relays
EN 50012	1978-01	Low voltage switchgear and controlgear for industrial use; Terminal marking and distinctive number for auxiliary contacts on particular contractors
EN 50013	1978-01	Low voltage switchgear and contolgear for industrial use; Terminal marking and distinctive number for particular control switches
EN 50015	1979-01	Electrical apparatus for potentially explosive atmospheres; Oil immersion "o"
EN 50016	1979-01	Electrical apparatus for potentially explosive atmospheres; Pressurised apparatus "p"

EN 50017	1979-01	Electrical apparatus for potentially explosive atmospheres; Powder filling "q"
EN 50018	1979-01	Electrical apparatus for potentially explosive atmospheres; Flameproof enclosure "d"
EN 50019	1979-01	Electrical apparatus for potentially explosive atmospheres; Increased safety "e"
EN 50022	1978-07	Low voltage switchgear and controlgear for industrial use; Mounting rails; Top hat rails 35mm wide for snap-on mounting of equipment
EN 50023	1978-07	Low voltage switchgear and controlgear for industrial use; Mounting rails; Top hat rails 75 mm wide for snap-on mounting of equipment
EN 50024	1980-07	Low voltage switchgear and controlgear for industrial use; Mounting rails C-profile and accessories for the mounting of equipment
EN 50025	1987-07	Low voltage switchgear and controlgear for industrial use; Inductive proximity switches, form C, for direct current, 3 or 4 terminals
EN 50026	1987-07	Low voltage switchgear and controlgear for industrial use; Inductive proximity switches, form D, for direct current, 3 or 4 terminals
EN 50027	1979-09	Low voltage switchgear and controlgear for industrial use; Terminal aperture sizes for unprepared round copper conductors
EN 50032	1982-07	Low voltage switchgear and controlgear for industrial use; Inductive proximity switches. Definitions, classification and designation
EN 50033	1985-09	Electrical apparatus for potentially explosive atmospheres. Caplamps for mines susceptible to firedamp
EN 50035	1980-07	Low voltage switchgear and controlgear for industrial use; Mountain rails, G-profile for the fixing of terminal blocks
EN 50036	1987-07	Low voltage switchgear and controlgear for industrial use; Inductive proximity switches, form A, for alternating current, 2 terminals
EN 50037	1987-07	Low voltage switchgear and controlgear for industrial use; Inductive proximity switches, form C, for alternating current, 2 terminals

EN 50038	1987-07	Low voltage switchgear and controlgear for industrial use; Inductive proximity switches, form D, for alternating current, 2 terminals
EN 50039	1982-01	Electrical apparatus for potentially explosive atmospheres; Intrinsic safety "i" systems
EN 50040	1987-07	Low voltage switchgear and controlgear for industrial use; Inductive proximity switches, form A, for direct current, 2 terminals
EN 50041	1981-01	Low voltage switchgear and controlgear for industrial use; Control switches. Position switches 42,5 x 80. Dimensions and characteristics
EN 50042	1982-01	Low voltage switchgear and controlgear for industrial use; Terminal marking. Terminals for external associated electronic circuit components and contacts
EN 50043	1984-09	Low voltage switchgear and controlgear for industrial use; Sizes numbers and gauges for flat connections
EN 50044	1982-01	Low voltage switchgear and controlgear for industrial use; Inductive proximity switches. Identification of connections
EN 50045	1982-01	Low voltage switchgear and controlgear for industrial use; Mounting rail. Top hat rail, 15 mm wide for the fixing of terminal blocks
EN 50047	1981-01	Low voltage switchgear and controlgear for industrial use; Control switches. Position switches 30 x 55. Dimensions and characteristics
EN 50049	1983-09	Domestic or similar electronic equipment interconnection requirements; peritelevision connector
EN 50050	1986-01	Electrical apparatus for potentially explosive atmospheres
EN 50052	1986-01	Cast aluminium alloy enclosure for gas-filled high-voltage switchgear and controlgear
EN 55014	1987-01	Limits and methods of measurement of radio interference characteristics of household electrical appliances, portable tools and similar electrical apparatus

EN 55015	1987-01	Limits and methods of measurement of radio interference characteristics of fluorescent and luminaries
EN 55022	1987-01	Limits and methods of measurement of radio interference characteristics of information technology equipment
EN 60238-1	1986-01	Edison screw lamp-holders
EN 60320-1	1987-06	Appliance couplers for household and similar general purposes
EN 60400	1985-01	Lamp-holders for tubular fluorescent lamps and starterholders
EN 60555-1	1987-07	Disturbances in supply systems caused by household appliance and similar electrical equipment. Part 1: Definitions
EN 60555-2	1987-07	Disturbances in supply systems caused by household appliance and similar electrical equipment. Part 2: Harmonics
EN 60555-3	1987-07	Disturbances in supply systems caused by household appliance and similar electrical equipment. Part 3: Voltage fluctuations
EN 60662	1987-01	High pressure sodium vapour lamps

APPENDIX 6

TABLE OF CASES

Case	Paragraph
Anns *v.* London Borough of Merton 1978 AC 728; 2 All ER 492	4.6
City of Edinburgh District Council *v.* Co-operative Wholesale Society 1986 SLT (Sh.Ct.) 57	2.19
Dunbarton County Council *v.* George Wimpey & Co. Ltd 1968 SLT 246	2.26
Galek's Curator Bonis *v.* Thomson (OH) 1991 SLT 29	4.4
Glasgow District Council *v.* Secretary of State (OH) 1981 SLT (Notes) 63	—
GUS Property Management *v.* City of Glasgow District Council (OH) 1989 SLT 390	2.22
Hadden *v.* City of Glasgow District Council (OH) 1986 SLT 557	4.7
Howard *v.* Hamilton District Council 1985 SLT (Sh.Ct.) 42	2.21
Manford *v.* George Leslie 1987 SCLR (Sh.Ct.) 232	4.8
McChlery *v.* J W Haran (OH) 1987 SLT 662	4.8
McWilliams *v.* Sir William Arrol & Co. Ltd and Lithgows Ltd 1962 SC (HL) 70	4.4
Millar *v.* Galashiels Gas Co. Ltd 1949 SC (HL) 31	4.2
Morrisons Associated Companies Ltd *v.* James Rome & Sons Ltd (OH) 1964 SLT (Notes) 75	4.2
Murphy *v.* Brentwood District Council 1991 1 AC 398	4.6
Pegg *v.* City of Glasgow District Council 1988 SLT (Sh.Ct.) 49	2.22
Sawyers *v.* Clackmannan County Buildings Authority 1973 SLT (Sh.Ct.) 4	—

NB. The following abbreviations have been used in listing the reports:

AC	Law Reports Appeal Case series	(England)
All ER	All England Law Reports	(England)
BLR	Building Law Reports	(England)
HL	House of Lords	(Scotland/England)
OH	Outer House, Court of Session	(Scotland)
SC	Session Cases	(Scotland)
SLR	Scottish Law Review	(Scotland)
Sh.Ct.	Sheriff Court	(Scotland)
SLT	Scots Law Times	(Scotland)

APPENDIX 7

BUILDING AND BUILDINGS
THE BUILDING (FORMS) (SCOTLAND) REGULATIONS 1991

Powers are conferred on the Secretary of State by s. 24(1)(a) of the act to make forms regulations. Forms which are suitable for use by buildings authorities are presented as prescribed forms within these regulations.

Each individual building authority then develops its own set of forms including the prescribed procedural arrangements which it then makes available to applicants for building work or uses (in terms of notices and orders) to control building work.

The Building (Scotland) Act, written in legalistic terms makes rather heavy reading whereas the various forms used for building control purposes are designed for ease of understanding of the requirements of the act by lay persons. It is recommended that a sensible approach to the understanding of the act is to read it in conjunction with the forms.

There follows a complete listing of the prescribed forms set out in the Building (Forms)(Scotland) Regulations 1991 and thereafter space only allows for a small selection of typical examples of forms which would be available from a building authority.

The forms selected are numbers 1, 4, 4B, 8, 9, 14 and 26.

SCHEDULE
INDEX OF PRESCRIBED FORMS

Form No.	Title
1	Application for Relaxation of Building Standards Regulations
2	Relaxation Direction
3	Relaxation of Building Standards Regulations – Appeal to S of S
4	Application for Building/Demolition/Change of Use Warrant
4A	Safety Design Certificate (New Building)
4B	Safety Design Certificate (Alterations/Extensions/Change of Use)
5	Building Warrant
6	Demolition Warrant
7	Application for Amendment of Building Warrant
8	Application for Certificate of Completion
9	Compliance Certificate for Electrical Installation
10	Certificate of Completion
11	Application for Temporary Occupation/Use
12	Authorisation for Temporary Occupation/Use
13	Application for Extension of Period of Use of Building Intended to have a Limited Life
14	Notice under s. 10 to Show Cause why Building Operations should not be Executed
15	Order under s. 10 requiring Execution of Building Operations
16	Notice under s. 11 Requiring Owner to Show Cause why Building Should not Conform to Building Standards Regulations
17	Order under s. 11 to Make Building Conform to Standards Regulations
18	Notice Requiring Operations on a Dangerous Building
19	Notice of Intention to Make Order Requiring Operations on a Dangerous Building
20	Order under s. 13 requiring Owner to Carry out Operations on a Dangerous Building
21	Notice to Remove from a Dangerous Building or Adjacent Building
22	Certificate that a Building is a Source of Immediate Danger
23	Certificate that Order has been made under s. 13 Requiring Building to be Demolished
24	Notice of Intention to Enter Premises
25	Authority to Enter Premises
26	Charging Order

Form 1

Example of the use of Form 1 (Application for a Relaxation) relates to a situation in an attic conversion where the applicant wishes to install an emergency window (escape velux) in the plane of the roof (which has a pitch of 35°) rather than use an upstand to install it at an angle of 40° which is the normal situation (see Regulation 9 as read with Technical Standard E2.60(b)). The building has some architectural merit and the argument is on aesthetic grounds.

FORM 1

EASTREN COUNCIL

APPLICATION FOR RELAXATION OF BUILDING STANDARDS REGULATIONS

Building (Scotland) Act 1959, section 4

Before completing this form please read the Notes at the back

PART A

This application should be sent to the Secretary of State/Eastren Council* (see note 1)

1. APPLICANT	2. AGENT (if any)
Name Jane Boyle	Name Tom Brown
Address 14 Windfall Lane Glasgow	Address 76 The Winds Glasgow
Post Code G44 3	Post Code G43 4
Tel No 0141 601 6929	Tel No 0141 621 6128

3. ADDRESS OF BUILDING
(in relation to which application is made)

14 Windfall Lane
Glasgow

4. NAME AND ADDRESS OF OWNER OF BUILDING
(if different from applicant)

As applicant

PART B

5. DETAILS OF APPLICATION

Regulation number and provisions of the TS from which dispensation or relaxation is being sought	Proposed dispensation or relaxation	Grounds for application
Regulation 9 as read with Technical Standard E2.60(b) diii	Relaxation of standard relating to angle of emergency window (velux) to permit fitting in plane of roof (34°)	Building is in area of buildings subject to Article Four planning controls

PART C (EXISTING BUILDINGS ONLY)

6. State present use of building Dwelling
7. State proposed use of building Dwelling

PART D (NEW BUILDINGS ONLY)

8. State proposed use of building Not relevant

PART E

9. FIRE AUTHORITY
Is the work being carried out at the request of the Fire Authority? ~~YES~~/NO*

10. SECTION 10 NOTICE
Has a notice under s. 10 of the Building (Scotland) Act 1959 been served on you? (see note 5) ~~YES~~/NO*

11. If so, give date of notice Not relevant

12. SECTION 11 NOTICE
Has a notice under s. 11 of the Building (Scotland) Act 1959 been served on you? (see note 5) ~~YES~~/NO*

13. If so, give date of notice — Not relevant

14. BUILDING WARRANT
Has an application for a building warrant been made? YES/~~NO~~*

If so, please give date of application and reference number — 29 February 1996 ZXZM 3459 96

15. LISTED BUILDING
Is the building listed as being of special architectural or historic interest, or in a conservation area? (If in doubt the planning authority can advise) YES/~~NO~~* — It is in a consevation area

16. If so, please state category — Article Four

PART F

17. I/~~We~~* apply for a direction dispensing with or relaxing those provisions of the Building Standards Regulations set out above (in accordance with any necessary plans (including drawings, specifications and other particulars) submitted with the application)*
(see note 6)

Signature of ~~applicant~~/Agent* Tom Brown Date 29/2/96

*Delete as appropriate

PART G

EASTREN COUNCIL	ENQUIRY POINT
Eastren Council	Name Mr Pat Roller
Eastren Road	Tel. 6943 6217
Eastren	Ext. 443
G48 4RT	

Notes

(1) Where on application it appears to the Secretary of State or local authority that it is unreasonable that any provision of the building standards regulations should apply to any particular building, the Secretary of State or the local authority may give a direction relaxing that provision in relation to that building. If you are in doubt to whom your application should be addressed, you should seek advice from your local authority. If your application is to be addressed to the Secretary of State, a copy must also be sent to the local authority.

(2) Applications to the local authority should be sent to the address shown at Part G of the form. The local authority can advise you where applications to the Secretary of State should be sent.

(3) If the local authority refuse to dispense with or relax the regulations or apply conditions to a relaxation which you find unacceptable you may appeal to the Secretary of State against their decision. You must appeal within 28 days from the local authority's decision by completing Form 3 and sending it to the Secretary of State (see note 2).

(4) If the local authority do not give a decision within two months of you making the application you may appeal to the Secretary of State following the procedure in note 3.

(5) Section 10 of the Building (Scotland) Act 1959 gives powers to the local authority to remove or make a building conform to the building standards regulations where the building was constructed without a warrant, in contravention of the conditions of the warrant, or where the life of a limited life building has expired. Section 11 gives power to the local authority to require buildings to conform to the building standards regulations.

(6) The local authority can advise you on what plans, drawings, specifications and other particulars should accompany your application.

The response from the building authority would be made on Form 2 either accepting (or accepting with conditions) or refusing the application for dispensation or relaxation.

If the applicant is unhappy about the local authority response then Form 3 can be used to appeal directly to the Secretary of State on the issue.

Form 4

The example of the use of Form 4 (warrant application form) relates to the attic conversion which was the subject of the relaxation application just dealt with.

FORM 4

EASTREN COUNCIL

APPLICATION FOR BUILDING/DEMOLITION/
CHANGE OF USE* WARRANT

*Delete as appropriate
Building (Scotland) Act 1959, Section 6

The local authority will grant a building warrant if they are satisfied that the building will be constructed in accordance with the building operations regulations and the building standards regulations. A warrant for demolition will be granted if the building operations regulations will be met

PART A (to be completed for all applications)

1. APPLICANT
Name Jane Boyle
Address 14 Windfall Lane
Glasgow
Post Code G44 3
Tel No 0141 601 6929

2. AGENT (if any)
Name Tom Brown
76 The Winds
Glasgow
Post Code G43 4
Tel No 0141 621 6128

3. PURPOSE FOR WHICH WARRANT IS SOUGHT
Please state the nature of the proposed operations, e.g. alteration, extension, change of use or demolition of house, shop or other building Attic conversion

4. ADDRESS OF BUILDING(S)
(for which application is being made)

14 Windfall Lane, Glasgow

5. LISTED BUILDINGS
Does this application concern buildings listed as being of special architectural or historic interest or in a conservation area? (If in doubt the planning authority can advise.)

Yes/~~No~~*

6. If so state which category Conservation area — subject to Article 4 Direction

7. USE(S) OF BUILDING(S)
The building(s) will be used as The proposal is to convert the attic into two additional bedrooms for this dwelling.

8. RELAXATION DIRECTIONS
The building(s) is/~~are~~* the subject of a relaxation direction given by the ~~Secretary of State~~/the local authority*
reference number ZXZM 3459 96 dated 29 Feb 1996

9. STAGES OF CONSTRUCTION
If a staged warrant is required, please indicate which stage(s) are requested Not relevant in this case

NB. You can arrange at the LA's discretion to be granted a warrant on condition that you provide further details before any of the following stages are commenced –

foundations
substructure
underground drainage system
superstructure, excluding the stages specified below
external wall cladding or internal walls or their linings
roof
installation of a lift, escalator or electrical, ventilation, heating or plumbing system

PART C (to be completed by all applicants)

10. COST OF OPERATIONS
The cost of the operations is estimated at £ 5,500
Please note that the local authority may wish to verify this figure.

PART D (to be completed where an existing building is to be used for a different purpose)

11. EXISTING USE(S) OF A BUILDING(S)
The building(s) is/~~are~~* at present used as a dwelling

12. PROPOSED USE(S) OF A BUILDING(S)
The building(s) will be used as a dwelling

PART E (complete this part only if the building is intended to have a life of less than 5 years – less onerous requirements may apply to buildings in this category)

13. LIMITED LIFE BUILDING(S)
The life of building(s) will be (not more than 5 years)
Not relevant

PART F (to be completed by all applicants)

14. DECLARATION
I/~~We~~* apply for a warrant in accordance with the details supplied above and with any necessary plans (including drawings, specifications, and other particulars) accompanying this application.
I/~~We~~* have/~~have not~~* included a certificate showing that the design complies with the structural requirements of the building standards regulations (see notes 2 and 3).

Signature of ~~applicant~~/agent Tom Brown
Date 29/2/96

*Delete as appropriate

PART G

ADDRESS TO WHICH YOU SHOULD SEND YOUR APPLICATION

Eastren Local Authority
Eastren Park
Glasgow G162 WF

Local Authority Enquiry Point – Building Control Department
Extension 445

Warning
A BUILDING WARRANT DOES NOT EXEMPT YOU FROM OBTAINING OTHER TYPES OF PERMISSION NECESSARY, E.G. – PLANNING PERMISSION OR LISTED BUILDING CONSENT. CONSULT THE LOCAL AUTHORITY IF IN DOUBT.

Notes

(1) If the local authority refuse to grant a warrant you may appeal to the sheriff within 21 days of the date of the decision.

(2) The local authority can advise you on what plans, drawings, specifications and other particulars should accompany your application.

(3) The Building (Self-Certification of Structural Design) (Scotland) Regulations 1992 enable an applicant for building warrant to submit with his application a certificate signed by a suitably qualified civil or structural engineer certifying that the design, including the specification of material used, complies with regulation 11 of the Building Standards (Scotland) Regulations 1990. Forms 4A and 4B refer.

In facilitating note 2 above, the typical information provided by the local authority would be as follows:

(1) The address to which the building warrant application should be sent.

(2) The list of items to be submitted, being as follows:
   (a) Application form (to be completed)
   (b) Fees form (to be completed)
   (c) Fees cheque (a table of fees would be provided to allow the applicant to determine in relation to the cost of the work, the fee to be rendered)
   (d) Structural design certificate if appropriate
   (e) One durable plan coloured + one paper copy of plan — where the estimated cost of the works is in excess of £10000

   or

   Two paper plans of which one must be coloured — where the estimated cost of the works is less than £10000

(3) All copies of the plan must be *signed* and *docqetted* by the applicant. This usually means stating on the drawing 'This is the plan referred to in my application dated .... and signed .... '

Plans should show the following:

	Minimum scale (metric)
Plans	1/100
Elevations	1/100
Sections	1/50
Block plans	1/500
Locality plan	1/1250
Details (if required)	1/20

A block plan is a line drawing of the building in relation to the boundaries, showing the position of any external alterations or erections.

(4) If central heating is to be installed and solid fuel is to be burned at a rate of 45.36 kg or more per hour, or liquid or gaseous fuel is to be burned at a rate equivalent to 0.37 MW or more per hour, an application form as prescribed in Regulations made under the Clean Air Act 1968 will require to be obtained from the Environmental Health Section of the local authority.

Structural design certificate
The structural design certificate referred to in these notes would be completed on Form 4A (for new buildings) or on Form 4B (for alterations and/or extensions to existing buildings).

Form 4B

The example given continues the theme of the attic conversion where the appropriate form is 4B and assurances may be required on such matters as depths and spans of attic joists, openings in roof for velux windows, openings in floor for staircase etc., as follows:

STRUCTURAL DESIGN CERTIFICATE FOR ALTERATIONS AND/OR EXTENSIONS TO EXISTING BUILDINGS

Building (Scotland) Act 1959, s. 6AA

This certificate is submitted with reference to the application by
Jane Boyle / Tom Brown (agent)
dated 29/2/96 for warrant to develop an attic in a terraced dwelling to create two additional bedrooms
at 14 Windfall Lane, Glasgow G44 3

*The design calculations for the new structural elements, i.e. ~~foundations~~, ~~in situ concrete~~, ~~pre-cast concrete~~, steel ✓, brick ✓, ~~masonry~~, ~~aluminium~~, timber ✓ .... (delete or add elements as necessary) having been prepared by me; or

*Being satisfied that the design calculations for the new structural elements, i.e. foundations, in situ concrete, pre-cast concrete, steel, brick, masonry, aluminium, timber ....N.R........ (delete or add elements as necessary) have been properly prepared.

I hereby certify that the design, including the specification of material to be used, complies with regulation 11 of the Building Standards (Scotland) Regulations 1990

SIGNED..Isgarth Greenview........ DATE.....15 March 1996.........

Name and address of practice
Greenview Structural Practice
Woodgrove Lane
Glasgow G88

Full Member of Institution of Civil Engineers ✓/Corporate Member ✓ of Institution of Structural Engineers*

Date of becoming a full/corporate member ..4...Oct....1965

or

Authorisation by Institution of Civil Engineers/Institution of Structural Engineers* to practise in the United Kingdom.

Date of Authorisation ....N.R.........

*Delete as appropriate

Having received Forms 4 (Warrant Application), 4B (Structural Design Certificate) and possibly Form 1 (Application for Relaxation) the local authority would check for compliance and when assured regarding compliance would issue the appropriate warrant on Forms 5, 6 or 7.

On completion of the building work the applicant (or agent) notifies the local authority by making application on Form 8 for a Certificate of Completion. This would be supported by Form 9 which provides similar assurances regarding electrical work, as do Forms 4A and B as regards structural matters. Again the attic conversion case is used to illustrate the use of Forms 8 and 9 as follows:

Form 8

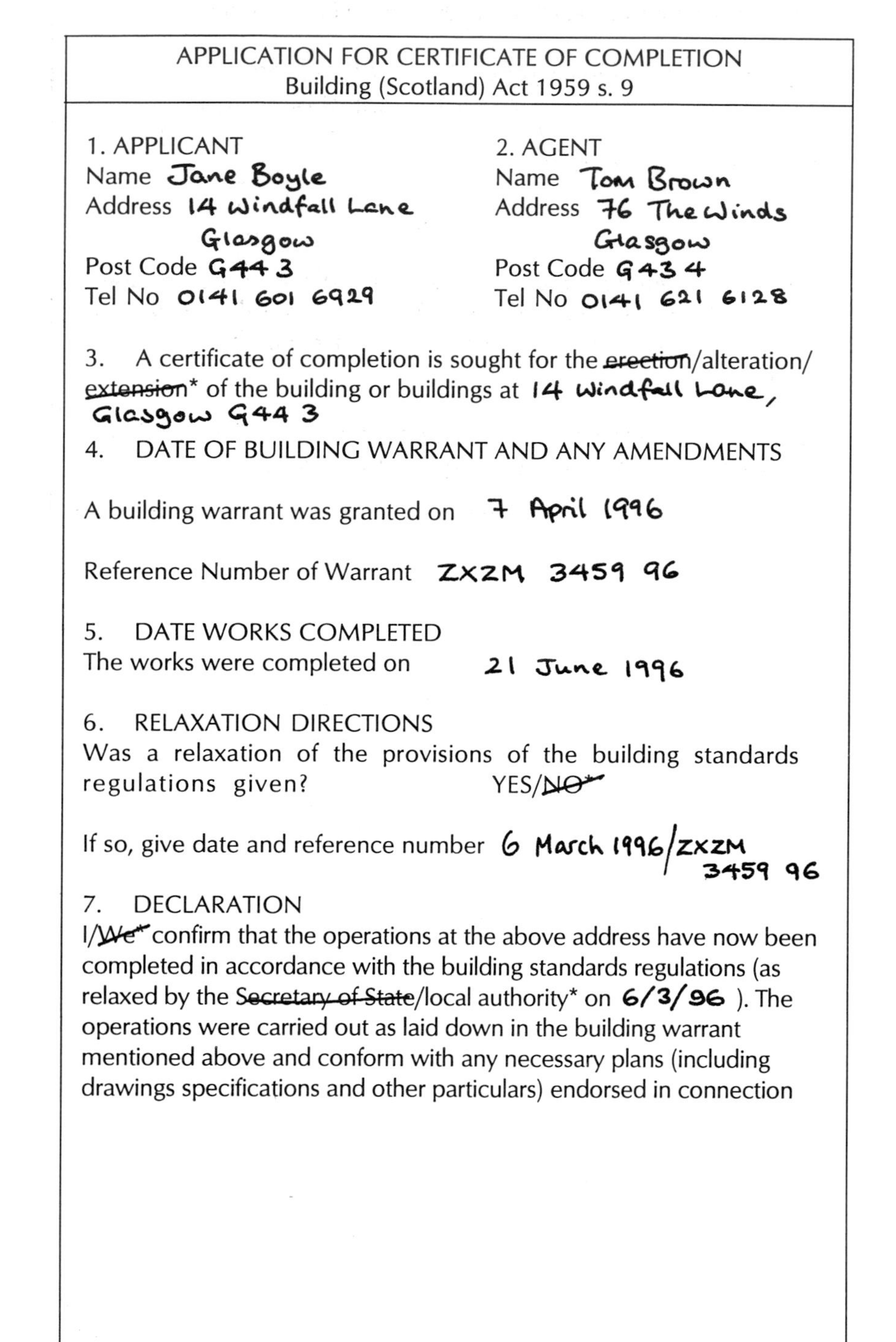

APPLICATION FOR CERTIFICATE OF COMPLETION
Building (Scotland) Act 1959 s. 9

1. APPLICANT
Name Jane Boyle
Address 14 Windfall Lane
Glasgow
Post Code G44 3
Tel No 0141 601 6929

2. AGENT
Name Tom Brown
Address 76 The Winds
Glasgow
Post Code G43 4
Tel No 0141 621 6128

3. A certificate of completion is sought for the ~~erection~~/alteration/~~extension~~* of the building or buildings at 14 Windfall Lane, Glasgow G44 3

4. DATE OF BUILDING WARRANT AND ANY AMENDMENTS

A building warrant was granted on 7 April 1996

Reference Number of Warrant ZXZM 3459 96

5. DATE WORKS COMPLETED
The works were completed on 21 June 1996

6. RELAXATION DIRECTIONS
Was a relaxation of the provisions of the building standards regulations given? YES/~~NO*~~

If so, give date and reference number 6 March 1996/ZXZM 3459 96

7. DECLARATION
I/~~We*~~ confirm that the operations at the above address have now been completed in accordance with the building standards regulations (as relaxed by the ~~Secretary of State~~/local authority* on 6/3/96). The operations were carried out as laid down in the building warrant mentioned above and conform with any necessary plans (including drawings specifications and other particulars) endorsed in connection

with the warrant. (I/~~We~~* also enclose a certificate showing that the electrical installation complies with the terms of the warrant)*

Signature of ~~applicant~~/agent* Tom Brown

Date 22 June 1996

*Delete as appropriate

Notes

(1) The local authority must either grant a certificate of completion or notify the applicant of their refusal to do so within 14 days of the date of receipt of the application for a certificate of completion.

(2) If the local authority refuse to grant a certificate of completion the applicant has the right of appeal to the sheriff.

Form 9

FORM 9

COMPLIANCE CERTIFICATE FOR ELECTRICAL INSTALLATION
Building (Scotland) Act 1959 s. 9

I certify that, to the best of my knowledge and belief, the electrical installation at 14 Windfall Lane carried out by me/~~under my supervision~~*, complies with the Building Standards (Scotland Regulations 1990 and the relevant conditions of the building warrant (reference number ZXZM 3459 96) granted by Eastren Council on 14 March 1996 for the ~~erection~~/alteration/~~extension~~* of the building.

Signature Bill Ohms
Date 10 June 1996
NAME Bill Ohms
ADDRESS 17 Riverway
Glasgow
Post Code G80 8UT
Tel No 0141 993 9356
Profession MIEE

*Delete as appropriate

THIS CERTIFICATE WHEN COMPLETED SHOULD BE SENT TO:

Eastren Council
Eastren Road
Eastren
G48 4RT

COUNCIL ENQUIRY POINT

Name Mr Pat Roller
Tel 6943 6217
Ext 443

Warning
IF A PERSON GRANTS OR PRODUCES A CERTIFICATE WHICH IS FALSE OR MISLEADING HE/SHE MAY BE GUILTY OF AN OFFENCE UNDER THE BUILDING (SCOTLAND) ACT 1959 AND LIABLE ON SUMMARY CONVICTION TO A FINE.

On satisfying itself that the work on site matches the plans and with possession of the Structural and Electrical Installation Certificates, the local authority would then proceed to the issue of a Certificate of Completion using Form 10.

The procedures discussed so far in this appendix can be seen as for facilitating building work. There are occasions when building control must use enforcing measures such as can occur where some person is undertaking building work which does not comply with the building standards regulations. Such maverick behaviour is often colloquially referred to as 'cowboy work'.

Earlier in this book, for instance, the use of s. 10 notices and orders was discussed at some length in association with circumstances where the local authority might be required to undertake the necessary remedial work and recover the cost of the work via the use of a charging order.

Forms 14 (s. 10 notice), 15 (s. 10 order) and 26 (charging order) would be the enforcement mechanisms used in such cases.

Suppose for example a sharp-eyed building control surveyor who works for Eastren Unitary Authority observes on his way to his work that two new dormer bedrooms have appeared on a single storey terrace dwelling at 776 Glibston Road. A check-up at the office reveals that there is no evidence of an application having been made for any alterations at this location. Over a frustrating period of time, no contact can be made with the perpetrators. A possible way forward would be to issue an s. 10 Notice by recorded delivery, along the following lines:

Form 14

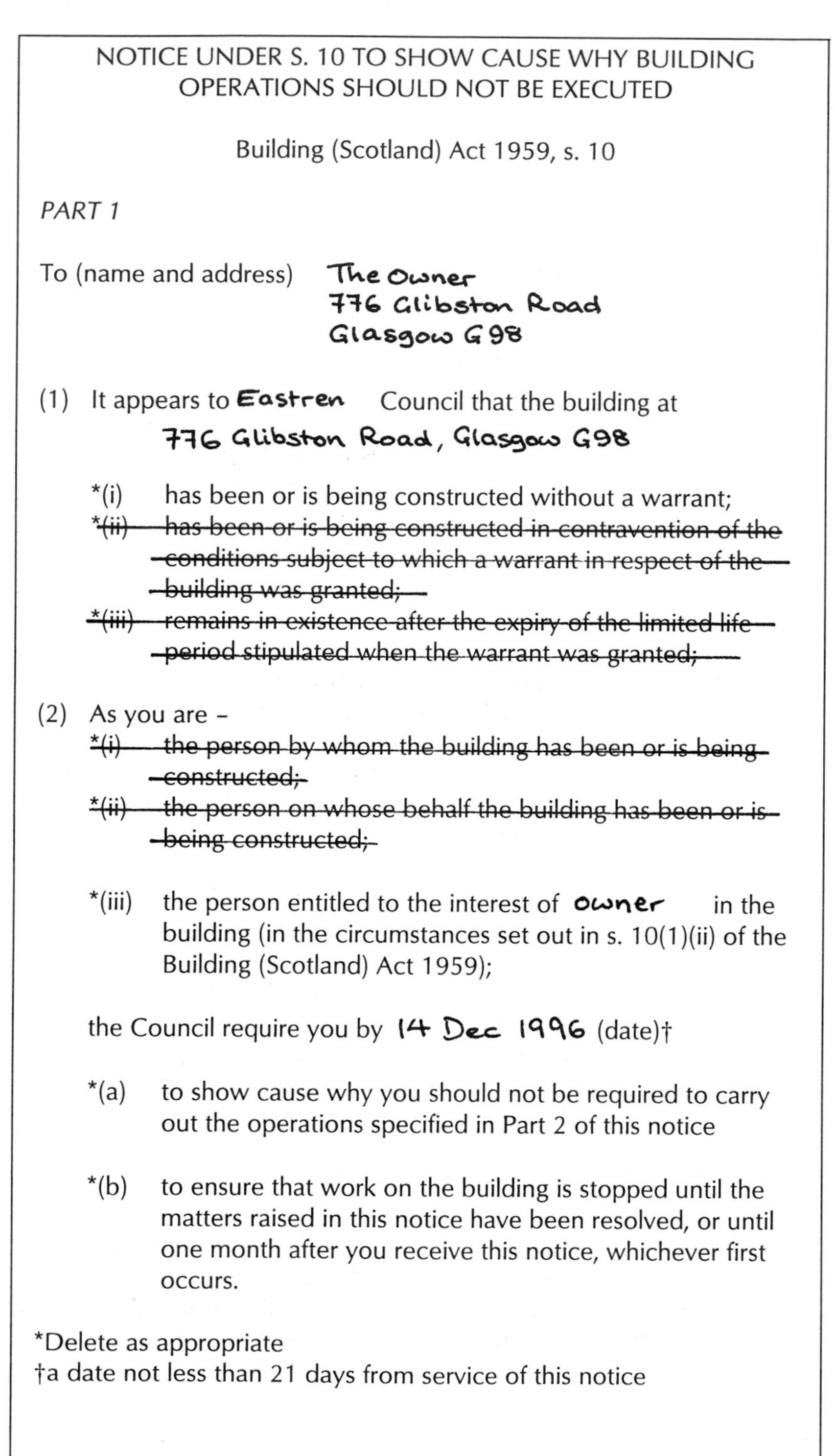

NOTICE UNDER S. 10 TO SHOW CAUSE WHY BUILDING OPERATIONS SHOULD NOT BE EXECUTED

Building (Scotland) Act 1959, s. 10

PART 1

To (name and address) The Owner
776 Glibston Road
Glasgow G98

(1) It appears to Eastren Council that the building at
776 Glibston Road, Glasgow G98

*(i) has been or is being constructed without a warrant;
~~*(ii) has been or is being constructed in contravention of the conditions subject to which a warrant in respect of the building was granted;~~
~~*(iii) remains in existence after the expiry of the limited life period stipulated when the warrant was granted;~~

(2) As you are –
~~*(i) the person by whom the building has been or is being constructed;~~
~~*(ii) the person on whose behalf the building has been or is being constructed;~~

*(iii) the person entitled to the interest of owner in the building (in the circumstances set out in s. 10(1)(ii) of the Building (Scotland) Act 1959);

the Council require you by 14 Dec 1996 (date)†

*(a) to show cause why you should not be required to carry out the operations specified in Part 2 of this notice

*(b) to ensure that work on the building is stopped until the matters raised in this notice have been resolved, or until one month after you receive this notice, whichever first occurs.

*Delete as appropriate
†a date not less than 21 days from service of this notice

PART 2 BUILDING OPERATIONS REQUIRED

The operations referred to in Part 1 are:

Change of use of building

Pat Roller (signed) for Eastren Council
15 Nov 1996 (date)

YOU SHOULD SEND YOUR REPLY TO:

Eastren Council	COUNCIL ENQUIRY POINT
Eastren Road	Name Mr Pat Roller
Eastren	Tel 6943 6217
Post Code G48 4RT	Ext 443

Notes

(1) In cases which fall under 1(i) or 1(ii) above, you may apply to the Secretary of State or the local authority for a relaxation direction under s. 4 of the Building (Scotland) Act 1959. You must apply by the date given in paragraph 2. The local authority can advise you to whom your application should be addressed. If you apply for a direction, the period of this notice will be extended.

(2) If you fail to show cause why the operations should not be carried out, the local authority may serve on you an Order requiring the work to be carried out. If you fail to carry out the operations in the Order you may be guilty of an offence under the Building (Scotland) Act 1959 and liable on summary conviction to a fine. In addition, the local authority may carry out the operations themselves and claim from you as a debt any expenses incurred.

Of course, in a situation where there was no positive response to the notice, the local authority could then issue Form 15 requiring the specified operations to be carried out (as spelled out in note 2 to Form 14). If this stronger persuasion still bore no fruit then the local authority would require to consider undertaking the work itself. This raises the difficulty of recovery of the cost of carrying out the work.

One possible recovery mechanism, as discussed earlier, is to use a charging order. This would be given to the person with an interest in the building on Form 26 as follows:

Form 26

CHARGING ORDER
Building (Scotland) Act 1959

The Eastren Council, in exercise of their powers under s. 6 to the Building (Scotland) Act 1959, hereby provide and declare that the subjects described in the Schedule below are charged and burdened with a payment to Eastren Council of an annuity of £400. 00 payable on the 3rd day of May in the year 1998 (insert date one year after the date of this Order) and in every year for the term of thirty years from the date of this Order.

A.B. Lawyer (signed) for Eastren Council
1/5/97 (date)

The Order should be signed in a manner in which the local authority's deeds are normally executed.

SCHEDULE

Description of subjects*	Name and designation of owner
Single storey terrace dwelling at 776 Glibston Road Glasgow G98	A. Strangefellow Owner

*This should, wherever possible, include reference to a recorded title or land certificate.

The reader is strongly recommended to develop the theme of this appendix by using the Building Forms (Scotland) Regulations 1991 in conjunction with section 2 of this book (The 1959 Act and its Regulations) to build up an understanding of the way in which administration of the building control process flows from the act.

Index

INDEX